MULTIPHASE POLYMER SYSTEMS

Micro- to Nanostructural Evolution in Advanced Technologies

MULTIPHASE POLYMER SYSTEMS

Micro- to Nanostructural Evolution in Advanced Technologies

edited by
Andreea Irina Barzic
Silvia Ioan

CRC Press
Taylor & Francis Group
Boca Raton London New York

CRC Press is an imprint of the
Taylor & Francis Group, an **informa** business

CRC Press
Taylor & Francis Group
6000 Broken Sound Parkway NW, Suite 300
Boca Raton, FL 33487-2742

First issued in paperback 2019

ISBN-13: 978-1-4987-5563-4 (hbk)
ISBN-13: 978-0-367-87653-1 (pbk)

Library of Congress Cataloging-in-Publication Data

Names: Barzic, Andreea Irina, editor. | Ioan, Silvia, editor.
Title: Multiphase polymer systems : micro- to nanostructural evolution in advanced technologies / edited by Andreea Irina Barzic and Silvia Ioan.
Description: Boca Raton : Taylor & Francis, CRC Press, 2017. | Includes bibliographical references and index.
Identifiers: LCCN 2016011001 | ISBN 9781498755634 (alk. paper)
Subjects: LCSH: Polymers. | Polymeric composites. | Phase rule and equilibrium.
Classification: LCC TP1120 .M85 2017 | DDC 668.9--dc23
LC record available at https://lccn.loc.gov/2016011001

Visit the Taylor & Francis Web site at
http://www.taylorandfrancis.com

and the CRC Press Web site at
http://www.crcpress.com

Dedication

This book is dedicated to the 150th anniversary of the Romanian Academy.

Contents

SECTION II Multiphase Polymer Systems in Solid Phase

Preface

Phase morphology in multicomponent, polymer-based systems represents the main physical characteristic that allows for a controlled design concomitantly with the development of new plastic materials. Many scientific publications and industrial patents are still focused on research regarding the interrelation of phase morphology with the physical and chemical features of the materials as a tool for enhancing the performance of polymer systems. This book, *Multiphase Polymer Systems: Micro- to Nanostructural Evolution in Advanced Technologies*, attempts to serve as a guide for specialists that work in this area with an emphasis on the properties of these promising materials in both solution and solid phase. This aspect is of major importance because, in most cases, polymers are processed from solutions or melts; thus, the physicochemical characteristics in these states must be clearly understood.

Research preoccupation with the field of multiphase materials covers a variety of topics, such as miscible and immiscible polymer blends, copolymer solution assemblies, porous polymer beads, polymer liquid crystal systems, composites, nanocomposites, and polymer/metal systems. This book reviews the most recent advances made in the areas of the morphological, flow behavior, interfacial, physical, fire-resistant, thermophysical, and biomedical properties of multiphase polymer systems, together with some basic investigation techniques that are sensitive in elucidating the features of each phase. The vast and complex scientific information presented in this manuscript contributes significantly to the basic knowledge of students and researchers working in the field of materials science, and it will also be of interest to the industrial media due to the synthesis and processing procedures outlined here that, through their novelty, introduce new trends in many applicable areas.

Andreea Irina Barzic and Silvia Ioan
"Petru Poni" Institute of Macromolecular Chemistry
Iasi, Romania

Introduction

Multiphase Polymer Systems: Micro- to Nanostructural Evolution in Advanced Technologies is a fundamental reference work on copolymers, polymer blends, polymer composites, interpenetrating polymers, and layered polymer/metal structures, covering aspects of science, engineering, technology, and application.

The book is divided into two sections. The first relates to multiphase polymer systems in solution phase. Thus, miscible and immiscible polymer blends in solution and the impact of different characterization methods in the development of commercial products are addressed. In addition, large macromolecular assemblies composed of different polymers in precise arrangements that carry out biological functions are also presented. The book emphasizes the different architectures of copolymer solutions that tend to form large aggregates, such as spherical or cylindrical micelles, under appropriate conditions. The micellization process is described based on hydrophobic interactions, electrostatic interactions, and hydrogen bonding of the constituent block copolymers. Changes in hydrophobic/hydrophilic balance, temperature, and pH conditions are analyzed in order to discuss the possibility of their application in regenerative medicine. Functionalization of polymeric micelle surfaces to modify physicochemical and biological properties is highlighted from the standpoint of designing high-performance medical devices, including scaffolds for tissue engineering and matrices for drug delivery systems. Various polymer structures can also be obtained from liquid–liquid two-phase systems as a result of phase separation during radical polymerization reactions. Different preparation methods for obtaining these polymer networks have been developed in recent years. Thus, the terminology and classification of liquid–liquid two-phase polymerization and crosslinking are described. By changing some of the independent variables—such as the amount of crosslinking agent, the diluent quantity, the initiator concentration, or the polymerization temperature and agitation speed—key factors involving kinetics and thermodynamics of reaction systems decisively influence the resulting macroporous architecture as well as the microspheres' size and porosity.

Some water-soluble polymers presented in this book have attracted significant interest in the preparation of different biomaterials as well as in other applications, including paper adhesives, textiles, coatings, thickener agents, and so on. From mixtures of natural and synthetic polymers, new materials can be obtained with a variety of properties and end uses. The final properties of polymer mixtures depend on the characteristics of their components, composition, and, mainly, on their miscibility. Based on the unique properties of these mixtures, such as controlled swelling and release capabilities, there is a great potential for their utilization in a wide range of biomedical applications (packaging, pharmaceutical preparations, surgical implants, controlled drug delivery systems, etc.). Also, a comprehensive discussion on the most recent developments in the area of polymer-based advanced materials is presented, with a special focus on their characteristics and applications. Flow domains of polymer solutions with liquid crystal phase as well as microstructure evolution during reinforcement of polymer solutions are important topics discussed in the first section of this book. Polymers with semirigid backbones present the ability to self-organize in solutions after a critical concentration. When these systems are subjected to shearing, a specific texture results after chain relaxation that can be further induced to other polymers.

This section of the book also presents aspects of the flow behavior of some polymer mixture solutions containing a lyotropic phase as a function of system composition, shear intensity, and temperature. The correlation between phase separation and the changes in rheological properties induced by the anisotropic component of the system is discussed. Also, viscoelastic solution properties are reviewed in electro-optics with regard to their practical uses. Finally, we discuss polymer composite materials, where mineral and organic fillers are used as reinforcing materials.

The second section of the book deals with aspects of multiphase polymer systems in solid phase. Well known for their applications in display manufacture, polymer dispersed liquid crystal composite materials return the researcher's attention to the possibilities for their use in a larger field of application including biomedicine. The newest contributions in the domain of polymer dispersed liquid crystals in terms of composite design, anchoring forces, properties, and the relationship between composite design and applications are also discussed. The surface morphological features of multiphase systems, including a wide range of materials such as composites, blends, and interpenetrating polymer networks, can be characterized at high magnification and resolution by atomic force microscopy (AFM). This method is an excellent complement to electron microscopy techniques, such as scanning electron microscopy (SEM) and transmission electron microscopy (TEM). Utilizing the phase contrast mode, all these microscopy techniques can be used for composition imaging of the different phases in multiphase polymer films. Moreover, Chapter 8 addresses the manner in which AFM can also provide additional quantitative information regarding the differences in local mechanical properties by measuring lateral forces (lateral force imaging) and adhesion forces (adhesion force mapping), and Young's modulus (modulus mapping and indentation mapping). Furthermore, a review of the current state of knowledge in the field of X-ray methods applied to multiphase polymer systems, depending on different factors (crosslinking, hydration, stretching, or temperature), is presented in Chapter 9.

Dynamic mechanical analysis in polymeric multiphase systems in solid state, as part of rheology, is associated with oscillatory tests that are employed to investigate all kinds of viscoelastic material from the point of view of flow and deformation behavior. In particular, it evaluates the molecular mobility in polymers, the pattern of which may be an indication of phase-separated systems. Although there are certain preferred tools for visual examination of phenomena for these kinds of systems, dynamic mechanical analysis has the advantage of examination in dynamic conditions and of the prediction of properties.

Chapter 11 also highlights the latest developments in multiphase polymeric materials and composites. Designing thermophysical behavior of polymers with nanometric inclusions for heat dissipation in electronic devices includes the transport abilities of polymers significantly enhanced by the embedding of nanometric inclusions with specific properties. Thermal conduction processes in polymer nanocomposites are described, starting from the matrix itself and continuing with the multiplication of conduction paths by inserting various types of nanofillers. Here, the most important theoretical models used for thermal conductivity of reinforced polymers are described in relation to the most important factors in designing polymer nanocomposites. Current trends in the manufacturing of such multiphase materials with enhanced thermal conduction, by controlling the nanofiller–matrix interactions and the composition, are presented for their heat dissipation purposes in electronics.

On the contrary, electroconductive materials have gained great importance in various industries. The continuous evolution of current technology has stimulated researchers to develop new and low-cost materials with improved electrical conductivity by reinforcement of commercial polymers with nanoparticles. Thus, Chapter 12 depicts the technology used to obtain advanced polymer nanocomposites and also includes the unsolved issues regarding the optimization of charge carrier transport. The performance of polymers containing at least one conductive nanophase is reviewed in accordance with the applications of these multiphase materials in electronics, opto-electronics, and other fields. Polymer composites based on maleimide compounds (addition polyimides) are materials that can be included in the multiphase systems used in various hi-tech applications, possessing an excellent balance of mechanical, thermal, and electrical properties. However, currently available systems have two major drawbacks—their limited processability and conditioned stability in the long term due to the aging phenomenon. This topic is included in Chapter 13, which aimed at assessing some of the latest achievements in improving the processability of bismaleimidic resins used to obtain high-performance composites for special applications. Epoxy resins, described in another chapter, represent a class of reactive prepolymers (or polymers with oxirane units) that can be converted into hard and infusible thermosets. These materials possess excellent properties, such as high strength and modulus, low shrinkage in cure, excellent chemical and solvent resistance, and so on. However, high flammability is one of the major drawbacks in extending their use as high-performance materials in applications that demand high flame retardancy. Organophosphorus compounds increase the flame retardancy of epoxy by acting in the condensed and/or gaseous phase. Also, they generate fewer toxic gases and less smoke during combustion. Chapter 14 also describes the flame retardant characteristics of epoxy thermosets containing some 9,10-dihydro-9-oxa-10-phosphaphenanthrene-10-oxide derivatives.

Chapter 15 outlines the current progress in fields such as consumer electronics, robotics, or biomedical engineering that require the development of sensing technologies and that also combine miniaturization of some composite materials with a high sensitivity at different stimuli and low power consumption. Advances in material integration and nanofabrication technologies have enabled new directions of research and development for microsensor systems. Examples include nanocomposite materials that combine the advantages of polymers—such as flexibility, elasticity, chemical resistance, or biocompatibility—with the unique properties of nanoparticles. Magnetic nanocomposites have been introduced as an alternative to the use of superparamagnetic particles, which require a highly magnetic field for magnetization, thereby increasing power consumption and limiting the possibility of integration and miniaturization. On the contrary, iron magnetic nanomaterials are highly magnetized in the absence of a magnetic field due to the strong anisotropic shape, and they can be easily fabricated using a cost-effective process. Incorporating these magnetic nanomaterials in a polymer induces the formation of permanent magnetic composites, with distinct features and reinforced mechanical properties. The magnetic properties of such a nanocomposite can be tailored for a wide range of possible uses by the choice of the material, the dimensions, and the concentration of the magnetic nanomaterials or by aligning the magnetic nanomaterials in the polymer. The magnetic field of the magnetic nanomaterials also generates a bias field for magnetic field sensors that can be used to detect the stray field of the nanocomposite for different sensitivities.

Chapter 16 is focused on imide-type polymers containing highly dispersed gold nanoparticles developed through electrospinning—an inexpensive, straightforward method for producing membranes with very large surface-to-volume ratios. Spinning condition parameters, including solution properties (solution concentration, solution viscosity, polymer molecular weight, surface tension), processing conditions (applied voltage, volume flow rate, spinning distance), and the effect of ambient conditions (temperature, humidity) on fiber morphology and size distribution are systematically studied in order to obtain nanofibers with low diameters and narrow polydispersity. The obtained nanofibrous hybrid structures are characterized from a morphological and functional point of view. The photo-catalytic performance of the hybrid fibers for degradation of formaldehyde, known to be a human carcinogen, is investigated as well.

Chapter 17 discusses the current state of the art in the field of physicochemical properties of metallic clusters or aggregates, which are of great interest due to the fact that they are a new type of catalyst that can be used in many organic syntheses. Moreover, they can be widely used for the production of different polymeric materials with extraordinarily different properties. Generally, clusters are aggregates of atoms, the sizes of which are larger than those of atoms and molecules but smaller than bulk materials. They do not have the same structure or atomic arrangements as a bulk solid (organic polymer substrate) and can change their structure with the addition of just one or few atoms. Clusters may be classified as metallic or nonmetallic according to the atoms from which they are made; in general, some metallic character persists from small clusters to bulk matter of the same composition. Growing clusters and layers on flat organic polymer substrates is a well-investigated process for solid–gas and solid–liquid surfaces. Usually, atoms diffuse and collide until they form a nucleus of a certain critical size, often with only a few atoms. Also, these clusters can be fabricated, starting from the bulk material by somehow reducing their dimension (the top-down approach).

On the contrary, metallization of polymers is of considerable technological importance and is also very interesting from a fundamental point of view (Chapter 18). Metalized polymers are used extensively in applications such as food packing, on-chip interconnection, reflectors for car lights, optical data storage (compact discs), electrically shielded computer cases, decorative coatings, optical ultrathin color filters, and substrates for biomolecules. In general, the properties of thin metal/polymer interfaces are determined by the interaction between the metal and the polymer and by the morphology of the interface. The cohesive energy of metals is typically two orders of magnitude higher than the cohesive energy of polymers.

Furthermore, the interaction between moderately reactive metals and polymers is generally very weak in comparison with the strong metal–metal binding forces. Therefore, noble metals and other metals of low reactivity do not wet untreated polymer surfaces but form three-dimensional spherical clusters (i.e., they exhibit a Volmer–Weber growth mode). Also, control of surface properties is very important for highly adhesive performance in thin metal layers, biomaterials, and paints. Modification of a hydrophobic polymer surface into a hydrophilic one can be achieved by wet (acid, alkali), dry (plasma), and radiation treatments (ultraviolet radiation and laser) without affecting the bulk properties. Plasma treatment of the polymer substrate with different gases represents one of the efficient methods for improving the wettability and metal adhesion of a polymer substrate.

The topics addressed in this book will be of interest to those working as researchers and students in doctoral or postdoctoral programs, as well as engineers; they are designed to be an information source for all aspects mentioned earlier. The performance of these materials depends on the properties of different components, as well as their spatial arrangements, controlled by thermodynamics and flow-imposed morphology. In this context, the aim of this book is to describe and interpret phase morphology and interface in multiphase complex materials.

Andreea Irina Barzic
Silvia Ioan

Editors

Andreea Irina Barzic has been a scientific researcher in the Department of Physical Chemistry of Polymers at the "Petru Poni" Institute of Macromolecular Chemistry, Iasi, Romania, since 2005. She received her PhD in chemistry from the Romanian Academy, Bucharest, Romania. Her major fields of interest include polymer blends, composites, and liquid crystal polymers. She has published more than 40 papers in peer-reviewed journals and has authored or coauthored several books. She is also involved in several research projects and is a referee (reviewer) for a number of prestigious journals in the field of polymer science.

Silvia Ioan is a senior researcher, with more than 50 years of scientific contributions, at the "Petru Poni" Institute of Macromolecular Chemistry, Iasi, Romania. Her areas of study have included interdisciplinary research for theoretical and experimental substantiations of the processes and technologies at nano-, micro-, and macroscales. These scientific contributions have focused on the participation of some polymers with special architectures in obtaining multifunctional materials including studies concerning the improvement of some properties for predetermined goals and applications. The direction of her research is reflected in many publications, such as books or book chapters and more than 300 scientific papers, as well as in patents, in several participations at the national and international scientific conferences, leader and member of numerous research projects. She also developed a specialized research group on the characterization of macromolecular compounds in solution and solid states at Romania's Institute of Macromolecular Chemistry. In addition, Dr. Ioan serves as an expert reviewer of several prestigious scientific journals, and she is the PhD supervisor in the materials science domain at the doctoral school of the Romanian Academy—"Petru Poni" Institute of Macromolecular Chemistry.

Contributors

Magdalena Aflori
Department of Polymer Materials Physics
"Petru Poni" Institute of Macromolecular Chemistry
Iasi, Romania

Raluca Marinica Albu
Department of Physical Chemistry of Polymers
"Petru Poni" Institute of Macromolecular Chemistry
Iasi, Romania

Andreea Irina Barzic
Department of Physical Chemistry of Polymers
"Petru Poni" Institute of Macromolecular Chemistry
Iasi, Romania

Maria Bercea
Department of Electroactive Polymers and Plasmochemistry
"Petru Poni" Institute of Macromolecular Chemistry
Iasi, Romania

Luminita Ioana Buruiana
Department of Physical Chemistry of Polymers
"Petru Poni" Institute of Macromolecular Chemistry
Iasi, Romania

Ionela-Daniela Carja
Department of Polycondensation
"Petru Poni" Institute of Macromolecular Chemistry
Iasi, Romania

Mariana Cristea
Department of Polymer Materials Physics
"Petru Poni" Institute of Macromolecular Chemistry
Iasi, Romania

Adina Maria Dobos
Department of Physical Chemistry of Polymers
"Petru Poni" Institute of Macromolecular Chemistry
Iasi, Romania

Vicente J. Forrat Pérez
AIDICO: Technological Institute of Construction
Novelda-Alicante, Spain

Giuseppino Fortunato
Empa
Swiss Federal Laboratories for Materials Science and Technology
St. Gallen, Switzerland

Constantin Gaina
Department of Polyaddition and Photochemistry
"Petru Poni" Institute of Macromolecular Chemistry
Iasi, Romania

Viorica Gaina
Department of Polyaddition and Photochemistry
"Petru Poni" Institute of Macromolecular Chemistry
Iasi, Romania

Corneliu Hamciuc
Department of Polycondensation
"Petru Poni" Institute of Macromolecular Chemistry
Iasi, Romania

Camelia Hulubei
Department of Polycondensation
"Petru Poni" Institute of Macromolecular Chemistry
Iasi, Romania

Silvia Ioan
Department of Physical Chemistry of Polymers
"Petru Poni" Institute of Macromolecular Chemistry
Iasi, Romania

Ridvan Karapinar
Department of Physics
Faculty of Science
Yuzuncu Yil University
Van, Turkey

Luminita Marin
Department of Polycondensation
"Petru Poni" Institute of Macromolecular Chemistry
Iasi, Romania

Simona Morariu
Department of Electroactive Polymers and Plasmochemistry
"Petru Poni" Institute of Macromolecular Chemistry
Iasi, Romania

Simona Luminita Nica
Department of Physical Chemistry of Polymers
"Petru Poni" Institute of Macromolecular Chemistry
Iasi, Romania

René Michel Rossi
Empa
Swiss Federal Laboratories for Materials Science and Technology
St. Gallen, Switzerland

Ion Sava
Department of Polycondensation
"Petru Poni" Institute of Macromolecular Chemistry
Iasi, Romania

Diana Serbezeanu
Department of Polycondensation
"Petru Poni" Institute of Macromolecular Chemistry
Iasi, Romania

Iuliana Stoica
Department of Polymer Materials Physics
"Petru Poni" Institute of Macromolecular Chemistry
Iasi, Romania

Fulga Tanasa
Department of Polyaddition and Photochemistry
"Petru Poni" Institute of Macromolecular Chemistry
Iasi, Romania

Mirela Teodorescu
Department of Electroactive Polymers and Plasmochemistry
"Petru Poni" Institute of Macromolecular Chemistry
Iasi, Romania

Oana Ursache
Department of Polyaddition and Photochemistry
"Petru Poni" Institute of Macromolecular Chemistry
Iasi, Romania

Cristina Doina Vlad
Department of Functional Polymers
"Petru Poni" Institute of Macromolecular Chemistry
Iasi, Romania

Tachita Vlad-Bubulac
Department of Polycondensation
"Petru Poni" Institute of Macromolecular Chemistry
Iasi, Romania

Section I

Multiphase Polymer Systems in Solution Phase

1 Miscible and Immiscible Polymer Blends

The Impact of Viscosity and Phase-Separation on the Development of Commercial Products

Adina Maria Dobos

CONTENTS

1.1 INTRODUCTION

The mixing of two or more polymers is an aspect of technology that lately has received special attention because blending represents a new, easy, and efficient method for obtaining advanced materials with special properties that are not found in individual components. In relation to polymer mixtures, the term *miscible* has been used to describe the multicomponent system behavior that is similar to that of a single-phase system (Krammer 1990). The term does not necessarily refer to polymers mixing at the segmental level, but it suggests that, at this level, the macroscopic properties are characteristics of a single-phase materials. By mixing at least two polymers, *miscible*, *immiscible*, or *partially miscible* blends can be obtained. In many cases, polymers are not miscible (i.e., compatible) because the thermodynamic parameters—namely the *entropy* and *enthalpy* of blending—are unfavorable. For this reason, studies on miscibility or on phase separation are essential. There are several methods for characterizing polymer blends, such as differential scanning calorimetry (DSC), dynamic mechanical thermal analysis (DMTA), dielectric measurements, cloud point measurements, and nuclear magnetic resonance spectroscopy (NMR). All of these allow the determination of polymer mixtures' compatibility only over a certain size range (Utracki 1990).

Thus, in obtaining polymer mixtures, not only should the thermodynamic aspects be considered but also the thermal and mechanical blending conditions that are involved. These factors influence the kinetic process and establish equilibrium in the system (Krammer 1990).

1.2 THERMODYNAMICS OF MISCIBLE AND IMMISCIBLE POLYMER BLENDS

The thermodynamic treatment of polymer/solvent systems is based on the Flory–Huggins parameter, χ, which measures polymer power dissolution and takes into account the specific interaction that occurs among the polymer segments and the solvent molecules (Schuld and Wolf 1999, 2001). Increasing polymer concentration causes preponderantly polymer/polymer interactions, and starting with a concentration value characteristic for each polymer/solvent system, the Flory–Huggins parameter presents values higher than 0.5 (Schneider et al. 2004).

Taking into account the interactions manifested in the system, miscible polymer blends can be classified into miscible polymer blends with *weak* or *strong* specific interactions. Based on the Flory–Huggins theory, when χ parameter has negative values ($\leq$0.001), it is assumed that weak interactions exist in the system. But, for values greater than 0.01, it is assumed that strong interactions occur among the components of polymer blends. However, this parameter is not the most suitable one to describe a miscible polymer blend in terms of its interactions. Intermolecular forces would be more appropriate (Huyskens et al. 1991; Israelachvili 1991). The most common interactions are *van der Waals* (1 KJ/mol energy), *dipole/dipole* (2–8 KJ/mol energy), *hydrogen bonding* (50–170 KJ/mol energy), and *ionic* (850–1,700 KJ/mol energy). Frequently, the chemical structure, molecular weight, concentration, and intermolecular interactions play an important role in the establishment of bonds among the blend components (i.e., the van der Waals or hydrogen bonding type). Because miscible polymer blends are of great practical interest, it is assumed that miscibility is always preferred. According to the literature (Schwahn 1994), perfect mixing complies with the following formula:

$$\Delta G_m = \Delta H_m - T \cdot \Delta S < 0 \tag{1.1}$$

where ΔG_m = Gibb's free energy, ΔH_m = mixing enthalpy, ΔS = mixing entropy, and T = temperature.

Miscible polymer blends act as single-phase systems that combine the properties of pure components so that the stability criteria for binary mixtures with composition φ, at fixed temperature T and pressure p, are expressed as

$$\Delta G_m < 0, \left(\frac{\partial^2 \Delta G_m}{\partial \Phi^2}\right)_{p,T} > 0 \tag{1.2}$$

Miscible polymer blends are characterized by homogeneity up to the molecular level, with negative values for ΔG_m and positive values for $T \cdot \Delta S$. At the same time, ΔG_m always depends on the value of ΔH_m; the pair of polymers tends to form a single phase when the entropic contribution to the free energy exceeds that of enthalpy and $\Delta H_m < T \cdot \Delta S$.

Nevertheless, there are situations when specific interactions may not exist in miscible polymer blends. In those cases, a two-phase behavior can be observed: an upper critical solution temperature (UCST) and a lower critical solution temperature (LCST). The exception being when the two phases are combined (Krammer et al. 1989; Ougizawa and Inoue 1986), as shown in Figure 1.1.

Generally, in miscible blends, phase separation occurs when one of the following parameters varies: composition, temperature, or pressure. Under these conditions, the system moves from the monophasic state to the metastable one, with phase separation occurring through nucleation (namely the appearance of crystallization germs in the amorphous phase and their growth in one or more directions). This is followed by an increase in the separated domains. When the system is forced to shift from a single phase into the spinodal region, phase separation occurs spontaneously—through spinodal decomposition. This is expressed thermodynamically as

$$spinodal\!: \left(\frac{\partial^2 \Delta Gm}{\partial \phi^2}\right)_{p,T} = 0 \tag{1.3}$$

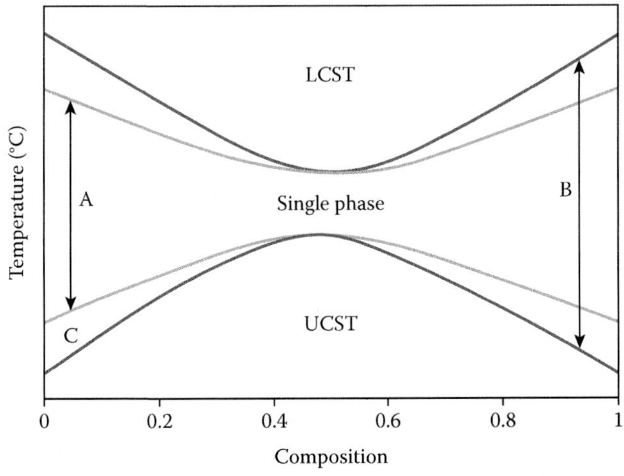

FIGURE 1.1 Phase behavior of polymer blends with the upper and the lower critical solution temperature UCST and LCST: (A) single-phase miscible region between two binodals, (B) two-phase separated regions of immiscibility, surrounded by spinodals, (C) four fragmented metastable regions between binodals and spinodals. (Adapted from Ougizawa Toshiaki and Inoue Takashi. *Polym. J.*, 18, no. 7, 521–527, 1986.)

and

$$\textit{critical point}: \left(\frac{\partial^2 \Delta Gm}{\partial \phi^2}\right)_{p,T} = \left(\frac{\partial^3 \Delta G_m}{\partial \phi^3}\right)_{p,T} = 0 \quad (1.4)$$

Usually, most binary polymer blends are immiscible. When two immiscible polymers are mixed, different phase morphologies may appear depending upon their nature, the amount of each blend component, and on their flow history (such as when polymers act like materials having a different flow). By mixing two immiscible polymers, materials with weak interfacial adhesion and, consequently, poor mechanical integrity result. For this reason, the goal has been to obtain as many miscible polymer blends as possible. The transition from immiscible blends to useful polymeric materials involves some processing at the interface, such as the reduction of interfacial tension (assuring adherence to the surface), which requires the utilization of a compatibilizer miscible in both media. Lately, the literature in this field has proposed rigid nanomaterials of silica nanoparticles as a type of compatibilizer (Elias et al. 2006; Sadiku-Agboola et al. 2011).

1.3 EFFECT OF THE COMPATIBILIZER

Various methods for enhancing the miscibility polymer blends have been proposed. These include the introduction of functional groups into the polymer molecular structure, crosslinking, and the formation of an interpenetrating polymer network (Zhang et al. 2004). However, the most common method for improving polymer miscibility is with the introduction of compatibilizers. A successful option is the utilization of nanofillers (Cao et al. 2011) and nanoclays (Xavier and Bose 2013; Zhang et al. 2011) that reinforce the polymer matrix of composites. Benefits from introducing these additives include low cost and the addition of strong mechanical properties to nanocomposites. The goals of this procedure are to reduce interfacial tension, stabilize the desired morphology, improve the interaction among solid-state phases, and obtain polymer alloys. The various morphological aspects attained depend not only on the level of mechanical mixing but, to a greater extent, on their control by the compatibilizer (Utracki 2000).

It has been shown that dispersion improves when the interfacial tension decreases (Paul 1978). However, Sundararaj and Macosko (1995) have challenged this hypothesis by claiming that the main role of a compatibilizer is not to reduce interfacial tension but rather to prevent the production of coalescence. The authors have observed that droplet diameter in a steady state increases significantly with an increase in a blend volume fraction where no compatibilizer has been added, becoming dependent on this for mixtures containing a compatibilizer. Therefore, prevention of coalescence is one of the most important aspects of compatibilization (Van Puyvelde et al. 2002). This can be explained by at least two physical mechanisms: the appearance of Marangoni stress (Figure 1.2a) and the wringing of the compatibilizer spread in the matrix phase, when droplets approach each other (Figure 1.2b)

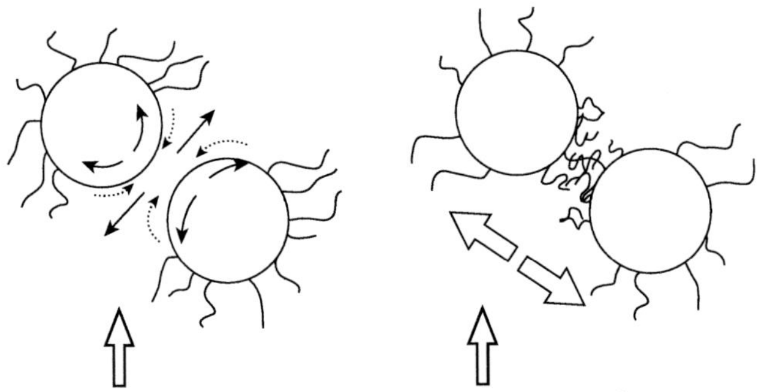

FIGURE 1.2 Schematic representation explaining the removal of coalescence. (a) Marangoni stress and (b) elastic repulsion. (Adapted from Van Puyvelde Peter, et al. *Polym. Eng. Sci.*, 42, 1956–1964, 2002.)

Although displaying different behaviors, both phenomena may occur simultaneously. For example, water-in-oil emulsions in which the adsorbed proteins reduce film drainage due to concentration gradients and stabilize the emulsion through steric hindrance (Walstra 1993). Fortelny (2004) has demonstrated that only when the Marangoni effect is insignificant can steric repulsion be manifested.

The presence of compatibilizer molecules at the interface determines a complex rheological behavior that cannot be explained by interfacial tension. For this reason, in recent years, researchers have attempted to correlate *interfacial rheology* with *morphology* development in polymer blends. Frequently, in both physically and chemically compatibilized blends, a slow relaxation process appears as a result of interfacial elasticity (Asthana and Jayaraman 1999; Jacobs et al. 1999; Moan et al. 2000; Riemann et al. 1996, 1997; Van Hemelrijck et al. 2004). Thus, the rheological behavior of components, along with the behavior at interface, will dictate the behavior of the entire system. Palierne (1990, 1991) has addressed these aspects from a mathematical perspective, correlating the oscillatory flow behavior with the basic morphology of blends. Thus, a *relaxation shoulder* in the linear dynamic moduli was predicted, this being a characteristic for the shape relaxation time of droplets that involves droplet radius and interfacial tension, as follows:

$$t_s = \frac{R_v \eta_m}{4\alpha} \cdot \frac{(19p+16)(2p+3-2\varphi(p-1))}{10(p+1)-2\varphi(5p+2)} \tag{1.5}$$

where t_s = the shape relaxation time of droplets, R_v = average radius of a droplet, p = viscosity ratio of the blend, φ = volume fraction of the dispersed phase, η_m = viscosity of the matrix phase, and α = interfacial tension.

Moreover, this model can also involve parameters that refer to the interfacial rheological properties, such as the *interfacial dilation modulus* and the *interfacial shear modulus* (Jacobs et al. 1999). Consequently, quantitative measurements of these parameters can predict the rheogical properties of compatibilized blends, especially for low molecular weight emulsions (Janssen et al. 1997; Peters et al. 2005). Exceptions are the molten polymers, whose measurement of surface properties is not simple because of their high viscosity. However, the Palierne model still offers an explanation for the additional low frequency relaxation process that has been observed in compatibilized blends. Riemann et al. (1996, 1997) reduced the full Palierne model to linear dynamic moduli in which the two relaxation processes are present, offering values for the shape relaxation time, t_s, as well as for the interfacial relaxation time, t_β.

The influence of compatibilizer concentration on the two relaxation times also was analyzed by Van Hemelrijck et al. (2004). They have fitted the Palierne model with an interfacial shear modulus by introducing a concentration gradient of the block copolymer along the interface. Figure 1.3 presents the dynamic moduli of compatibilized blends—polydimethylsiloxane (PDMS)/polyisoprene (PI)—at various ratios of compatibilizers versus angular frequency (Van Hemelrijck et al. 2005).

As one can observe, for uncompatibilized blends (dotted lines), a single shoulder appears, corresponding to t_s and representing the shape relaxation time of droplets. However, for compatibilized blends within the domain of lower frequencies, the second relaxation mechanism, t_β, which depends considerably on compatibilizer concentration, becomes obvious.

The literature (Van Hemelrijck et al. 2004) provides examples of the evolution of the t_s and t_β time functions on strain. Moreover, the shape relaxation time of droplets, t_s, which depends on their radius, displays evidence of the capacity of the compatibilizer to prevent the emergence of coalescence. The effects of molecular weight and

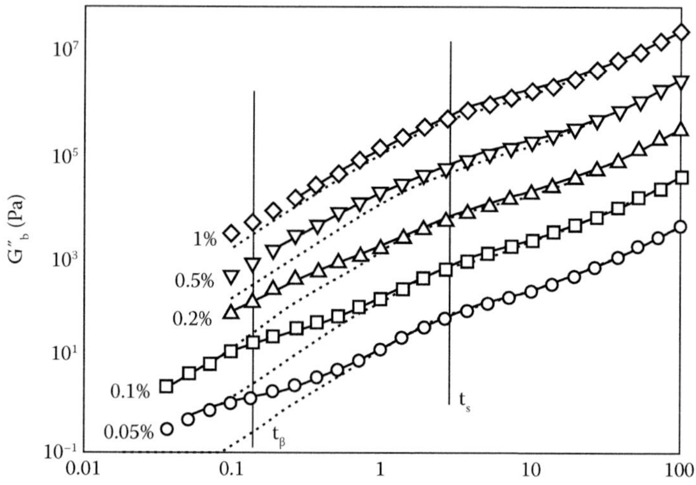

FIGURE 1.3 Storage moduli versus angular frequency for compatibilized blends (10% PDMS in PI) with compatibilizer concentrations ranging from 0.05 to 1 wt%. (Adapted from Van Hemelrijck Ellen, et al. *J. Rheol.*, 49, no. 3, 783–798, 2005.)

the asymmetry of block copolymers on t_s and t_β were also studied (Van Hemelrijck et al. 2005), with the conclusion being that a high molecular weight of block copolymers, at the same concentration, determines a visible suppression of coalescence.

The effect of the compatibilizer is not only to suppress the coalescence process but also to break up the drops; it can reduce the interfacial tension, and thus the hydrodynamic stress, leading to drop destruction. Another phenomenon, which can be mathematically expressed (Eggleton and Stebe 1998; Li and Pozrikidis 1997; Milliken et al. 1993; Pawar and Stebe 1996; Stone and Leal 1990), refers to the development of a large gradient within interfacial tension that affects local deformation and breakup of the droplets. By measuring the steady drop size, Ramic et al. (2000) have shown that when the compatibilizer is present in small amounts, breakup of the droplets does not occur. The compatibilizer succeeds in stabilizing drops with a viscosity greater than one. Figure 1.4 illustrates the possibilities of breaking up drops with the addition of compatibilizers (Hu et al. 2000).

Viscosity plays an important role in polymer mixing. Several papers have been devoted to the impact of the compatibilizer upon the rheological behavior of blends. Some data in the literature (Iza et al. 1999) showed that the addition of a compatibilizer did not affect the steady shear properties but that it significantly modified behavior during the start-up of the shear flow. For example, in uncompatibilized blends with Newtonian components, an overshoot in the first normal stress difference is observed that is then followed by a minimum (Vinckier et al. 1997). The variations are explained by the modification of surface geometry: droplet deformation, fibril formation, and, finally, fibril breakup. Van Hemelrijck et al. (2002) have observed that, for blends with a small content of compatibilizer, variation of the first normal stress difference was similar to that of uncompatibilized blends, in that this occurred at lower time values. In contrast, with a greater compatibilizer content, there is no variation in the first normal stress difference, which means that no substantial deformation or breakup of the compatibilized droplets appears in the system. In this case, droplet behavior is similar to that of a suspension. Figure 1.5 presents the variation of the first normal stress difference function on the strain for polydimethylsiloxane (PDMS)/polyisoprene (PI) blends; the blue color representing the droplet images was obtained by small-angle light scattering (SALS) for both uncompatibilized and compatibilized samples (Van Hemelrijck 2005).

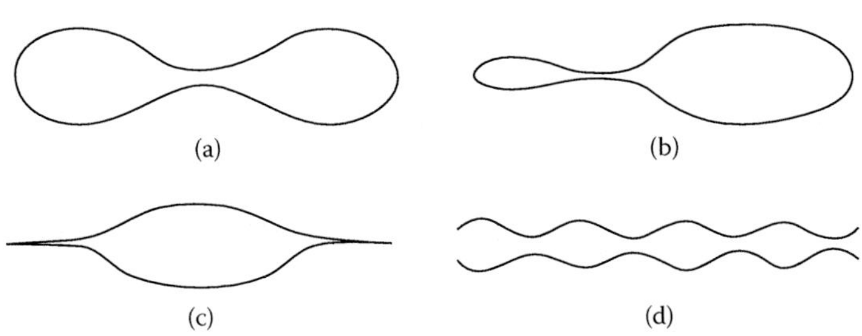

FIGURE 1.4 Representation of droplet breaking modes. (a) Simple breakup by waist thinning; (b) asymmetric breakup by waist thinning; (c) tip streaming; and (d) capillary breakup. (Adapted from Hu Thomas et al. *Phys. Fluid.*, 12, no 3, 484–489, 2000.)

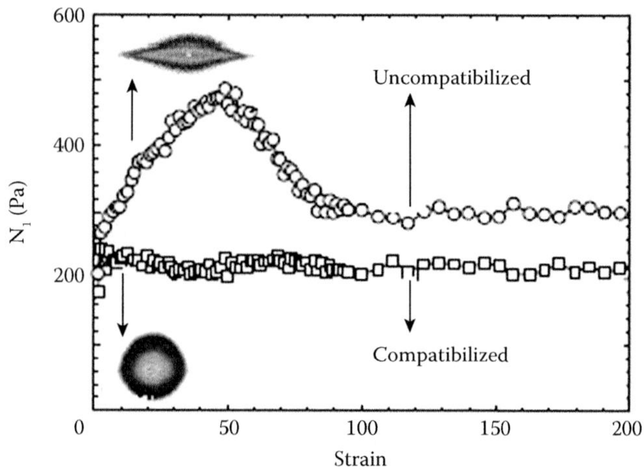

FIGURE 1.5 Variation of the first normal stress difference versus strain for (○) uncompatibilized blends of polydimethylsiloxane (PDMS) in polyisoprene (PI) and for (□) 10% compatibilized blends of polydimethylsiloxane (PDMS) with polyisoprene (PI). (Adapted from Van Hemelrijck Ellen. Effect of physical compatibilization on the morphology of immiscible polymer blends. PhD thesis, K U Leuven, 2005.)

The SALS images indicate anisotropy for uncompatibilized blends, although for 10% compatibilized blends, an isotropic SALS pattern can be observed, meaning that under such conditions, there is neither breakup nor deformity.

Regarding the steady-state viscosity of compatibilized blends, Moan et al. (2000) have found that the viscosity curves showed two plateau regions at low and at moderate frequencies, respectively. The occurrence of the plateau at low frequencies has been attributed to interfacial rheology, according to the Palierne pattern. For compatibilized blends, which contain a Newtonian component, Velankar has observed that the steady shear viscosity was higher than for corresponding uncompatibilized blends (Velankar et al. 2004). In the case of immiscible blends, abnormally low viscosity values were found (Utracki 1983; Utracki and Kamal 1982); these were sometimes even lower than the viscosity of pure components. The literature (Zhao and Macosko 2002) shows that, for incompatible systems with a high Flory–Huggins interaction parameter, viscosity displays high negative deviations. This can be explained by the fact that, at the interface, the polymer chains are less entangled and, consequently, their slip is facilitated. However, the slip process can be avoided by adding the premade block copolymer, which is able to remove the interfacial slip, allowing shifting of the compatibilizer at the interface.

In conclusion, obtaining a perfect blend—namely a combination of the components on the basis of structure dispersion—is required. The composition range is very narrow because low values of interfacial tension lead to different continuous structures. The addition of a compatibilizer determines interfacial tension reductions as well as coalescence suppression, preventing the formation of a continuous phase. Therefore, the order in which the phenomena will occur cannot be predicted due to the competitive effect of the compatibilizer.

1.4 MISCIBLE/IMMISCIBLE POLYMER BLEND RHEOLOGY

As has already been discussed, the most common method for analysis of multicomponent systems is rheology. The relationship between polymer structure and rheological behavior is of great interest from two viewpoints:

- The susceptibility of the rheological properties to various structural aspects, with rheology being a much easier method to apply than nuclear magnetic resonance.
- The appearance of the properties that govern the rheological behavior of polymers during the mixing process. Therefore, at the microstructural level, the system will present a complex rheology due to the modification that occurs during the flow process (Van Puyvelde and Moldenaers 2005).

For *homogeneous* polymer systems, special attention has focused on phase separation, where the rheological and thermodynamic phenomena interact (Takahashi et al. 1994). The molecular weight blends with different chain lengths for the same species are the most easily studied mixtures (Struglinski and Graessley 1985). For these systems, the Flory–Huggins parameters equal zero, regardless of chain length and temperature. The viscoelastic and diffusional behavior is characterized by the monomeric friction coefficient (ζ_0), the entanglement molecular weight (M_e), and the glass transition temperature (T_g). These parameters remain equal for all components and are not affected by mixing. Gell et al. (1997) have observed that, for the miscible blends of saturated hydrocarbon polymers, there is only one glass transition temperature and that differences appear exclusively in the apparent glass temperature T* of individual components. Unlike other authors (Zawada et al. 2013), Gell et al. have found no significant deviation from the time–temperature superposition, this principle being suitably applied so that, by representing the dynamic shear storage ($\log G'$) versus loss ($\log G''$), a single curve moduli was obtained (Han et al. 1995).

In the case of miscible polymer blends with weak interactions, such as polystyrene (PS)/poly(2,6-dimethylphenylene ether) (PPE) (Kim et al. 1992), poly(methyl methacrylate) (PMMA)/poly(styrene-coacrylonitrile) (PSAN) (Aoki and Tanaka 1999; Pathak et al. 1998), polystyrene/poly(vinyl methyl ether) (PVME) (Kapnistos et al. 1996; Kitade et al. 1994; Pathak et al. 1999), and poly isoprene (PI)/poly(vinyl ethylene) (PVE) (Alegria et al. 1995; Friedrich et al. 1996; Watanabe and Urakawa 2009; Zawada et al. 2013), a single glass transition temperature was also observed, confirming that, thermodynamically, the blends are miscible. However, a large single glass transition temperature, T_g, may appear, especially when the difference among the glass transition temperatures (ΔT_g) of each component is higher. But, when lower, this difference is lower than a critical value (Alegria et al. 1995; Friedrich et al. 1996; Pathak et al. 1998). Scientists have intensively analyzed the glass transition temperature of miscible polymer blends (with weak or very weak interactions) by varying the polymer solution concentration, which determines the distribution of segmental relaxation times (Kapnistos et al. 1996; Miller et al. 1990; Wendorff 1980) and is influenced, in turn, by blend composition and temperature. For such systems, the time–temperature superposition principle fails and is considered thermorheologically complex (Colby 1989; Kapnistos et al. 1996; Ngai and Plazek 1990). When determining polymer miscibility, it is assumed that hydrogen bondings play an important role (Coleman et al. 1991). In this context, the presence of weak or very weak interactions in systems such as poly (4-vinylphenol) (PVPh)/poly (vinyl methyl ether) (PVME), poly (4-vinylphenol) (PVPh)/poly (vinyl acetate) (PVAc), poly (4-vinylphenol) (PVPh)/poly (2-vinyl pyridine) (P2VP), or poly (4-vinylphenol) (PVPh/poly (4-vinyl pyridine) [P4VP]) was analyzed (Yang 2007).

By representing the viscoelastic moduli (G′ and G″) for different temperatures as a function of angular frequency, ω, it was found that, in the terminal region, the slope for G″ is close to 1 and for G′ is 2 for all temperatures at which the experiments were conducted (Figure 1.6).

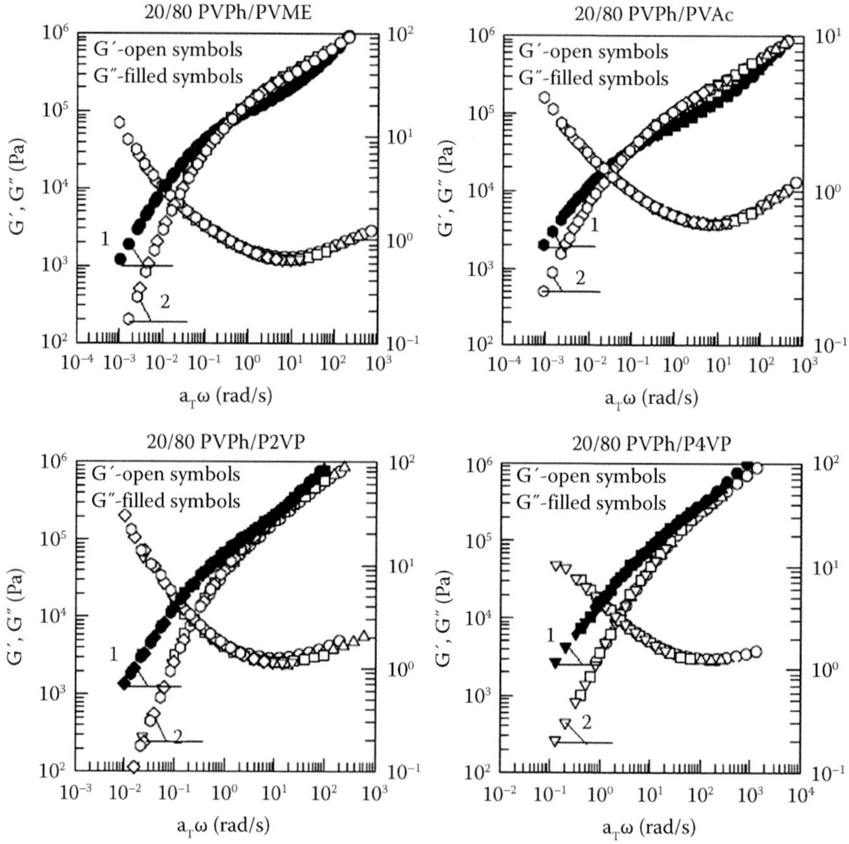

FIGURE 1.6 Logarithmic plot of G′ and G″ versus $a_T\omega$ for some PVPh blends at different temperatures: ○ = 74°C, Δ = 84°C, □ = 94°C, ∇ = 104°C, ◊ = 114°C, and ⬡ = 124°C. (Adapted from Yang Zhiyi. Rheology of miscible polymers blends with hydrogen bonding. Dissertation, University of Akron, Ohio 2007.)

Also, the logarithmic plot of G′ versus G″ confirms that the slope values approach 2 because they are independent of temperature (Figure 1.7). This suggests that the blends are homogeneous and also that the effects of self-association, crosslinking, or entanglement processes, if they exist, are very low, which follows the time–temperature superposition principle.

In addition, the *partially miscible* polymer blends are thermorheologically simple, so that all considerations available for homogeneous systems can be applied in their case as well (Utracki 1990). The composition of the two phases from the vicinity of phase separation is not the same as in the case of immiscible mixtures where it is supposed that, if droplets are present in their corresponding morphology, they are composed from one of the mixture components and that their total volume is equal to that of the blend composition. This theory is not applicable to partially

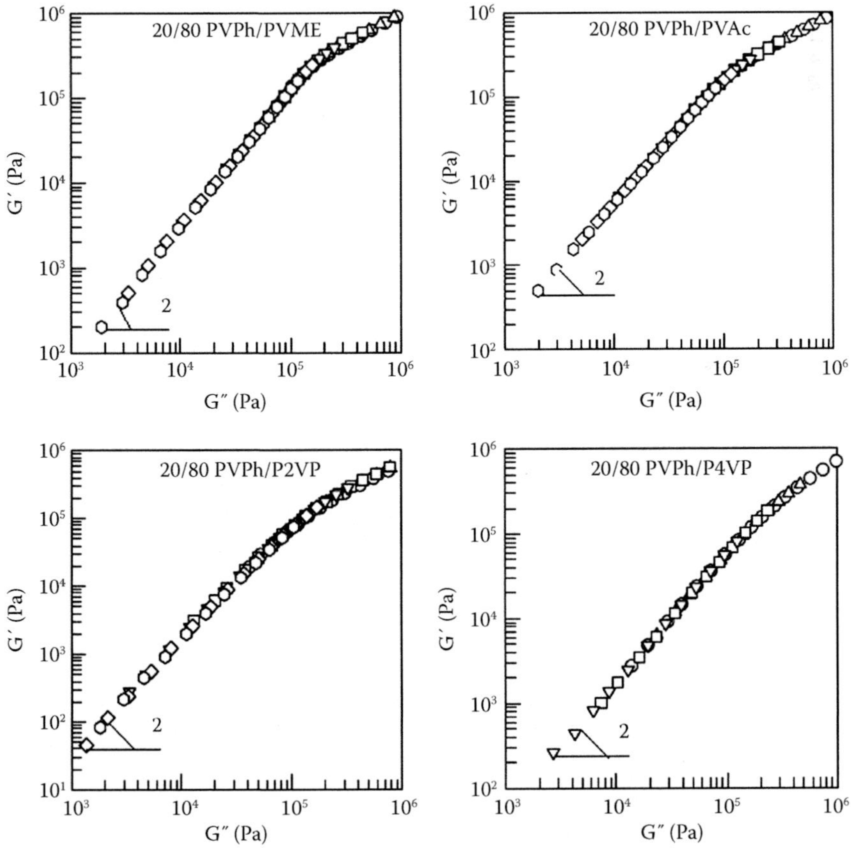

FIGURE 1.7 Logarithmic plot of G′ versus G″ for some PVPh blends at different temperatures: ○ = 74°C, Δ = 84°C, □ = 94°C, ∇ = 104°C, ◊ = 114°C, and ⬡ = 124°C. (Adapted from Yang Zhiyi. Rheology of miscible polymers blends with hydrogen bonding. Dissertation, University of Akron, Ohio 2007.)

miscible blends where droplet composition is obtained from the bimodal curves and their total volume fraction by applying the lever rule as follows (Vlassopoulos 1996):

$$\phi_{dr} = (\phi'_A - \phi_A)/(\phi'_A - \phi''_A) \tag{1.6}$$

where ϕ_{dr} = the volume fraction of droplets, ϕ_A = the blend composition, ϕ'_A = the matrix composition in the region of phase separation, and ϕ''_A = the droplet composition in the region of phase separation.

The viscoelastic properties of partially miscible mixtures can be analyzed by applying incompressible emulsion models (Palierne 1990). Various studies (Graebling et al. 1993; Lacroix et al. 1998; Vinckier and Launn 1999) have confirmed the success of this model, providing information on the complex, typically bimodal, terminal relaxation of two immiscible or partially miscible polymers subjected to phase separation and capable of attaining dynamic equilibrium. Assuming that the droplet size is uniform, the complex modulus $G^*(\omega)$ can be written as

$$G^*(\omega) = G_M \cdot \frac{1 + 3\phi_{dr} H_i(\omega)}{1 - 2\phi_{dr} H_i(\omega)} \tag{1.7}$$

where

$$H_i(\omega) = \frac{4\left(\frac{\alpha}{R}\right)\left[2G^*_m(\omega) + 5G^*_i(\omega)\right] + \left[G^*_i(\omega) - G^*_m(\omega)\right]\left[16G^*_m(\omega) + 19G^*_i(\omega)\right]}{40\left(\frac{\alpha}{R}\right)\left[G^*_m(\omega) + G^*_i(\omega)\right] + \left[2G^*_i(\omega) + 3G^*_m(\omega)\right]\left[16G^*_m(\omega) + 19G^*_i(\omega)\right]} \tag{1.8}$$

where m and i refer to the matrix and disperse phase (inclusions), α is the interfacial tension between the matrix and the dispersed phase, and R is the radius of the droplet.

The literature has shown that when the blends are nearly monodisperse and the increase in droplet size is insignificant with respect to time and temperature, utilization of the mean radius is recommended (Lacroix et al. 1998). Knowledge and modeling of viscoelastic properties are necessary for understanding blend morphology and the compatibility degree of polymers.

In this respect, a theory that takes into account the deformation of one droplet (Doi and Ohta 1991) can be applied to describe the shear and normal stress transients. According to this model, blend morphology is characterized by a *scalar* (referring to a specific interfacial area) and a *tensor* (characterizing interface anisotropy). These parameters may be expressed in two equations—one describing the stresses of the interfacial structures and the other for the evaluation of the scalar and interface tensor. For immiscible blends with Newtonian or weakly viscoelastic fluids and an increase in shear, the droplets deform into fibrils while maintaining their initial diameter, d. In comparison, in a highly elastic matrix where droplet shape is approximated by a cylinder, after their breakup the stresses are expressed by the following equations (Iza et al. 1999; Levitt et al. 1996):

$$(\gamma,\dot{\gamma},d)=\frac{\left(\sum \phi_i \eta_i\right)\dot{\gamma}+2\alpha\phi_{dr}\left(1+\frac{\gamma^2}{2}+\frac{\gamma}{2}\sqrt{\gamma^{2+4}}\right)^{1/2}}{d\sqrt{\gamma^2+4}} \tag{1.9}$$

$$N_1(\gamma,\dot{\gamma},d)=\left(\sum \phi_i N_{1,i}\right)+\frac{2\gamma\alpha\phi_{dr}\left(1+\frac{\gamma^2}{2}+\frac{\gamma}{2}\sqrt{\gamma^2+4}\right)}{d\sqrt{\gamma^2+4}} \tag{1.10}$$

where $\dot{\gamma}$ is the shear rate, γ is the total strain, and ϕ_{dr} is the volume fraction of inclusions.

In these conditions, it has been reported that at low shear rates, a globular structure appears, while at a high shear rate (or elongational flow), the stratified or fibrillar morphology is predominant (Grizzuti et al. 2000). In addition, the viscoelastic properties of the concentrated binary polymer blends can be modeled by the Frankel and Acrivos (1970) approximation. The literature mentions a few such systems, including polydimethylsiloxane/polyethylmethylsiloxane (PDMS/PEMS) with a high content of polyethylmethylsiloxane (Chopra 2002). These components are considered dynamically symmetric and were selected because of their extremely low glass transition temperature. Analysis of these systems showed that polymer miscibility is highly sensitive to the molecular weight, M_W, of components. Moreover, the activation energy of blends is influenced by blend composition and level of shear. In the phase-separated regime, the optical microscopy images show droplet–matrix morphology, droplet size, and interfacial tension increasing with greater PDMS content in PDMS/PEMS blends. The absence of the interaction among blend components results from optical and rheological measurements as well as from the Flory–Huggins theory. Poly(styrene-*co*-maleic anhydride)/poly(methyl methacrylate) (SMA/PMMA) blends also were investigated because they are industrially relevant and exhibit high glass transition temperature and low dynamic asymmetry, that is, differences among the glass transition of components (Chopra et al. 2000). It was also noted that the viscoelastic properties are susceptible to phase behavior, in which case, the time–temperature superposition fails, marking the beginning of phase separation. The activation energy for these two regions is lower than for single-phase systems, although the additional relaxation mechanism (not present in the homogeneous region) can be a marker of phase separation. The nonlinear viscoelastic properties' dependence on the shear rate and hysteresis observed in the thixotropic loops also confirms the presence of inclusions in the region of phase separation.

As a conclusion to the aforementioned observations, it can be asserted that the rheological features of complex *partially miscible polymer blends* are only partially elucidated in terms of their thermodynamics, based on the behavior of immiscible blends and if their phase state is well known.

Recently, similar studies also have been conducted for *immiscible polymer blends*, in view of their technological application. For these systems, elasticity is pronounced and the relaxation time is great, and the structure obtained during spinodal decomposition has been discussed often in literature (Laeuger et al. 1995). Although phase separation is responsible for the thermorheological complexity of biphasic polymer systems, this behavior has not been fully elucidated. Establishment of a connection between nonlinear viscoelastic behavior and phase separation represents a challenge for researchers, the main concern being the study of shear viscosity and normal stress impact on phase separation.

1.5 IMPACT OF POLYMER BLEND RHEOLOGY ON THE DEVELOPMENT OF COMMERCIAL PRODUCTS

Polymer blending offers the advantage of developing new products with commercializable properties, thereby reducing the time and costs necessary for synthesizing new monomers/polymers with the same properties. Moreover, blending products with special properties can be obtained through a less complex manner than by the direct synthesis of monomers/polymers.

A series of analytical and characterization methods has been developed for a better understanding of polymer blend miscibility and phase separation behavior. Study of polymer blend miscibility is of particular interest because of its thermodynamic implications, while phase separation behavior greatly influences the mixing process of polymers (Bousmina et al. 2002). Mechanical compatibility, representing an average property of blend constituents among the final mechanical properties of the mixture, is attained through mixing. Rheology, rupture strength, lack of transparency, flammability, film characteristics, and adherence and performance of coatings are properties that can be obtained by specific combinations of immiscible polymers. As already mentioned, to obtain mechanical compatibility in phase-separated systems, several techniques have been used, such as adding a graft or block copolymer, interpenetrating a polymer network, reactivating extrusion, adding polymeric interfacial agents, or adding a compatibilizer. For example, the reactive extrusion technology initially begins with super tough polyamides, where maleic anhydride grafted onto an ethylene propylene rubber makes possible the graft formation with a terminal $-NH_2$ groups of polyamide which will stabilize the interface between the ungrafted constituents. Recently, several technological and commercial improvements have taken place in polymer blend engineering, with many of the commercial polymers being produced by a combination of those already in existence. In this respect, a special place is held by polyester and polycarbonate, namely polybutylene terephthalate/polycarbonate, cyclohexane dimenthanol-based polyesters/polycarbonate, and polyethylene terephthalate/polycarbonate. The use of polymeric materials for auto panels resulted in a large number of potential candidates based on polymer blends (Utracki 2002). Ethylene-carbon monoxide alternating copolymers, initially introduced by Shell, are a novelty in this field. The presence of small amounts of propylene in commercial polymers supposedly with impact a decrease in melting point, thus allowing their easier processing.

Other fields studying the incorporation of polymer blends include molecular composites, blends based on biodegradable or electrically conductive polymers, and theoretical evaluations implying phase separation behavior. In addition, there are computation model predictions of polymer miscibility. Due to the biphasic/multiphasic nature of polymer blends, a study of their rheological behavior should begin with phenomenological considerations by applying a molecular approach. Studying the phase morphology of polymer blends presents a challenge for researchers involved in the design of polymeric materials. Blend morphology is the main parameter that dictates the final properties of immiscible polymer blends, and their analysis represents another step in the development of new performant materials. Therefore, it is obvious that the rheology of polymer blends depends on a full understanding of the structure–properties relationship of materials, requiring a multiscale model to predict material properties. The actual research is limited at the time and length scale, and thus must be completed with the development of functions in dynamic environments across a wide range of temperatures and frequencies. These parameters tend to accelerate blend degradation and, consequently, to alter their functional response through electromechanical couplings. New technology, even without detailed scientific understanding, plays an important role in the field of polymer blends especially with reference to nanocomposites with exceptional properties, conductive plastics for electronics, self-assembly processes for the creation of special polymeric structures, and fabrication of biomaterials and polymers for tissue engineering.

Thus, progress in the field of smart polymers will depend not only on researchers' focus on developing rheological systems with enhanced adaptive capabilities but also on the establishment of theoretical models to simulate such rheological behaviors.

1.6 CONCLUSIONS

The chapter summarizes the current knowledge concerning polymer blends, with special emphasis on thermodynamic and rheological aspects. Present developments focus on rheological properties, structure, and morphology of polymer blends and play an important role in the rapid advances in polymer blend engineering. Moreover, future research is focused on developing mathematical models for simulating the rheological properties of polymer blends to advance technological progress. The theoretical and experimental studies presented in this chapter are useful for students and scientists, from academic as well as industrial backgrounds, in the study of producing new polymeric materials through blending of those already in existence and also for expanding their areas of application.

REFERENCES

Alegria Ángel, Elizetxea Cristina, Cendoya Ione, and Colmenero Juan De Dios. Anomalous dynamical homogeneity of the dielectric a-relaxation in miscible polymer blends of poly(epichlorohydrin) and poly(vinyl methyl ether). *Macromolecules.* 28 no. 26 (1995): 8819–8823.

Aoki Yuji, and Tanaka Takeshi. Viscoelastic properties of miscible poly(methyl methacrylate)/poly(styrene-co-acrylonitrile) blends in the molten state. *Macromolecules.* 32 no. 25 (1999): 8560–8565.

Asthana Himanshu, and Jayaraman Krishnamurthy. Rheology of reactively compatibilized polymer blends with varying extent of interfacial reaction. *Macromolecules*. 32 no. 10 (1999): 3412–3419.

Bousmina Mustapha, Lavoie Paul-André, and Riedl Bernard. Phase segregation in SAN/PMMA blends probed by rheology, microscopy, and inverse gas chromatography techniques. *Macromolecules*. 35 no. 16 (2002): 6274–6283.

Cao Yewen, Zhang Jing, Feng Jiachun, and Wu Peiyi. Compatibilization of immiscible polymer blends using graphene oxide sheets. *ACS Nano*. 5 no. 7 (2011): 5920–5927.

Chopra Divya. Thermodynamics and rheology of partially miscible polymer blends. A dissertation presented to the graduate faculty of the university of British Columbia, April 2002.

Chopra Divya, Vlassopoulos Dimitris, and Hatzikiriakos Savvas. Nonlinear rheological response of phase separating polymer blends: Poly(styrene-co-maleic anhydride)/poly(methyl methacrylate). *J. Rheol.* 44 no. 1 (2000): 27–45.

Colby Ralph. Breakdown of time-temperature superposition in miscible polymer blends. *Polymer.* 30 no. 7 (1989): 1275–1278.

Coleman Michael, Graf John, and Painter Paul. *Specific interactions and the miscibility of polymer blends*. Lancaster, PA: Technomic Publishing, 1991.

Doi Masao, and Ohta Takao. Dynamics and rheology of complex interfaces. *J. Chem Phys.* 95 no. 2 (1991): 1242–1248.

Eggleton Charles Dionisio, and Stebe Kathleen. An adsorption-desorption controlled surfactant on a deforming droplet. *J. Colloid. Interface Sci.* 208 no. 1 (1998): 68–80.

Elias Liju, Fenouillot Francoise, Jean-Charles Majeste, and Philippe Cassagnau. Morphology and rheology of polymer blends filled with silica nano particles. *34th Journees d'Etude des Polymers*, Lorient, September 2006, pp. 24–29.

Fortelny Ivan. An analysis of the origin of coalescence suppression in compatibilized polymer blends. *Eur. Polym. J.* 40 no. 9 (2004): 2161–2166.

Frankel Neil, and Acrivos Andreas. The constitutive equation for a dilute emulsion. *J. Fluid. Mech.* 44 (1970): 65–78.

Friedrich Christian, Schwarzwaelder Kerstin, and Riemann Randolf. Rheological and thermodynamic study of the miscible blend polystyrene/poly(cyclohexyl methacrylate). *Polymer.* 37 no. 12 (1996): 2499–2507.

Gell B. Carol, Krishnamoorti Ramanan, Kim Eugene, Graessley W. William, and Fetters J. Lewis. Viscoelasticity and diffusion in miscible blends of saturated hydrocarbon polymers. *Rheol. Acta.* 36 (1997): 217–228.

Graebling Didier, Muller René, and Palierne Jean-Francois. Linear viscoelastic behavior of some incompatibile polymner blends in the melt. Interpretation of data with a model of emulsion of viscoelastic liquids. *Macromolecules*. 26 no. 2 (1993): 320–329.

Grizzuti Nino, Buonocore Giovanna Giuliana, and Iorio Giulia. Viscous behavior and mixing rules for an immiscible model polymer blends. *J. Rheol.* 44 no. 1 (2000): 149–164.

Han Han, Choi-Feng Chin, Li De-Jie, and Han Chang Dae. Effect of flow geometry on the rheology of dispersed two-phase blends of polystyrene and poly(methyl methacrylate). *Polymer.* 36 no. 12 (1995): 2451–2462.

Hu Thomas, Pine David, and Leal Gary. Drop deformation, breakup, and coalescence with compatibilizer. *Phys. Fluid.* 12 no 3 (2000): 484–489.

Huyskens Pierre, Luck Werner, and Zeegers-Huyskens Therese. *Intermolecular forces*. Berlin: Springer-Verlag, 1991.

Israelachvili Nissim. *Jacob Intermolecular and surface forces*. 2nd ed. New York: Academic Press, 1991.

Iza Mustapha, Bousmina Mosto, and Jérôme Robert. Rheology of compatibilized immiscible viscoelastic polymer blends. *Rheol. Acta.* 40 no. 1 (1999): 10–22.

Jacobs Ursula, Fahrlander Michael, Winterhalter Judy, and Friedrich Carsten. Analysis of Palierne's emulsion in the case of viscoelastic interfacial properties. *J. Rheol.* 43 no. 6 (1999): 1495–1509.

Janssen J. M. Jo, Boon Ai-Ching, and Agterof Wim. Influence of dynamic interfacial properties on droplet breakup in simple shear flow. *AIChE. J.* 43 no. 6 (1997): 1436–1447.

Kapnistos Michael, Hinrichs Axel, Vlassopoulos Dimitris, Anastasiadis Spiros, Stammer Andreas, and Wolf Bernhard Anton. Rheology of a lower critical solution temperature binary polymer lend in the homogeneous, phase-separated, and transitional regimes. *Macromolecules.* 29 no. 22 (1996): 7155–7163.

Kim Jin Kon, Han Chang Dae, and Lee Young Ju. Rheological behavior of miscible blends of polystyrene and poly(2,6-dimethyl-1,4-phenylene ether). *Polym. J.* 24 (1992): 205–213.

Kitade Shinichi, Takahashi Yoshiaki, and Noda Ichiro. Viscoelastic properties of homogeneous polymer blends in the terminal region. *Macromolecules.* 27 no. 25 (1994): 7397–7401.

Krammer Hans Werner. Phase behavior of polymer blends. *J. Macromol. Sci. Chem. A.* 27 no 13–14 (1990): 1713–1732.

Krammer Hans Werner, Inoue Takashi, and Ougizawa Toshiaki. Upper and lower critical solution temperature behavior in polymer blends and its thermodynamic interpretation. *Polymer.* 30 no. 5 (1989): 888–892.

Lacroix Christophe, Grmela Miroslav, and Carreau Pierre. Relationship between rheology and morphology for immiscible molten blends of polypropylene and ethylene copolymers under shear flow. *J. Rheol.* 42 no. 1 (1998): 41–62.

Laeuger Joerg, Lay Rainer, Maas Swen, and Gronski Wolfram. Structure development of a polybutadiene/polyisoprene blend during spinodal decomposition. Comparison between light scattering and optical microscopy. *Macromolecules.* 28 no. 20 (1995): 7010–7015.

Levitt Leon, Macosko Christopher, and Pearson Scott. Influence of normal stress difference on polymer drop deformation. *Polym. Eng. Sci.* 36 no. 12 (1996): 1647–1655.

Li Xiaofan, and Pozrikidis Constantine. The effect of surfactants on drop deformation and on the rheology of dilute emulsions in stokes flow. *J. Fluid. Mech.* 341 no. 25 (1997): 165–194.

Miller Joel, McGrath Kenneth, Roland Mike, Trask Craig, and Garroway Allen. Nuclear magnetic resonance study of polyisoprene/poly(vinylethylene) miscible blends. *Macromolecules.* 23 no. 21 (1990): 4543–4547.

Milliken William, Stone Howard, and Leal Gary. The effect of surfactant on the transient motion of newtonian drops. *Phys. Fluid.* 5 no. 1 (1993): 69–79.

Moan Michel, Huitric Jacques, Mederic Pascal, and Jarrin Jose Marin. Rheological properties and reactive compatibilization of immiscible polymer blends. *J. Rheol.* 44 no. 6 (2000): 1227–1245.

Ngai Kialing, and Plazek Donald. Breakdown of time-temperature superposition in miscible polymer blends and the coupling scheme. *Macromolecules.* 23 no. 19 (1990): 4282–4287.

Ougizawa Toshiaki, and Inoue Takashi. UCST, LCST behavior in polymer blends and its thermodynamic interpretation. *Polym. J.* 18 no. 7 (1986): 521–527.

Palierne Jean-Francois. Linear rheology of viscoelastic emulsions with interfacial tension. *Rheol. Acta.* 29 no. 3 (1990): 204–214.

Palierne Jean-Francois. Erratum. *Rheol. Acta.* 30 (1991): 497–497.

Pathak Jai, Colby Ralph, Floudas George, and Jerome Robert. Dynamics in miscible blends of polystyrene and poly(vinyl methyl ether). *Macromolecules.* 32 no. 8 (1999): 2553–2561.

Pathak Jai, Colby Ralph, Klamath Sudesh, Kumar Sanat, and Staler Reimund. Rheology of miscible blends: SAN and PMMA. *Macromolecules.* 31 no. 25 (1998): 8988–8997.

Paul Donald. *Polymer blends*. Vol. 2. New York: Academic Press, 1978.

Pawar Yashodhara, and Stebe Kathleen. Marangoni effects on drop deformation in an extensional flow: The role of surfactant physical chemistry, I. Insoluble surfactants. *Phys. Fluids*. 8 no. 7 (1996): 1738–1751.

Peters Gerrit, Zdravkov Alexander, and Meijer Han. Transient interfacial tension and dilatational rheology of diffuse polymer-polymer interfaces. *J. Chem. Phys.* 122 no. 10 (2005): 104901.

Ramic Anthony, Stehlin Julia, and Hudson Steven, Jamieson Alexander, Manas-Zloczower Ica. Influence of block copolymer on droplet breakup and coalescence in model immiscible polymer blends. *Macromolecules*. 33 no. 2 (2000): 371–374.

Riemann Randolf, Cantow Hans-Joachim, and Friedrich Carsten. Rheological investigation of form relaxation and interface relaxation processes in polymer blends. *Polym. Bull.* 36 no. 5 (1996): 637–643.

Riemann Randolf, Cantow Hans-Joachim, and Friedrich Carsten. Interpretation of a new interfacegoverned relaxation process in compatibilized polymer blends. *Macromolecules*. 30 no. 18 (1997): 5476–5484.

Sadiku-Agboola Oluranti, Sadiku Emmauel Rotimi, Adegbola Adesola Taoreed, and Olusesan Frank Biotidara. Rheological properties of polymers: Structure and morphology of molten polymer blends. *Mater. Sci. Appl.* 2 (2011): 30–41.

Schneider Andre, Schuld Norbert, Bercea Maria, and Wolf Bernhard Anton. On what terms and why the thermodynamic properties of polymer solutions depend on chain length up to the melt. *J. Polym. Sci. Part B. Polym. Phys.* 42 no. 9 (2004): 1601–1609.

Schuld Norbert, and Wolf Bernhard Anton. *Polymer-solvent interaction parameters. Polymer handbook*. 4th ed. Edited by J. Brandrup, E. H. Immergut. New York: Wiley, 1999.

Schuld Norbert, and Wolf Bernhard Anton. Solvent quality as reflected in concentration and temperature-dependent Flory–Huggins interaction parameters. *J. Polym. Sci. Part B: Polym. Phys.* 39 no. 6 (2001): 651–662.

Schwahn Dietmar. *Polymer blends and phase separation. Neutron and synchrotron radiation for condensed matter studies*. Vol. III. New York: Springer Verlag, 1994.

Stone Howard, and Leal Gary. The effects of surfactants on drop deformation and breakup. *J. Fluid. Mech.* 220 (1990): 161–186.

Struglinski Mark, and Graessley William. Effects of polydispersity on the linear viscoelastic properties of entangled polymers. 1. Experimental observations for binary mixtures of linear polybutadiene. *Macromolecules*. 18 no. 12 (1985): 2630–2643.

Sundararaj Uttandaraman, and Macosko Christopher. Drop breakup and coalescence in polymer blends—The effects of concentration and compatibilization. *Macromolecules*. 28 no. 8 (1995): 2647–2657.

Takahashi Yoshiaki, Suzuki Hirokazu, Nakagawa Yoshiki, and Noda Ichiro. Effects of shear flow on viscoelastic properties of polystyrene/poly(vinyl methyl ether) blends near the phase separation temperature. *Macromolecules*. 27 no. 22 (1994): 6476–6481.

Utracki Leszek. Melt flow of polymer blends. *Polym. Eng. Sci.* 23 no. 11 (1983): 602–609.

Utracki Leszek. *Polymer alloys and blends: Thermodynamics and rheology*. New York: Hanser, 1990.

Utracki Leszek. Polymer blends. *RAPRA Rev. Rep.* 11 no. 3 (2000): 1–170.

Utracki Leszek. *Polymer blends handbook*. Dordrecht: Kluwer, Academic, 2002.

Utracki Leszek, and Kamal Musa. Melt rheology of polymer blends. *Polym. Eng. Sci.* 22 no. 2 (1982): 96–114.

Van Hemelrijck Ellen. Effect of physical compatibilization on the morphology of immiscible polymer blends. PhD thesis, K U Leuven, 2005.

Van Hemelrijck Ellen, Van Puyvelde Peter, and Moldenaers Paula. Rheology and morphology of compatibilized and uncompatibilized polymer blends. *Proceedings of the Meeting of the European Polymer Processing Society*, Guimaraes, Portugal, 2002.

Van Hemelrijck Ellen, Van Puyvelde Peter, Macosko Christopher, and Moldenaers Paula. The effect of block copolymer architecture on the coalescence and interfacial elasticity in compatibilized polymer blends. *J. Rheol.* 49 no. 3 (2005): 783–798.

Van Hemelrijck Ellen, Van Puyvelde Peter, Velankar Sachin, Macosko Christopher, and Moldenaers Paula. Interfacial elasticity and coalescence suppression in compatibilized polymer blends. *J. Rheol.* 48 no. 1 (2004): 143–158.

Van Puyvelde Peter, and Moldenaers Paula. Rheology and morphology development in immiscible polymer blends. *Rheol. Rev.* (2005): 101–145.

Van Puyvelde Peter, Velankar S. Sachin, Moldenaers Paula, and Mewis Jan. Effect of Marangoni stresses on the deformation and coalescence in compatibilized immiscible polymer blends. *Polym. Eng. Sci.* 42 (2002): 1956–1964.

Velankar Sachin, Van Puyvelde Peter, Mewis Jan, and Moldenaers Paula. Steady-shear rheological properties of model compatibilized blends. *J. Rheol.* 48 no. 4 (2004): 725–744.

Vinckier Inge, and Launn Hans Martin. Manifestation of phase separation processes in oscillatory shear: Droplets matrix systems versus co-continuous morphologies. *Rheol. Acta.* 38 no. 4 (1999): 274–286.

Vinckier Inge, Moldenaers Paula, and Mewis Jan. Transient rheologcial response and morphology evolution of immiscible polymer blends. *J. Rheol.* 41 no. 3 (1997): 705–718.

Vlassopoulos Dimitris. Rheology of LCST polymer blends: Poly(styrene-co-maleic anhydrite)/poly(methyl methacrylate). *Rheol. Acta.* 35 no. 6 (1996): 556–566.

Walstra Pieter. Principles of emulsion formation. *Chem. Eng. Sci.* 48 no. 2 (1993): 333–349.

Watanabe Hiroshi, and Urakawa Osamu. Component dynamics in miscible polymer blends: A review of recent findings. Korea-Aust. *Rheol. J.* 21 no. 4 (2009): 235–244.

Wendorff Joachim. Concentration fluctuations in poly(vinylidene fluoride)—Poly(methyl methacrylate) mixtures. *J. Polym. Sci. Polym. Lett. Ed.* 18 no. 6 (1980): 439–445.

Xavier Xavier, and Bose Suryasarathi. Multiwalled-carbon-nanotube-induced miscibility in near-critical PS/PVME blends: Assessment through concentration fluctuations and segmental relaxation. *J. Phys. Chem. B.* 117 no. 28 (2013): 8633–8646.

Yang Zhiyi. Rheology of miscible polymers blends with hydrogen bonding. A dissertation presented to the graduate faculty of the university of Akron, August 2007.

Zawada Jeffrey, Fuller Gerry, Colby Ralph, Fetters Lewis, and Roovers Jacques. Component Dynamics in Miscible Blends of 1,4-Polyisoprene and 1,2-Polybutadiene. *Macromolecules.* 27 no. 23 (2013): 6861–6870.

Zhang Qin, Yang Hong, and Fu Qiang. Kinetics-controlled compatibilization of immiscible polypropylene/polystyrene blends using nano-SiO_2 particles. *Polymer.* 45 no. 6 (2004): 1913–1922.

Zhang Wengong, Lin Minyue, Winesett Werner Ade, et al. The use of functionalized nanoparticles as non-specific compatibilizers for polymer blends. *Polym. Adv. Technol.* 22 no. 1 (2011): 65–71.

Zhao Rui, and Macosko Christopher. Slip at polymer-polymer interfaces: Rheological measurements on coextruded multilayers. *J. Rheol.* 46 no. 1 (2002): 145–167.

2 Complex Macromolecular Architectures of Copolymer Solution Assemblies for Regenerative Medicine and Drug Delivery

Luminita Ioana Buruiana

CONTENTS

2.1 INTRODUCTION

Self-assembly provides a simple approach for obtaining well-defined nanostructures from designer macromolecules in bulk, thin film, and solution environments. The variety of available macromolecules gives rise to different applications that rely on the ability of amphiphilic block copolymers to self-assemble in solution and to adsorb at surfaces or interfaces. Depending on the morphologies obtained (size, shape, periodicity), these self-assembled systems—that can be included in multiphase polymer systems used in various advanced technologies—have already been shown to be suitable for a number of applications in nanotechnology, reusable elastomeric materials, electronics, drug delivery, paints, cosmetics, lubricants, and in detergents (Disher and Eisenberg 2002; Yang 2002; Lazzari and Lopez-Quintela 2003; Kelley et al. 2013).

Amphiphilic macromolecules, like block copolymers, are promising for applications as molecular design, offering chemical versatility and exquisite control over the size and shape of solution assemblies. These macromolecules form various nanostructures in aqueous solutions, including spherical/cylindrical micelles and vesicle bilayers (Smart et al. 2008; Blanazs et al. 2009; Mai and Eisenberg 2012). Among various aggregation processes, the most studied ones belong to the self-assembly of block or graft copolymers. As a consequence of their molecular structure, block copolymers in the solid state form a variety of superlattices in sizes of only a few nanometers. Copolymers that self-organize in a selective solvent (for one of the two blocks) have found industrial applications, but one point of concern is the relatively poor stability of the aggregates formed (Wooley 2000).

In an attempt to find many more potential applications, micellar systems have been designed adaptable to their environment and able to respond in a controlled manner to external stimuli. The design of synthetic structures that approach proteins in their complexity, functionality, and performance allows their use in different domains of medicine. In order to obtain a sufficiently strong attraction in solution and at the surface, micelles form surface and bulk micelles, some of the latter being bonded to the surface by connectivity of the chains and through the interacting hydrophobic monomers (Thurmond et al. 1999).

Surface micelles are formed below the bulk critical concentration, and they become more organized as the surface concentration increases. For sufficiently long copolymers, a secondary layer of micelles develops that is adjacent and connected to the layer of surface micelles. The features of these systems can be used for its patterning (formation of surface or interfacial organized layer of micelles) and for its rheological properties (elasticity of the surface network connected to a secondary dense layer of bulk micelles) at the same time (Hayward and Pochan 2010; Taco et al. 2010; Hugouvieux et al. 2011).

2.2 PREPARATION OF MICELLES/SELF-ASSEMBLIES: TYPES AND CLASSIFICATION

The literature mentions many procedures that can be followed to prepare micelles. One approach consists of the introduction of the copolymer in a nonselective solvent that represents the common solvent for both polymer blocks. The desired micellar structure is obtained by the selection of certain solution parameters such as temperature of a cosolvent or the subsequent addition of a selective solvent into a previously prepared solution from a common solvent; after that, the common solvent is removed from the solution by dialysis. Another approach in the preparation of micelles involves the addition of the dry copolymer powder into a selective solvent. The preference for one method or another depends on the system investigated (Rodriguez-Hernandez et al. 2005).

One important class of block copolymers are amphiphilic block copolymers that have affinities for two different environments. These two blocks interact very differently with the environment due to their chemical nature, and they behave distinctively in solution. The aforementioned differences can induce microphase separation of an amphiphilic block both in aqueous media and in organic solvents. Two basic processes can be distinguished for block copolymers in solvent media, namely micellization and gelation. Micellization occurs when the block copolymer is dissolved in a large amount of a selective solvent for one of the blocks. In this case, the polymer chains tend to organize themselves in a diversity of structures, from micelles to vesicles to cylinders. The soluble block will be oriented toward the continuous solvent medium and will become the micelle "shell," whereas the insoluble part will be coated by the solvent in the "core" of the structure. Conversely, gelation takes place from the semidilute to high concentration regime of block copolymer solutions that are obtained from an arrangement of ordered micelles.

One can classify micelles into different types with morphologies varying from spherical to vesicular or other structures, such as bilayers, cylinders, and inverse micelles (Figure 2.1).

Taking into consideration the latest research interest in biological applications and green chemistry, the amphiphilic block copolymers in aqueous solution have been studied more often.

Spherical micelles ("core-shell") are prepared by self-assembly of diblock copolymers, directed by an entropically driven association mechanism. In recent years, aqueous micellar systems have attracted special attention, which has been motivated especially by their applications in biomedicine (as stabilizing agents in dermatological creams or lotions) as emulsifiers, foam stabilizers, or detergents. Worth mentioning in this category is the micellar system based on poly(ethylene oxide)-*b*-poly(b-benzylaspartate) (PEO-PBLA) block copolymers with poly(ethylene oxide) (PEO) as corona. The interest in the use of micelles as potential drug carriers demonstrates that micelles can act as ideal vehicles for delivery due to their diminished side effects; the drug is protected from degradation and is better targeted at the deposition site. Advantages of this system are the small micellar size and long-term stability that is required for a prolonged circulation time (Disher et al. 2000). Another system is formed by α-acetoxypoly(ethylene oxide)-*b*-poly(D,L-lactide) block copolymers. The α-acetal group in PEO was transformed into an aldehyde group and conjugated with a peptide segment of phenylalanine (Phe) or tyrosil-glutamic acid (Tyr-Glu). Within this system, the influence of the peptidic receptor on drug delivery has been analyzed as well as the effect of the surface charge on pharmacokinetic behavior.

Spherical micelles were prepared by solubilization/encapsulation from rod-coil block copolymers of poly(phenylquinoline)-*b*-poly(styrene) (PPQ-PS) using binary solvents selective for the rod-type block copolymers. The resulting assemblies could be used to encapsulate large molecules, such as C_{60} and C_{70} fullerenes (Chen and Jenekhe 1999).

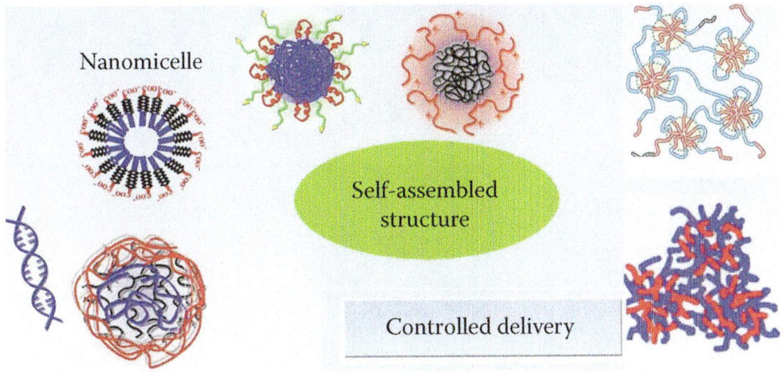

FIGURE 2.1 Examples of different self-assembled structures. (Adapted from Tian Jia, et al. *J. Mater. Chem.*, 21, no. 42, 16928–16934, 2011.)

The encapsulation process has consequences for the micellization and self-assembly of the polymers; with this method, regular hollow spheres that increase in size and aggregation number have been obtained. Materials prepared in this way present advantages for use in electronic, optoelectronic, or photonic applications.

On the contrary, amphiphilic block copolymers can form various vesicular architectures in solution—a nanometer-sized structure with a double-layer outer membrane that encloses an inner volume, known as polymer-based liposomes (polymersomes). These include uniform common vesicles, large polydisperse vesicles, entrapped vesicles, or hollow concentric vesicles. Examples in this context are vesicles based on poly(ethylene oxide)-*b*-poly(ethylene) (PEO-*b*-PE) and poly(ethylene oxide)-*b*-polybutadiene (PEO-*b*-PB) block copolymers, with various block compositions (Lee et al. 2001).

Self-assembly of rod-coil-rod block copolymers based on novel poly(quinoline-styrene-quinoline) triblocks have been prepared as well. The presence of a poly(phenyl quinoline) block, which is a π-conjugated system, makes these materials suitable for electronic, optoelectronic, and photonic applications (Chen and Jenekhe 2000).

A unique pH-dependent phase behavior from a copolymer micellar solution to a hydrogel with micelles ordered in a hexagonal phase was presented in the literature (Determan et al. 2006). Thus, between 3.0 and 7.4 pH, one could observe the presence of charged spherical micelles. Once the pH increased above 7.4, the size of the micelles was enhanced due to the increasing hydrophobicity of the copolymer blocks above their pKa of 7.6. This size increase resulted in a transition to a cylindrical micelle morphology in the 8.1–10.5 pH range; at pH greater than 11, the copolymer solution undergoes macroscopic phase separation. A representation of the different shapes of micelles is presented in Figure 2.2. Further self-assembly of

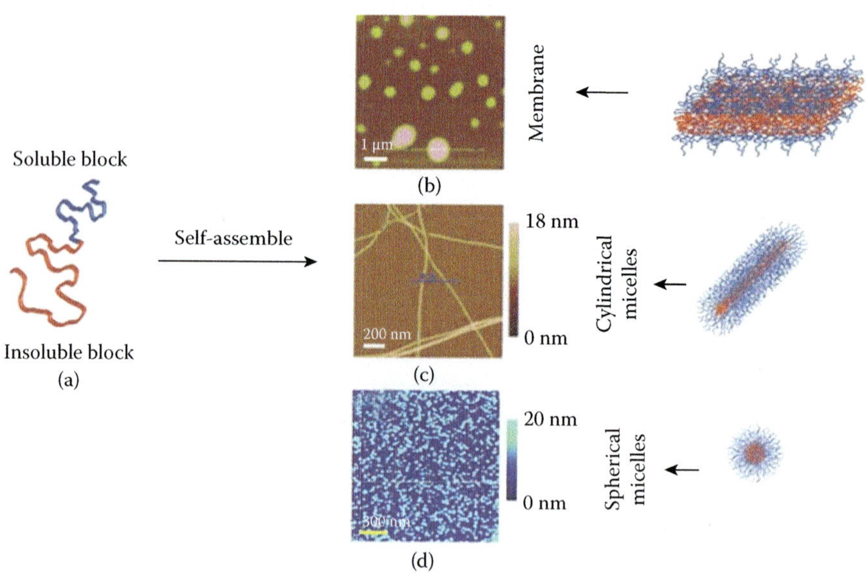

FIGURE 2.2 Micelles with various shapes: (a) soluble/insoluble block, (b) membrane, (c) cylindrical micelles, and (d) spherical micelles.

the cylindrical micelles into an ordered hexagonal phase occurs at higher temperatures due to the decreased solubility of the polymer blocks in water above 70°C. This special characteristic—the ability to change the reversible self-assembly at multiple length scales (ranging from nanoscale to macroscale) by changing pH and temperature conditions—offers interesting possibilities for material design and fabrication (Ding et al. 2007; Iatridi and Tsitsilianis 2011).

The solubility of polymers dictates the type of block copolymer obtained. Thus, if the copolymer is water soluble, micelles can be prepared by two methods:

1. *Direct dissolution method*—This consists of adding the copolymer to the aqueous media at a level above the critical micelle concentration; in this method, the drug is allowed to partition into the core of the micelles (Kabanov and Alakhov 2002).
2. *Film casting method*—This involves the dissolution of the copolymer and drug in a volatile solvent; then, the solvent is evaporated, leaving the polymer film that will be used in the bottom of a vial. Afterward, a warm buffer (or water) is added under agitation to dissolve the polymer film (Burt et al. 1999; Letchford and Burt 2007). If the copolymer is not soluble in water, the literature mentions dialysis or oil in water emulsion as micelles preparation procedures:
   a. In the *dialysis method*, the copolymer and drug are eluted in a water miscible organic solvent. Micelles are formed by the addition of water or, alternatively, by the addition of the copolymer/drug/solvent solution under water stirring, followed by dialysis against aqueous media to remove the solvent (Kim et al. 1998; Park et al. 2005).
   b. The *oil in water emulsion procedure* involves the addition of a solution formed by the copolymer drug and a volatile, non-water-miscible organic solvent into a rapidly stirring aqueous media, with or without a surfactant; then, the evaporation of the solvent takes place (Sant et al. 2004).

2.3 SPECIFIC METHODS OF MICELLES CHARACTERIZATION

Different parameters describing the principal properties of self-assemblies are mentioned in the literature. Critical micellization concentration (CMC) is described using the pyrene fluorescence method. From a stock solution of pyrene dissolved in chloroform, an aliquot is transferred with a micropipette into a series of dry test tubes; the solvent is allowed to evaporate under vacuum by protecting it from light to obtain dry pyrene. Then, a series of polymer solutions in buffer solution (pH 7.4) are added to the pyrene. Mixtures are stirred in the dark for 24 h at 25°C and then filtered through a membrane filter for separation of undissolved pyrene crystals. The concentration of solubilized pyrene in the micellar phase is determined spectrofluorometrically at 339 nm wavelengths of excitation and 390 nm of emission.

Ultraviolet-absorption spectroscopy has also been used as an important technique for the determination of the CMC. This method is based on the tautomerism of 1-phenyl-1,3-butadione between the keto and enol forms, which possess different absorption maxima: 312 nm for the enolic form and 250 nm for the keto form. The former structure appears in nonpolar solvents (e.g., cyclohexane) and the latter in polar solvents, where H-bonding is destabilized in favor of the keto configuration (Dominguez et al. 1997).

Static light scattering (SLS) is a powerful technique for estimating the *average molar masses* of self-assembled structures (and their CMC). In addition, if scattering from the core and the corona of the micellar system is not very different, the radius of gyration (R_G) can be also calculated. Dynamic (or quasi-elastic) light scattering (DLS) can be used to estimate the *hydrodynamic radius* (R_H) of a block copolymer micellar system from the determination of its diffusion coefficient. Moreover, the sensitivity and versatility of DLS allow changes in the micelle equilibrium due to variations of temperature, pH, or other monitored parameters.

Small-angle X-ray scattering (SAXS) has been used in the analysis of micellar solutions to obtain overall and internal sizes from differences in electron density of the solvent and the solute. Small-angle neutron scattering (SANS) gives information not only about the shape but also about the cross section of the samples (Rodriguez-Hernandez et al. 2005).

Micelle hydrodynamic diameters are measured using dynamic light scattering with a Coulter N4 nanosizer. The analyses are performed at scattering angle of 90° at 20°C (Zhang et al. 2004). Mean diameters are calculated according to the formula:

$$M_0 = L_{M0} + \left(\frac{d_1}{d_1 + d_2} \right) w \tag{2.1}$$

where L_{M0} is the lower limit of mode class; d_1 represents the frequency of the modal class-frequency of the previous class; d_2 is the frequency of the modal class-frequency of the following class; and w is the width of the interval.

Other nonscattering methods, such as transmission electron microscopy (TEM) and atomic force microscopy (AFM), provide images whereby size, shape, or internal structure of the micelles can be confirmed. As an example, the morphology of the micelles is determined by atomic force microscopy using a commercial microscope MFP-3D Asylum Research type, with triangular cantilevers (DNP C, Bruker), each of them having a nominal resonance frequency of 56 KHz and a spring constant of 0.24 N/m. Images are recorded using contact mode in an ambient environment after drying the sample with nitrogen gas; also, images using amplitude modulation mode have yielded similar results (Satyam et al. 2014).

Zeta potential of the micelles is measured using a phase analysis light scattering zeta potential analyzer. Before the measurements, the micelle formulations are dialyzed against buffer solution to remove NaCl ions. Zeta potential is automatically calculated by the analyzer using the Smoluchowski equation:

$$m = \frac{e \cdot z}{h} \tag{2.2}$$

where z is the zeta potential, m is the mobility, e is the dielectric constant, and h is the absolute viscosity of electrolyte solution (Sezgin et al. 2006).

Further methodologies include dilute solution capillary viscometry, membrane osmometry, ultracentrifugation, size exclusion chromatography, and typical spectroscopic methods (e.g., nuclear magnetic resonance) (Forster and Plantenberg 2002).

2.4 EXAMPLES OF MICELLES/SELF-ASSEMBLIES AND APPLICATIONS

2.4.1 Therapy and Drug Delivery

Recent progress in nanotechnology has a positive effect on the field of drug delivery. In particular, self-assembled nanostructures are extremely important for therapeutic and delivery applications. Their size, high loading capacity, and reversible character are major assets for the controlled release of therapeutics.

Polymeric micelles used as drug delivery systems offer many advantages, which are summarized here as follows:

- Colloidal stability with low CMC
- Tunable sizes with narrow size distribution
- Ability to preserve the activities of drugs during circulation in blood
- High physical loading efficiency of drugs
- Bioconjugation for active targeting to specific diseased cells
- Controllable release of drugs upon external stimuli and degradation

A drug is administered in a high dose at a given time, for repeating that dose several hours or days later. As a consequence, increasing attention has been paid to methods of giving drugs continually for prolonged time periods and in a controlled way. In the pharmaceutical field, in addition to the importance of polymers, the understanding of physiological barriers in the human body is essential for developing appropriate controlled release systems. In addition, biomolecules and biopolymers can be efficiently used to design nanostructured materials that are of high interest and that can confer those biocompatibility and bioactivity properties (Lim et al. 2009).

In recent years, several classes of biomolecules (peptides, lipids, nucleic acids, and sugars) have been studied as building blocks for self-assembled nanostructures, yielding complex therapeutic agents and nanoscale vectors (Kuzuya and Komiyama 2010; Namiki et al. 2011; Matson and Stupp 2012; Reichardt et al. 2013).

Recently, novel polymeric and peptide self-assembling systems have been developed for drug delivery. These carriers mimic the characteristics of conventional lipid systems and, in some cases, demonstrate improved drug delivery qualities. In this case, diblock copolymers are commonly used to construct micellar assemblies. Self-assembly of block copolymers is often spontaneous and can be performed in the presence of small molecules that form micelles encapsulated with drugs, with diameters up to hundreds of nanometers (Lee and Yuk 2007; Branco and Schneider 2009). In addition to micelles, amphiphilic block copolymers can self-assemble into vesicles in water. Polymersomes have been used for delivery of small hydrophobic drugs and DNA/RNA molecules and possess drug delivery characteristics similar to micelles (Hosseinkhani et al. 2013).

Macroscopically stable gels are useful for delivery applications because their porous structures allow the controlled release of a therapeutic agent over an extended period of time. In particular, self-assembled gelators are of great interest because they can be formed by in situ gelation of the blocks. This could translate into clinical applications in which an injection of soluble blocks results in the gelation of the drug delivery materials by the formation of self-assembled nanostructures. In this context, gels formed by the self-assembly of peptides or polypeptides are well known due to their degradation and biocompatible capacity. Recent trends in this area can be illustrated by the encapsulation of *curcumin* in self-assembling β-hairpin peptide hydrogels. Curcumin is an antioxidant with anti-inflammatory and antitumor properties that has also poor water solubility and a relatively low bioavailability that limits its therapeutic use (Altunbas et al. 2011). Taking those factors into consideration, the authors have demonstrated multiple modes of interactions between the hydrophobic curcumin and the valine rich β-hairpin MAX8. They were able to control the release rate and toxicity against cancer cells because of the bioactive principle, by changing the concentration of the gelating peptide. Drug release from gels prepared by the network formation of self-assembled peptide nanofibers conjugated to drugs by reversible covalent bonds has also been described. The curcumin–hydrogel complex is prepared in situ, while curcumin encapsulation within the hydrogel network is accomplished with peptide self-assembly. Physical and in vitro biological studies have demonstrated the efficiency of curcumin-loaded β-hairpin hydrogels as injectable agents for localized curcumin delivery. In vitro experiments have proved that encapsulation of the curcumin within the hydrogel do not have an adverse effect on its bioactivity. Moreover, the rate of curcumin release and its therapeutic efficacy can be modulated as a function of the peptide concentration (Altunbas et al. 2011).

In recent years, self-assembly of nucleic acids has appeared as the most important tool for aqueous complex nanostructures. The fact that nucleic acids are also biocompatible, degradable, and bioactive makes them ideal building blocks for the development of a drug delivery vehicle of nanoscale dimensions (Bhatia et al. 2011). An essential advantage of DNA acids that allow the preparation of self-assembled nanostructures with a perfect control over the size and shape of the constructs represents the bioactivity of these nanostructures.

In order to design an ideal nanoscale drug carrier, one would envision a three-dimensional structure that could be filled with a therapeutic agent, closed, and then selectively reopened at the targeted delivery site. The first attempt to produce a DNA nanobox that potentially could be used as a delivery vector was the design of a DNA origami nanocube with a controllable lid; this lid could be held in a closed position by a set of two locks composed of DNA duplexes with sticky end extensions, allowing it to be opened by the dissociation of the duplex upon an externally added "key" (Andersen et al. 2009).

Recently, DNA aptamers have been used as gates to control the opening and closing of self-assembly targeted nanorobots. The keys that trigger the opening of a self-assembled DNA barrel are two orthogonal aptamers (Douglas et al. 2012).

Bottlebrush polymers are a type of branched or graft polymer with polymeric side chains attached to a linear backbone. The special architecture of bottlebrushes ensures several unique and potentially useful properties. These include a high entanglement molecular weight that enables rapid self-assembly of bottlebrush block copolymers into large domain structures as well as functionalization of their side chains for recognition, imaging, or drug delivery in aqueous environments (Verduzco et al. 2015). Bottlebrush block copolymers form micelles in a selective solvent but have a much lower critical micelle concentration in comparison with linear diblock copolymers and surfactants; this makes them suitable for applications as detectors or sensors in biological media that require dilute conditions. The resulting micelles typically are larger than those for linear diblock copolymers, and subsequently some functional groups can be incorporated into the bottlebrush side chains. An amphiphilic bottlebrush block copolymer with poly(acrylic acid) and polystyrene side chains was studied for its ability to self-assemble into micelles. The bottlebrush block copolymer was fully soluble in DMF and formed micelles in water. A micellar diameter of 48 nm and an aggregation number of 60 was estimated using atomic force microscopy, transmission electron microscopy, and dynamic light scattering; these were smaller than typically observed for linear block copolymer micelles (Li et al. 2010).

The literature has detailed several studies concerning bottlebrush polymers that were used in drug delivery applications. A series of bottlebrush star polymers and copolymers as degradable drug carriers were developed using a two-step synthesis approach (Liu et al. 2012; Burts et al. 2014a; Liao et al. 2014). This approach involves the preparation of living, end-functional bottlebrush arms by ring-opening metathesis polymerization. These are then coupled in a second reaction step with bis-norbornene crosslinkers. In this type of polymerization, there is no residual unreacted macromonomer after the first step, and the bottlebrush star molecular weight can be controlled by varying the length of the bottlebrush arms and the molar ratio of the crosslinkers. Subsequently, an ultraviolet (UV)-responsive nitrobenzyloxycarbonyl crosslinker was incorporated that led to bottlebrush star polymer degradation upon UV irradiation (Liu et al. 2012). Another study focused on the anticancer drug, doxorubicin (DOX), covalently linked to the bottlebrush star polymers through a copper-catalyzed azide-alkyne click coupling reaction. Drug-loaded star polymers were nontoxic and, on exposure to 365 nm light, DOX was released from the star polymers—this was confirmed by FTIR and cell viability measurements (Burts et al. 2014b). In other research, the potential for bottlebrush star polymers to deliver multiple drugs at a desired dosage in response to distinct stimuli was demonstrated. Two different norbornenyl macromonomers were synthesized, containing DOX or camptothecin (CPT) as drug carriers. These macromonomers were mixed at a desired ratio and copolymerized with polyethylene glycol macromonomer in the first step of the "brush-first" synthesis approach. Then, a $Pt_{(IV)}$ bis-norbornene complex was coupled with the brush arms. Hydrolysis of the bottlebrush star polymer crosslinkers thus resulted in release of DOX and CPT and, after reduction of the $Pt_{(IV)}$ complex, cytotoxic cisplatin. Confocal microscopy measurements and in vitro cellular viability studies were carried out to confirm internalization and cytotoxicity of drug-loaded bottlebrush star polymers (Liao et al. 2014).

A number of opportunities still remain to be developed for both fundamental and applied work regarding bottlebrush polymers. One specific area is a more quantitative understanding of bottlebrush block copolymer self-assembly in bulk and thin films. Also, deeper understanding leading to a predictive model of bottlebrush copolymer self-assembly would enable the design of bottlebrush copolymers to target specific self-assembled structures with desired phase behavior.

However, the anticancer drug DOX was covalently conjugated to m-polyethylene glycol-*b*-polylactide-*co*-dihydroxymethylpropylene carbonate by carbamate or acid-labile hydrazone linkage. The OH groups of the structure were activated with 4-nitrophenyl chloroformate and then reacted with amino groups of DOX. The resulting multifunctional mixed micelles exhibited pH-dependent release behavior—the release of DOX upon cleavage of carbamate and hydrazone bonds was proved to be enhanced in acidic pH (Hu et al. 2010; Oh 2011).

Cellular uptake has been studied using laser confocal fluorescence microscopy; nanoparticles of polylactide-based amphiphilic block copolymers are encapsulated with fluorescence dyes and incubated with living cells, such as HeLa and 3T3 cells. The obtained results imply that nanoparticles are uniformly distributed inside cells. Active targeting is a promising approach through the increase of nanoparticles' local accumulation in diseased tissue. This method requires the design and preparation of functional nanoparticles that are further conjugated with biomolecules targeting specific cells by specific receptor–ligand or antigen–antibody interactions. Biomolecules attached to polylactide-based nanoparticles include folate, peptide, carbohydrate (galactosamine), and aptamer (Byrne et al. 2008; Veiseh et al. 2010).

2.4.2 Regenerative Medicine and Tissue Engineering

Exciting developments in the field of regenerative medicine, based on the application of self-assembled biocompatible polymeric scaffolds for regeneration of tissues and organs, have taken place in recent years. Figure 2.3 illustrates this important domain in which the study of tissue engineering and regenerative medicine is essential.

Self-assembled nanomaterials show additional advantages because of their kinetic lability by providing properties such as self-construction, degradability, and adaptivity (Aida et al. 2012). Self-construction implies that their assembly can be achieved in situ, in the presence of cells, or in vivo, after injections of their molecular components at a desired site. Their reversible nature facilitates the dissociation into their molecular constituents. Another important property of these systems is their ability to adapt to their environment. One example would be a cell surface coated with receptors that can potentially induce the clustering of its ligands on the surface of a self-assembled structure to maximize the bond between the two surfaces. Among all examples of nanomaterials for tissue engineering and regenerative medicine, self-assembled peptide nanofibers are the most common. They are formed by the self-assembly of simple peptidic building blocks obtained through a fast synthesis that facilitates the introduction of peptidic or nonpeptidic epitope for specific applications (Cui et al. 2010; Luo and Zhang 2012; Wu et al. 2012; Stephanopoulos et al. 2013). In addition, their peptidic structure provides unique biocompatibility and degradability for use as nutrients by cells. The design of self-assembling peptides usually involves short

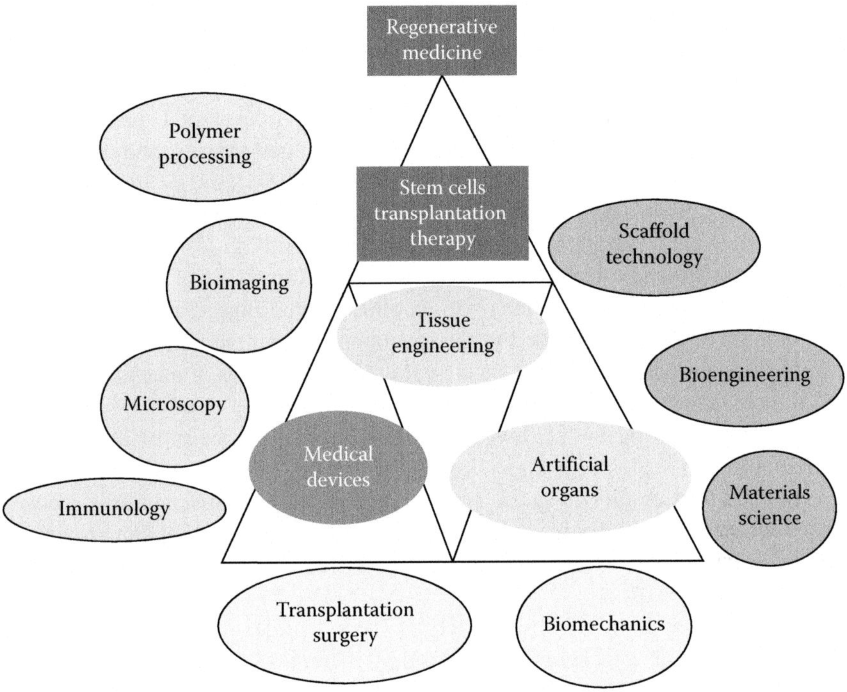

FIGURE 2.3 Important applications in tissue engineering and regenerative medicine.

peptide sequences known to facilitate the formation of secondary structures such as β-sheets or α-helices, and their conjugation to hydrophobic chains, in order to promote aggregation in water.

Another important issue in regenerative medicine is protecting grafts and cells after transplantation. In this context, there is great interest in the design of biomaterials that could encapsulate cells, protecting them from the host immune system while at the same time allowing nutrients to diffuse in and out of the capsule (Capito et al. 2008).

More recent literature describes a new method for recreating collagen fibers and gels from the self-assembly of simple peptides with sequences derived from the collagen's praline–hydroxyproline–glycine repeating unit. These peptides self-assemble into triple helices having electrostatic sticky ends, which promote the elongation of triple helices involving salt-bridged hydrogen bonds between lysine and aspartate (O'Leary et al. 2011). With the major role of collagen matrices in tissue engineering, this strategy is promising for the preparation of novel collagen analogs in regenerative medicine.

Supramolecular self-assembly can combine signals and control their concentration in a specific nanostructure as well as mediating interactions among nanostructures to create more complex objects. Besides the use of peptides, the notion of introducing saccharides and nucleic acids in the design of nanostructures for signaling or genetic manipulations is also of interest. Different nanostructures have been developed and tested over the years, including β-hairpins that create networks, ionic self-complementary peptides that form nanoscale fibrils, and nanostructures based on molecules known as peptide amphiphiles with spherical, planar, and cylindrical shapes (Figure 2.2) (Hartgerink et al. 2001; Bitton et al. 2005).

The literature mentions the cell signaling platforms developed for regenerative medicine involving the spinal cord, cartilage, bone, blood vessels, and the heart, in which peptide amphiphiles molecules are being used. Specific peptide amphiphiles' structures are formed by self-assembling cylindrically shaped one-dimensional nanostructures that contain a peptide sequence covalently grafted to a nonpeptidic hydrophobic segment, typically an alkyl segment (Stupp 2010). These molecules combine the hydrophobic segments with peptides containing at least two domains. One domain is covalently bonded to the hydrophobic segment named the β domain because of its property to form β sheets with other molecules; the terminal domain carries biological information for signaling.

Future directions in this field will consider how to organize rationally bioactive nanostructures at scales of a larger length. This would create systems with much more potent functionality to guide cells in regeneration. Future systems incorporating nanoscale design could include complex nanostructures that can repair cells through their ability to interact specifically with intracellular sites in organelles or the nucleus (Stupp 2010).

Biomaterials made from self-assembling, short peptides and peptide derivatives have great potential to generate powerful new therapies in regenerative medicine. The high signaling capacity and therapeutic efficacy of peptidic scaffolds represent the important issues in this area. Different classes of self-assembling peptide-based materials, including peptide amphiphiles, self-complementary ionic peptides, hairpin peptides, and others, are detailed in the latest research. Peptide amphiphiles—short peptide sequences attached to a hydrophobic tail—have been studied extensively. The key component of this peptide is represented by a β-sheet region of 4–8 amino acids directly adjacent to the palmitic acid tail, inducing the formation of cylindrical nanofiber. Peptide amphiphiles' structure also is formed by the same charged residues that were added to improve solubility and gelation. Self-assembly of molecules into cylindrical nanofibers takes place by hydrophobic collapse along with the formation of a hydrogen bonding network down the long axis of the nanofiber. Gelation of the nanofibers into networks can be triggered by charge screening through the addition of electrolytes or a change in pH (Greenfield et al. 2010; Matson and Stupp 2012). Also, the self-assembly designs, bioactive signaling strategies, and cell signaling capabilities of these bioactive materials have been reported.

In the next decade, new targets include simple hierarchically structured materials that display order parameters over scales of several lengths. Moreover, there are novel materials that interface peptide-based supramolecular structures with naturally occurring biopolymers or specifically designed synthetic polymers. In the interim, several advantages having a major impact on regenerative medicine are forecast. Most prevalent among these is the development of dynamic scaffolds that respond to their environment through internal or external stimuli. Further developments on enzyme-responsive dynamic scaffolds in cell culture and regenerative medicine can be expected. Another area that needs to advance is the integration of bioactive peptide materials with solid implants that could become a useful strategy for better outcomes at tissue–implant interfaces. Future challenges involving long-term goals such as immune system programming, subcellular targeting, and the development of highly integrated scaffold systems are being addressed (Missirlis et al. 2010; Matson and Stupp 2012).

Dendritic polymers, including dendrimers and hyperbranched polymers, are new polymer architectures that follow linear, branched, and crosslinking polymers. Hyperbranched polymers are highly branched macromolecules that have demonstrated great potential as excellent precursors in supramolecular self-assembly. Their special topological structures and properties provide unique advantages in supramolecular self-assembly and biomedical applications in comparison with the linear analogues, which have stimulated interest in this domain. The most commonly known types of hyperbranched polymers are polyphenylenes, polysiloxanes, polyethers, polyesters, polyamides, polyimides, polyurethanes, poly(sulfone-amine)s, poly(ester-amine)s, or poly(amido-amine)s (Zhou et al. 2010). These supramolecular structures have shown promising applications in biomedical areas such as drug delivery, protein purification/detection/delivery, gene transfection, with antibacterial/antifouling materials, and in cytomimetic chemistry.

Hyperbranched polymer self-assemblies are used as drug delivery systems, with five drug-loaded routes being developed:

- For Type 1 (drug complexes) and Type 2 (unimolecular micelles)—the drugs are attached to the supramolecular structures by noncovalent bonding or are entrapped inside the nanocavities of dendritic structures.
- For Types 3 and 4 (multimolecular micelles)—the drugs are loaded in the micelles cores (or shells).
- For Type 5 (drug conjugates)—the drugs are covalently conjugated to the terminal groups of hyperbranched polymers.

From these, only zero-dimensional micelles and two-dimensional membranes self-assembled from hyperbranched polymers have been reported in the literature to have biomedical applications; knowledge of interactions with cells and other biomacromolecules, in vivo transport and metabolism routes, and the cytotoxicity mechanism of the supramolecular structures is still limited. However, these drug delivery systems do display some advantages, such as enhancing the aqueous solubility and bioavailability of drugs, prolonging the circulation time, inducing the preferential accumulation at the tumor sites through the enhanced permeability and retention effect, and increasing the systemic side effects (Haag 2004; Calderón et al. 2010; Wilms et al. 2010). In the future, one can expect that these new materials will prove their efficiency in medicine therapy and reduce the pain of patients, and they may even ultimately defeat some fatal diseases, such as cancer.

2.5 FUTURE OUTLOOK

In years to come, the question of how to address self-organized systems will become more and more important. Regenerative medicine, drug delivery, and tissue engineering each represent a critical need today and for the future in order to raise quality of life and reduce the cost of health care in the context of an increasing human lifespan. It is clear that this field offers an extraordinary opportunity for chemists to design supramolecular structures that can promote regeneration in the human body. Future developments in these areas will include materials that can deliver a specific signal to a designated organelle. It will be very exciting to see how these recent trends will affect the future of drug delivery systems. On the contrary, the development of self-organized information structures, such as signal transmission and storage in the human brain, represents a challenge—but also a point of inspiration. All future applications promote the necessary training of scientists in interdisciplinary fields of chemistry, physics, biology, medicine, and materials research, as multidisciplinarity is a characteristic feature of this kind of research. It has become evident that new routes in traditional specific fields of applications can be very promising because they often represent the starting point for novel and unconventional approaches.

REFERENCES

Aida Takuzo, Meijer W. Egbert, and Stupp I. Samuel. Functional supramolecular polymers. *Science*. 335 (2012): 813–817.

Altunbas Aysegul, Lee J. Seung, Rajasekaran A. Sigrid, Schneider P. Joel, and Pochan J. Darrin. Encapsulation of curcumin in self-assembling peptide hydrogels as injectable drug delivery vehicles. *Biomaterials*. 32 no. 25 (2011): 5906–5914.

Andersen S. Ebbe, Dong Mingdong, Nielsen M. Morten, et al. Self-assembly of a nanoscale DNA box with a controllable lid. *Nature*. 459 no. 7243 (2009): 73–76.

Bhatia Dhiraj, Sharma Suruchi, and Krishnan Yamuna. Synthetic, biofunctional nucleic acid-based molecular devices. *Curr. Opin. Biotechnol.* 22 no. 4 (2011): 475–484.

Bitton Ronit, Schmidt Judith, Biesalski Markus, et al. Self-assembly of model DNA-binding peptide amphiphiles. *Langmuir.* 21 no. 25 (2005): 11888–11895.

Blanazs Adam, Armes P. Steven, and Ryan J. Anthony. Self-assembled block copolymer aggregates: From micelles to vesicles and their biological applications. *Macromol. Rapid Commun.* 30 no. 4–5 (2009): 267–277.

Branco C. Monica and Schneider P. Joel. Self-assembling materials for therapeutic delivery. *Acta Biomater.* 5 no. 3 (2009): 817–831.

Burt M. Helen, Zhang Xichen, Toleikis Philip, Embree Leanne, and Hunter L. William. Development of copolymers of poly(D, L-lactide) and methoxypolyethylene glycol as micellar carriers of paclitaxel. *Colloids Surf. B.* 16 no. 1–4 (1999): 161–171.

Burts O. Alan, Gao X. Angela, and Johnson A. Jeremiah. Brush-first synthesis of core-photodegradable miktoarm star polymers via ROMP: Towards photoresponsive self-assemblies. *Macromol. Rapid Commun.* 35 no. 2 (2014a): 168–173.

Burts O. Alan, Liao Longyan, Lu Y. Ying, et al. Brush-first and click: Efficient synthesis of nanoparticles that degrade and release doxorubicin in response to light. *Photochem. Photobiol.* 90 no. 2 (2014b): 380–385.

Byrne D. James, Betancourt Tania, and Brannon-Peppas Lisa. Active targeting schemes for nanoparticle systems in cancer therapeutics. *Adv. Drug Delivery Rev.* 60 no. 15 (2008): 1615–1626.

Calderón Marcelo, Quadir Mohiuddin Abdul, Sharma Sunil Kumar, and Rainer Haag. Dendritic polyglycerols for biomedical applications. *Adv. Mater.* 22 no. 2 (2010): 190–218.

Capito M. Ramille, Azevedo S. Helena, Velichko S. Yuri, Mata Alvaro, and Stupp I. Samuel. Self-assembly of large and small molecules into hierarchically ordered sacs and membranes science. *Science*. 319 no. 5871 (2008): 1812–1816.

Chen X. Linda and Jenekhe A. Samson. Solubilization and encapsulation of fullerenes by amphiphilic block copolymers. *Langmuir.* 15 no. 23 (1999): 8007–8017.

Chen X. Linda and Jenekhe A. Samson. Supramolecular self-assembly of three-dimensional nanostructures and microstructures: Microcapsules from electroactive and photoactive rod-coil-rod triblock copolymers. *Macromolecules.* 33 no. 13 (2000): 4610–4612.

Cui Honggang, Webber J. Matthew, and Stupp I. Samuel. Self-assembly of peptide amphiphiles: From molecules to nanostructures to biomaterials. *Biopolymers.* 94 no. 1 (2010): 1–18.

Determan D. Michael, Guo Liang, Pradeep Thiyagarajan, and Mallapragada K. Surya. Supramolecular self-assembly of multiblock copolymers in aqueous solution. *Langmuir.* 22 no. 4 (2006): 1469–1473.

Ding Ke, Alemdaroglu E. Fikri, Börsch Michael, Berger Rüdiger, and Herrmann Andreas. Engineering the structural properties of DNA block copolymer micelles by molecular recognition. *Angew. Chem. Int. Ed. Engl.* 46 no. 7 (2007): 1172–1175.

Disher M. Bohdana, Hammer A. Daniel, Bates S. Frank, and Disher E. Dennis. Polymer vesicles in various media. *Curr. Opin. Colloid Interface Sci.* 5 no. 1–2 (2000): 125–131.

Disher E. Dennis, and Eisenberg Adi. Polymer vesicles. *Science.* 297 no. 5583 (2002): 967–973.

Dominguez Ana, Fernandez Aurora, Gonzalez Noemi, Iglesias Emilia, and Montenegro Luis. Determination of critical micelle concentration of some surfactants by three techniques. *J. Chem. Educ.* 74 no. 10 (1997): 1227–1231.

Douglas M. Shawn, Bachelet Ido, and Church M. George. A logic-gated nanorobot for targeted transport of molecular payloads. *Science.* 335 no. 6070 (2012): 831–834.

Forster Stephan and Plantenberg Thomas. From self-organizing polymers to nanohybrid and biomaterials. *Angew. Chem. Int. Ed. Engl.* 41 no. 5 (2002): 688–714.

Greenfield A. Megan, Hoffman R. Jessica, Olvera de la Cruz Monica, and Stupp I. Samuel. Tunable mechanics of peptide nanofiber gels. *Langmuir.* 26 no. 5 (2010): 3641–3647.

Haag Rainer. Supramolecular drug-delivery systems based on polymeric core–shell architectures. *Angew. Chem.Int. Ed. Engl.* 43 no. 3 (2004): 278–282.

Hartgerink D. Jeffrey, Beniash Elia, and Stupp I. Samuel. Self-assembly and mineralization of peptide-amphiphile nanofibers. *Science.* 294 no. 5547 (2001): 1684–1688.

Hayward Ryan C. and Pochan J. Darrin. Tailored assemblies of block copolymers in solution: It is all about the process. *Macromolecules.* 43 no. 8 (2010): 3577–3584.

Hosseinkhani Hossein, Hong Po-Da, and Yu Dah-Shyong. Self-assembled proteins and peptides for regenerative medicine. *Chem. Rev.* 113, no. 7 (2013): 4837–4861.

Hugouvieux Virginie, Axelos A. V. Monique, and Kolb Max. Micelle formation, gelation and phase separation of amphiphilic multiblock copolymers. *Soft Matter.* 7 no. 6 (2011): 2580–2591.

Hu Xiuli, Liu Shi, Huang Yubin, Chen Xuesi, and Jing Xiabin. Biodegradable block copolymer-doxorubicin conjugates via different linkages: Preparation, characterization, and in vitro evaluation. *Biomacromolecules.* 11 no. 8 (2010): 2094–2102.

Iatridi Zacharoula and Tsitsilianis Constantinos. Water-soluble stimuli responsive star-shaped segmented macromolecules. *Polymers.* 3 no. 4 (2011): 1911–1933.

Kabanov V. Alexander and Alakhov Yu Valeri. Pluronic® block copolymers in drug delivery: From micellar nanocontainers to biological response modifiers. *Crit. Rev. Ther. Drug Carr. Syst.* 19 no. 1 (2002): 1–72.

Kelley Elizabeth G., Albert N. L. Julie, Sullivan O. Millicent, and Epps H. Thomas. Stimuli-responsive copolymer solution and surface assemblies for biomedical applications. *Chem. Soc. Rev.* 42 no. 17 (2013): 7057–7071.

Kim So Yeon, Shin I. L. Gyun, Lee Young Moo, Cho Chong Su, and Sung Yong Kiel. Methoxy poly(ethylene glycol) and epsilon-caprolactone amphiphilic block copolymeric micelle containing indomethacin. II. Micelle formation and drug release behaviours. *J. Controlled Release.* 51 no. 1 (1998): 13–22.

Kuzuya Akinori and Komiyama Makoto. DNA origami: Fold, stick, and beyond. *Nanoscale*. 2 (2010): 309–321.

Lazzari Massimo and Lopez-Quintela Manuel Arturo. Block copolymers as a tool for nanomaterial fabrication. *Adv. Mater.* 15 no. 19 (2003): 1583–1594.

Lee C.-M. James, Bermudez Harry, Discher M. Bohdana, et al. Preparation, stability, and in vitro performance of vesicles made with diblock copolymers. *Biotechnol. Bioeng.* 73 no. 2 (2001): 135–145.

Lee K. Yong and Yuk H. Soon. Polymeric protein delivery systems. *Prog. Polym. Sci.* 32 no. 7 (2007): 669–697.

Letchford Kevin and Burt Helen. A review of the formation and classification of amphiphilic block copolymer nanoparticulate structures: Micelles, nanospheres, nanocapsules and polymersomes. *Eur. J. Pharm. Biopharm.* 65 no. 3 (2007): 259–269.

Li Zhou, Ma Jun, Cheng Chong, Zhang Ke, and Wooley L. Karen. Synthesis of hetero-grafted amphiphilic diblock molecular brushes and their self-assembly in aqueous medium. *Macromolecules.* 43 no. 3 (2010): 1182–1184.

Liao Longyan, Liu Jenny, Dreaden C. Erik, et al. A convergent synthetic platform for single-nanoparticle combination cancer therapy: Ratiometric loading and controlled release of cisplatin, doxorubicin, and camptothecin. *J. Am. Chem. Soc.* 136 no. 16 (2014): 5896–5899.

Lim Yong Beom, Moon Kyung Soo, and Lee Myongsoo. Recent advances in functional supramolecular nanostructures assembled from bioactive building blocks. *Chem. Soc. Rev.* 38 no. 4 (2009): 925–934.

Liu Jenny, Burts O. Alan, Li Yongjun, et al. Brush-first method for the parallel synthesis of photocleavable, nitroxide-labeled poly(ethylene glycol) star polymers. *J. Am. Chem. Soc.* 134 no. 39 (2012): 16337–16344.

Luo Zhongli and Zhang Shuguang. Designer nanomaterials using chiral self-assembling peptide systems and their emerging benefit for society. *Chem. Soc. Rev.* 41 no. 13 (2012): 4736–4754.

Mai Yi Yong and Eisenberg Adi. Self-assembly of block copolymers. *Chem. Soc. Rev.* 41 no. 18 (2012): 5969–5985.

Matson B. John and Stupp I. Samuel. Self-assembling peptide scaffolds for regenerative medicine. *Chem. Commun.* 48 no. 1 (2012): 26–33.

Missirlis Dimitris, Krogstad V. Daniel, and Tirrell Matthew. Internalization of $p53_{14-29}$ peptide amphiphiles and subsequent endosomal disruption results in SJSA-1 cell death. *Mol. Pharmaceutics.* 7 no. 6 (2010): 2173–2184.

Namiki Yoshihisa, Fuchigami Teruaki, Tada Norio, et al. Nanomedicine for cancer: Lipid-based nanostructures for drug delivery and monitoring. *Acc. Chem. Res.* 44 no. 10 (2011): 1080–1093.

Oh Jung Kwon. Polylactide (PLA)-based amphiphilic block copolymers: Synthesis, self-assembly, and biomedical applications. *Soft Matter.* 7 no. 11 (2011): 5096–5108.

O'Leary E. R. Lesley, Fallas A. Jorge, Bakota L. Erica, Kang K. Marci, and Hartgerink D. Jeffrey. Multi-hierarchical self-assembly of a collagen mimetic peptide from triple helix to nanofibre and hydrogel. *Nat. Chem.* 3 no. 10 (2011): 821–828.

Park Eun Kyoung, Kim So Yeon, Lee Sang Bong, and Lee Young Moo. Folate-conjugated methoxy poly(ethylene glycol)/poly(ε-caprolactone) amphiphilic block copolymeric micelles for tumor-targeted drug delivery. *J. Control. Release.* 109 no. 1–3 (2005): 158–168.

Reichardt Niels, Martín-Lomas Manuel, and Penadés Soledad. Glyconanotechnology. *Chem. Soc. Rev.* 10 (2013): 4358–4376.

Rodriguez-Hernandez Juan, Chécot Frédéric, Gnanou Yves, and Lecommandoux Sébastien. Toward 'smart' nano-objects by self-assembly of block copolymers in solution. *Prog. Polym. Sci.* 30 no. 7 (2005): 691–724.

Sant P. Vinayak, Smith Damon, and Leroux Jean-Christophe. Novel pH-sensitive supramolecular assemblies for oral delivery of poorly water soluble drugs: Preparation and characterization. *J. Control. Release.* 97 no. 2 (2004): 301–312.

Satyam Abhigyan, Kumar Pramod, Fan Xingliang, et al. Macromolecular crowding meets tissue engineering by self-assembly: A paradigm shift in regenerative medicine. *Adv. Mater.* 26 no. 19 (2014): 3024–3034.

Sezgin Zerrin, Yuksel Nilufer, and Baykara Tamer. Preparation and characterization of polymeric micelles for solubilization of poorly soluble anticancer drugs. *Eur. J. Pharm. Biopharm.* 64 no. 3 (2006): 261–268.

Smart Thomas, Lomas Hannah, Massignani Marzia, Flores-Merino Miriam V., Ruiz Perez Lorena, and Giuseppe Battaglia. Block copolymer nanostructures. *Nano Today.* 3 no. 3 (2008): 38–46.

Stephanopoulos Nicholas, Ortony H. Julia, and Stupp I. Samuel. Self-assembly for the synthesis of functional biomaterials. *Acta Mater.* 61 no. 3 (2013): 912–930.

Stupp I. Samuel. Self-assembly and biomaterials. *Nano Lett.* 10 no. 12 (2010): 4783–4786.

Taco Nicolai, Colombani Olivier, and Chassenieux Christophe. Dynamic polymeric micelles versus frozen nanoparticles formed by block copolymers. *Soft Matter.* 6 no. 14 (2010): 3111–3118.

Thurmond Bruce K., Huang Haiyong, Clark Christopher G., Jr., Kowalewski Tomasz, and Wooley Karen L. Shell cross-linked polymer micelles: Stabilized assemblies with great versatility and potential. *Colloid Surf B: Biointerfaces.* 16 no. 1 (1999): 45–54.

Tian Jia, Zheng Fan, Duan Qingjiao, and Zhao Hanying. Self-assembly of polystyrene with pendant hydrophilic gold nanoparticles: The influence of the hydrophilicity of the hybrid polymers. *J. Mater. Chem.* 21 no. 42 (2011): 16928–16934.

Veiseh Omid, Gunn W. Jonathan, and Zhang Miqin. Design and fabrication of magnetic nanoparticles for targeted drug delivery and imaging. *Adv. Drug Deliv. Rev.* 62 no. 3 (2010): 284–304.

Verduzco Rafael, Li Xianyu, Pesek L. Stacy, and Stein E. Gila. Structure, function, self-assembly, and applications of bottlebrush copolymers. *Chem. Soc. Rev.* 44 no. 8 (2015): 2405–2420.

Wilms Daniel, Stiriba Salah-Eddine, and Frey Holger. Hyperbranched polyglycerols: From the controlled synthesis of biocompatible polyether polyols to multipurpose applications. *Acc. Chem. Res.* 43 no. 1 (2010): 129–141.

Wooley L. Karen. Shell crosslinked polymer assemblies: Nanoscale constructs inspired from biological systems. *J. Polym. Sci. Part A: Polym. Chem.* 38 no. 9 (2000): 1397–1407.

Wu C. Elisabeth, Zhang Shuguang, and Hauser A. E. Charlotte. Self-assembling peptides as cell-interactive scaffolds. *Adv. Funct. Mater.* 22 no. 3 (2012): 456–468.

Yang Jiang. Viscoelastic wormlike micelles and their applications. *Curr. Opin. Colloid Interface Sci.* 7 no. 5–6 (2002): 276–281.

Zhang Can, Qineng Ping, and Zhang Hongjuan. Self-assembly and characterization of paclitaxel-loaded N-octyl-O-sulfate chitosan micellar system. *Coll. Surf. B Biointerf.* 39 no. 1–2 (2004): 69–75.

Zhou Yongfeng, Huang Wei, Liu Jinyao, Zhu Xinyuan, and Yan Deyue. Self-assembly of hyperbranched polymers and its biomedical applications. *Adv. Mater.* 22 no. 41 (2010): 4567–4590.

3 Porous Polymer Structures by Synthesis from Liquid Two-Phase Systems

Camelia Hulubei and Cristina Doina Vlad

CONTENTS

3.1 INTRODUCTION

There are a large number of materials that include polymeric multiphase systems. These refer to structured copolymers, various types of blends with flexible linear chains, weakly crosslinked linear chains (elastomers), interpenetrating polymer networks, composites, three-dimensional crosslinked networks, high thermoresistant materials, and so on. Among the polymer structures obtained from liquid–liquid two-phase systems, porous crosslinked polymers occupy an important position due to their special properties (high chemical and mechanical stability) and structural variety, both of which generate a large number of usage possibilities. Particle morphology is an important characteristic of porous polymers, with pore size a fundamental factor in specific high tech applications.

The term *crosslinked* refers to a small region in a macromolecule (from a polymer chain) where there at least one point is tetra-connected. Covalent polymer networks can be considered among the largest known molecules. Such networks usually are referred to as macronetworks. Conversely, the dimensions of micronetworks range from only a few nanometers to several millimeters (Mester et al. 2011).

The terminology and classification of porous polymers have been extensively discussed by specialists. Different classifications—based on pore size (Becher and Birkmeier 1964), level of particle agglomeration (Jacobeli et al. 1979), crosslinker content, solvent retention (Rabelo and Coutinho 1994), or porosity type (Sederel and De Jong 1973)—have been made. But the existence of distinct high and low crosslinking regions within the same polymer, as a consequence of the specific polymerization mechanisms, has not been taken into consideration in these classifications.

Macroporosity, usually related to the permanent porosity in a dry polymer, signifies the network density due to voids from the polymer matrix. *Gel porosity* refers to the materials that have no pores in a dry state (Kun and Kunin 1964, 1968). At present, we refer to the classification made by the International Union of Pure and Applied Chemistry (IUPAC) in Table 3.1 (Rouquerol et al. 1994).

The internal structure of porous beads consists of many interconnected cavities (pores) of different sizes, the beads' rigidity being due to the extensive crosslinks. These structures are obtained in presence of porogen which is a substance inert to monomers and is included within beads. Finally, the pore-forming is removed in order to create the granule pores.

3.2 CROSSLINKED (CO)POLYMERS

Network formation is possible if at least one of the initial components has a functionality greater than two. There are many synthetic methodologies that allow the preparation of porous polymeric materials (Barbetta et al. 2009; Weigang et al. 2010).

TABLE 3.1
Pore Classification by the International Union of Pure and Applied Chemistry

Name	Width (nm)	
Micropores	<2	• Ultramicropores below 0.7 nm
Mesopores	2–50	• Supramicropores 0.7–2 nm
Macropores	>50	

Taking into account the chemical formation mechanism, networks can be obtained by polyaddition, polycondensation, or by crosslinking (co)polymerization of polyunsaturated monomers or of existing primary chains. When written without the parentheses, this term refers exclusively to copolymerization (Dusek 1993).

Porous crosslinked (co)polymers in a bead shape can be prepared by using a mixture of two or more liquids and various polymerization techniques such as suspension, dispersion, emulsification, precipitation, seeded polymerization, and so on (Dragan and Vlad 2006; Mohamed and Wilson 2012).

Considering the large range of porous polymer structures that are synthesized in heterogeneous systems, this chapter presents some aspects of the suspension polymerization technique that is performed by a radical polymerization mechanism (Gokmen et al. 2012; Falk et al. 2004). Radical crosslinking (co)polymerization (Degirmenci et al. 2007) is a more complex method of polymer network preparation. This is due to the kinetic reaction mechanism and the strong spatial correlations that cause big differences in a polymer chain's branching mode (Dusek and MacKnight 1988; Gridnev et al. 2001).

3.2.1 Suspension Polymerization Technique

Suspension polymerization (SpP) can be considered the least complicated heterogeneous polymerization technique in terms of its mechanism.

Generally, we can discuss two types of suspension polymerization, "bead" and "powder" SpP (Kotoulas et al. 2006). In "bead" SpP, the polymer is soluble in its monomers and smooth spherical particles are produced. In the "powder" SpP process, the polymer is insoluble in its monomers; thus, it precipitates out, which leads to the formation of irregular grains of particles.

In "bead" SpP, the monomers are dispersed (as droplets) in the continuous phase (commonly water) by the combined action of surface agents (inorganic and/or organic water soluble polymers) and agitation. All the reactants (monomers, initiators, etc.) reside in the organic or "oil" phase. The droplets adopt a spherical shape for minimum interfacial energy (maximum stability), their average size primarily being controlled by a number of factors. Among these, the stirring speed and the stabilizing agent quality and the quantity are very important. The correct prediction of droplet size, mixing effects, and polymer sedimentation require particular attention (Silva et al. 2004). Polymerization occurs in the monomer droplets that are progressively transformed into sticky, viscoelastic monomer–polymer droplets and, finally, into rigid and spherical polymer particles. Particle size and internal structure affect the use of resins in their applications.

3.2.2 Bead Formation Mechanism

The mechanism affecting the appearance and development of crosslinked polymer particles during suspension polymerization may be described usually by different schemes, depending on the monomer systems used. Starting from the suspension of water-insoluble monomers, the formation mechanism of the crosslinked polymers might be schematically depicted such as in Figure 3.1.

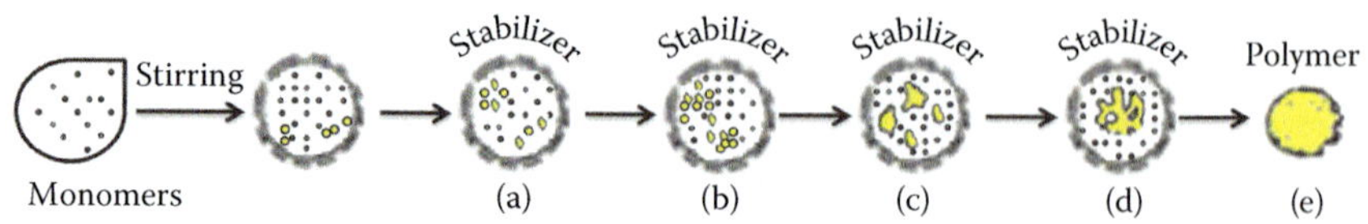

FIGURE 3.1 The suspension polymerization mechanism of the water-insoluble monomers. (Adapted from Vlad Cristina Doina. Suspension polymerization. In *Crosslinking (co)polymers obtained by suspension polymerization*, pp. 7–42. Iasi: PIM, 2008.)

Covering the monomer droplets with a stabilizer layer (Figure 3.1a) controls and reduces the coalescence process, helping to stabilize droplet size. Radical crosslinking (co)polymerization (CRR) has some specific behaviors, and the processes of chain initiation and growth are fundamental factors in the morphological differences of the resulting crosslinked structures. When considering the copolymers obtained from mono- and poly-unsaturated monomers by use of the CRR mechanism, we must assume that at the beginning of the (co)polymerization reaction, polyfunctional linear macromolecules—which remain dissolved in monomers (Figure 3.1a)—are formed. As the reaction progresses, the insoluble polymer granules that constitute "centers" or "nuclei" of polymerization appear (Figure 3.1b) and increase in number; the granules become sticky, bind to each other (Figure 3.1c), begin to contract, and polymerization continues according to the mechanism required by the type of monomers. Finally, a solid phase is formed (Figure 3.1d), which—under certain conditions—can contain pores.

As the polymerization progresses, the monomer is consumed, and the pores from the surface clog while the pores from inside close, resulting in a bead-shape polymer (Figure 3.1e). The SpP kinetics and mechanism depend on the nature, structure, and quantity of monomers; the water to monomers ratio; initiators and stabilizers; temperature; the form of reactor and agitator; the hydrodynamic regime; and so on. All these factors are essential for the products' properties (molecular weight distribution, the amount of the residual monomers, the particle morphology, side reactions, etc.) and their performance.

3.2.3 Factors Influencing the Formation of Crosslinked Beads

Bead formation imposes some conditions that are essential for obtaining uniform suspension of monomer droplets in aqueous phase and for reducing coalescence during polymerization (Hulubei et al. 2014).

3.2.3.1 The Stabilizer

To a great extent, the morphology of the resulting crosslinked (co)polymers is determined by the type of suspending agent used because it protects the droplets from collision during polymerization. Throughout the entire suspension polymerization process, the two phases (organic and water) remain separated (Guerrero et al. 2006). Polymerization takes place in the suspended droplets, which act like isolated microreactors.

In the continuous phase, a stable dispersion of droplets is ensured by stabilizing and stirring. The action mechanism of the stabilizer depends on its nature. Chemical structure is of prime importance. The balance between the polar groups (that ensure the dissolution of the stabilizer in water) and the hydrophobic groups (that physically interact with the monomer droplets' surface) can help locate efficient stabilizers for each suspension polymerization system.

It is well known that the final particle size and particle size distribution are decisively impacted by the initial drop size distribution of the monomer/water dispersion and the controlled breakage/coalescence processes, even from the earliest stages of polymerization.

Following are the features of a good stabilizer:

- The stabilizer does not inhibit or delay polymerization.
- It has an optimum ratio of hydrophilic/hydrophobic groups (promoting the solubilization in water and physically interacting with the monomer).
- It ensures the formation of a mechanically resistant layer around the protected droplets.
- The stabilizer ensures optimal growth of the environment viscosity to protect against the good monomers dispersing.

3.2.3.2 Stirring

Nonuniform agitation cannot generate a monodisperse system because there is a direct relationship between the dispersion velocity and the particle average size. Optimal stirring is ensured by the coalescence/dispersion ratio.

Some of the basic aspects for calculating stirring conditions include the ratio between the stirrer and the liquid level, the stirrer type, the viscosity ratio of the initial monomer and of the aqueous phase, the density of the two phases, the fluid flow parameters, and the Reynolds and Weber criteria (Leon et al. 2011; Vivaldo-Lima et al. 1997).

Agitation should be sufficient to prevent particle separation caused by the differences between the specific densities of the two phases. The monomer must be insoluble in water, with a surface tension lower than water. When the stirring is stopped, the system separates into two phases. Coalescence is prevented by a continuous control of the combined effect that is induced by the surfactant/stirring system. If there are no stabilizers in the system, an unstable dispersion occurs because of the breaking/coalescence ratio of the droplets.

3.2.3.3 Initiation

The radical initiators (equal to 0.1% of the monomer weight) are organic peroxides (benzoyl or lauroyl peroxide) or azo compounds (AIBN); these agents, insoluble in the aqueous phase, are usually solubilized in an organic solvent or in monomers after their introduction into the reactor (Vlad et al. 1994).

3.2.3.4 The Temperature

In the suspension polymerization process, the temperature usually is in the range 40°C–90°C, depending on the specific initiation type, kinetics, and on the system's reaction mechanism. The temperature value is selected to ensure that the initiator has a high decomposition rate (resulting in a great number of free radicals and, implicitly more nuclei and microspheres, as well as the diluents having a high solvating power). These factors contribute to smaller pore size because the polymerization rate increases with temperature (Okay 2000; Svec and Frechet 1995).

Polymer solubility in monomers is characterized by the solubility parameter. A slight difference between the solubility parameter of the monomer and the polymer implies that the mixing enthalpy of these two components is small. In such systems, sample agglomeration is avoided and the lack of defects suggests that the union of two particles and their adherence happens for a short period of time. Even though there is a density difference between the monomer and polymer, the droplets do not collapse because crosslinking occurs. The solubility parameter is a convenient measure that estimates polymer system solubility and influences bead morphology. The solvent–polymer compatibility can be semiquantitatively approximated using the solubility parameter (Tang et al. 2015; Vlad et al. 1997).

Bulk and suspension polymerization follow the same principles when all the system components are soluble in the monomer phase (initiators, macro radical, chain transfer agents). The rate of polymerization is not influenced by particle size and stabilizer type.

3.2.4 Free-Radical Crosslinking (Co)Polymerization Mechanism

In suspension polymerization, the crosslinked polymer formation from mono- and poly-unsaturated monomers takes place by free-radical crosslinking of the chains (Elliott and Bowman 2002). At the beginning of (co)polymerization, only a few linear polyfunctional macromolecules (that remain dissolved in monomers) are formed.

Radical crosslinking (co)polymerization (CRR) has several specific behaviors. The primary feature is the initiation process of the growing chains, a process that causes morphological differences in the resulting crosslinked structures.

From the standpoint of elementary reaction steps, free-radical crosslinking (co)polymerization starts with initiation, proceeds by propagation (in which the monomers are added to the active ends of the growing chains accompanied by chain transfer), and ends with termination reactions (when the active chain ends are deactivated).

Differences arise because of the unsaturated monomers' presence. This can be incorporated into the growing chain by a single double bond. Others, that remain pendant (free), are able to participate in (co)polymerization so that, gradually, the chains are united by the pendant double bonds, and thus forming crosslinks.

It often happens that chemical reactivity of the pendant bonds is equal to that of the polyunsaturated monomer; in this case, the copolymerization scheme is identical to that of the linear binary copolymerization.

The macroradical has different possibilities for growth by reacting with the following to form covalent bonds:

- A double bond of the polyunsaturated monomer
- The other pendant bonds

- An active end of the same chain (intramolecular cyclization crosslinked)
- An active end of a growing free chain (intermolecular crosslinking)

During the crosslinking process, two types of cyclization reactions take place: (1) cyclization with a pendant double bond from the same molecule and (2) cyclization with a double bond from another existing chain in the same molecule, the so-called "multiple cyclization."

At the beginning of the polymerization process, there are only a few pendant double bonds in the reaction system that can participate in the intermolecular crosslinking. The intramolecular cyclization reaction can occur if the cycle closing is supported by the chain flexibility.

For unequal reactivity of the vinyl groups, the scheme corresponds to that of ternary copolymerization. Therefore, there is exclusively intramolecular crosslinking at the beginning of process. Cycles of various sizes are formed, with the smallest cycles having the highest probability of formation. A very important cyclization can be observed when the reaction system contains a large amount of divinyl monomer (Vlad and Vasiliu 2010). However, a sharp decline in the pendant double bonds' reactivity appears. This feature of crosslinking (co)polymerization is explainable because many of these links are trapped inside cycles. For a complete conversion of monomers, the presence of double bonds in the reaction system supports the inaccessibility of these links.

Internal crosslinking leads to a very compact microgel structure. The insides of the pendant groups are extremely immobile and, although they are reactive, they do not react with monomers because the monomers cannot diffuse them.

Only the mobile pendant bonds from the periphery of this species (microgel) can react with macro radicals. Therefore, the trend of microgel formation depends on the length and flexibility of the primary chain; the possibility for accessing the unreacted double bonds from inside (their "apparent reactivity") depends on the microgel size, a fact reflected in the molecular weight value of the polymer chain. The decrease in the apparent reactivity of the pendant double bonds that produces crosslinking is due to the steric factors leading to the formation of inter- and intramolecular crosslinks. These factors strongly control diffusion in the polymerization reaction.

The specific behavior of the formation of a crosslinked (co)polymer structure by radical mechanism for a high crosslinker concentration is determined by: (1) an increase in the unreacted double bonds' concentration from the unsaturated monomer units (at low conversion)and (2) an increase in the number of pendant double bonds in the newly formed structure.

In the reaction system, a high concentration of polyunsaturated monomer involves a large number of double bonds. At low conversions, crosslinked (co)polymer structures with many unreacted double bonds are formed. When the conversion increases, the crosslinking efficiency decreases (the reaction is very difficult for both double bonds). Even at a low conversion, a decrease in the mobility of the formed structures takes place.

From the perspective of tridimensional structure evolution in crosslinking polymerization, the pendant double bonds' reactivity is strongly affected by the spatial correlations, resulting in an increase in the forming probability for cycles of different sizes and a decrease in double bonds' reactivity due to steric effects induced by the newly formed structure.

An increase in crosslinker concentration leads to the formation of compact macromolecules by internal crosslinking (the particles of the formed microgel bind together via the peripheral vinyl groups' participation during the growth of the macroradical in the liquid phase).

The pendant double bonds' reactivity is strongly affected by the spatial correlations induced by steric effects and, as consequence, the probability of cycle formation increases.

Possible reactions in the free radical crosslinking (co)polymerization of mono- and divinyl monomers are shown in the work of Vlad (2008).

Dilution favors contact among particles, and the macroradical is more likely to attack a peripheral vinyl link in the same particle. Paradoxically, the conversion increase does not increase the system heterogeneity. The spaces between the particles are gradually filled with the new polymer formed from monomers (by the peripheral vinyl groups participating from the all existing particles) so that the density fluctuation flattens (heterogeneity fades).

3.3 POROUS CROSSLINKED STRUCTURES BY SUSPENSION POLYMERIZATION

Adding a diluent in the polymerization system leads to the formation of heterogeneous permanent structures (with pores, after drying). To obtain rigid porous material, the dispersed liquid of the polymerizable mixture must contain not only monomers but also a diluent medium. The diluent medium is called "porogen" and must not react during the polymerization; it must remain within the newly formed beads, surrounded by the polymerized material. Finally, the porogen is removed and the places it occupied become the pores of the respective structure. Several special features of free-radical crosslinking (co)polymerization make the modeling of the network formation very difficult.

3.3.1 Role of Solvents

Because the monomers themselves act as diluents in the crosslinking (co)polymerization process and the monomers and (co)polymer concentration depend on the duration of polymerization, the question of how to find the critical conversion at which phase separation takes place (described as function: phase equilibrium versus conversion) becomes very important. Apart from the effect of phase separation, the diluent produces other effects such as lowering the glass transition temperature and the reaction rate, promoting the cyclization reaction, and changing the conformation of chains when it is removed from the network.

There are three types of inert media (and/or mixtures): solvents, nonsolvent, and linear polymers (Krueger et al. 1985).

When *inert swelling media* (solvents) are used, the macromolecular chains of the crosslinked polymer are completely solvated during all of the synthesis process, and the packing of nuclei becomes more ordered. With increasing dilution, the size of the chains among the microgel particles (nuclei) increases, and the solvent removal causes the reversible chains to collapse and return to the form that they had during polymerization. Beads with a large specific surface area and small pore diameter may be obtained. The resulting polymers become "porous by solvent."

An inert, ideal *nonsolvent medium* is difficult to find because, in the liquid phase, the reaction system always contains monomers—the nonsolvent character being exacerbated with an increase in the degree of transformation.

The precipitant character of the inert medium depends on (1) the initial concentration of monomers that are the proper solvent for the (co)polymer and (2) the inert medium/(co)polymer interaction parameter. This, in turn, determines an increase in the degree of permanent packing chains and, hence, an increase in the degree of effective crosslinking, which is manifested by the reduction in the volume. The difference between the volume of the copolymer and the initial monomers will increase as the amount of precipitant increases (resulting in an increase in porosity over the values corresponding to the quantity of inert medium from synthesis) (Bradford et al. 1955).

Linear polymers are part of the macromolecular inert media category. They bring a large number of variables to the reaction system because changes in chemical composition as well as in molecular weight and its distribution affect the porosity. When inert polymer media are used, the precipitant character of these macromolecular compounds depends on their degree of incompatibility with the newly formed (co)polymer. This incompatibility is responsible for phase separation at a much lower concentration for this type of inert medium as compared to others.

A mixture of linear polymer and good solvent as an inert medium can be used to obtain copolymers with a porous structure in more advantageous conditions than with the use of a single component as an inert medium. The presence of a solvent in addition to a polymer will contribute to a more uniform distribution of the inert polymer medium in the network, thus increasing the copolymer dispersion degree. The mixture can affect the size and distribution of pores because it acts as a unit of inert medium—the two components are inseparable and unable to react independently of one another.

A mixture of an inert swelling medium together with an inert precipitation medium with small molecules is frequently used to obtain copolymers having specific structures. The ratio of the two inert media depends on the polymer destination. In this case, the softer character of phase separation and the high degree of polymer dispersion lead to appreciable specific surfaces (Cheng et al. 1992).

Mixtures consisting of two inert precipitation media, one with small molecules and the other with a macromolecular structure, were studied theoretically. In that case, the size of the agglomerates greatly increased, resulting in products with very low mechanical strength.

3.3.2 Porous Polymers Obtained from Two Liquid Phase Systems

To understand pore formation during suspension polymerization, imagine that every drop is a spherical microreactor, the spherical form being ensured for the final polymer by stirring and interfacial tension. As a result of the crosslinking and solubility changes associated with increases in polymer chain length, the macromolecules formed inside this "microreactor" precipitate in the environment that surrounds them (porogen/monomer mixture). This *phase separation* occurs right at the beginning of polymerization, resulting in the formation of microscopic globular entities that begin to grow and are not coalescent because of crosslinking. Eventually, these come in contact with one another and, in an associated form, lead to clusters composed of interconnected cells and voids (pores). In essence, each drop is transformed into a microporous bead, the free volume being well correlated with the used porogen percentage.

The macroporous structure formation is determined exclusively by the behavior of the crosslinked chains obtained in the (co)polymerization reaction. The processes that occur in this stage can be described as follows: the initially formed microgel particles (nuclei) are linked together by chains that contain very few crosslinker molecules. This area of low crosslinking is very swollen in the diluent/monomer mixture. At drying, the new structure does not totally collapse, and the pores formation corresponds to the spaces between particles. The change in diluent thermodynamic quality reflects a different degree of association for the nuclei.

The inhomogeneous character of crosslinking polymerization is not a sufficient condition for the formation of the two-phase structure (the second phase refers to diluent or voids after drying). Therefore, the mechanism of phase separation (Hsiao et al. 1995) that appears near the limit of the thermodynamic stability is very important.

The phase separation in the presence of an inert medium occurring during the crosslinked formation was studied using Flory's thermodynamic theory (Dusek and Duskova-Smrckova 2000). Because the monomers themselves act as diluents, phase separation will be strongly influenced by the polymerization time, the phase equilibrium being described by the conversion function.

The model scheme for the formation of three-dimensional macroporous structures (Scheme 3.1) shows the steps that occur in the suspension polymerization process. Each particle (that can be seen as a "microreactor") is composed of monomer solution, initiator, and diluent (I). This organic phase is suspended in an aqueous phase containing the suspension stabilizer.

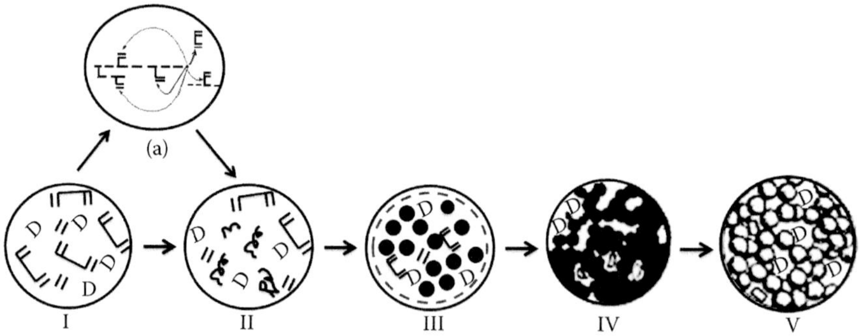

SCHEME 3.1 Model scheme for the formation of three-dimensional porous structures. (Adapted from Vlad Cristina Doina. Suspension polymerization. In *Crosslinking (co)polymers obtained by suspension polymerization*, pp. 7–42. Iasi: PIM, 2008.)

The polymerization reaction results in (co)polymer molecules rich in bis- or polyunsaturated components that still contain a large number of unreacted double bonds. The propagation, cyclization, and crosslinking reactions lead to weak intermolecular crosslinked primary nuclei (due to the diluent presence), intramolecular crosslinked microgels, and linear chains soluble in the unreacted monomers.

The presence of a low number of intermolecular bonds among the newly formed nuclei gives rise to a low crosslinking area that swells in the diluent/monomer mixture. The crosslinked polymer formation causes phase separation between the crosslinked polymer and the inert medium, with the monomers being distributed between the two phases (II). Because the weak crosslinked polymer is swollen in the rest of monomers, it behaves like liquids that (due to interfacial tensions) take the most stable form in terms of energy. In this way, mass separation of microspheres takes place (III). During the formation of this microgel, the minimum droplet energy is achieved based on the network deformation (observed as volume change). At a high monomer conversion, the formation of macro gelation and microsphere agglomerates takes place (IV). The first stage of the porous structure formation is the production and aggregation of highly crosslinked microspheres. The second stage is the binding and fixing of the microspheres and of their agglomerates. This is achieved by advancing the crosslinking polymerization reaction when the formation of new nuclei takes place. Based on the monomers' consumption, these nuclei bind the existing structures in the system (V). The inert medium is now in the spaces between microspheres and their agglomerates. These places occupied by the inert medium, after its removal, become the network pores, and they determine the macroporous structure properties (shape, size distribution, etc.).

3.4 THE MECHANISM OF PHASE SEPARATION

All the porosity formation theories are based primarily on phase separation (PS) and the solvating power of the porogen. A large amount of crosslinker is needed for producing a macroporous resin, to facilitate the phase separation between polymer and porogen during synthesis. A variety of phase morphologies are obtained and decided by the competition between the kinetics of phase separation and the crosslinking chemical reaction. The physical properties of a crosslinked polymer are strongly dependent on its phase-separation morphologies. Among the methods used to investigate phase separation (PS) are as follows:

- Thermally induced phase separation (TIPS): an increased reaction temperature, or a high concentration, produces a high-connectivity phase structure so that the polymerization surpasses the phase separation; using a low-temperature stage, a moderate decrease in phase separation is observed at a lower nucleation temperature (Kim et al. 2016; Padilla et al. 2011; Stieger et al. 2003).
- Nonsolvent-induced phase separation (NIPS).
- Combined NIPS-TIPS (N-TIPS).
- Polymerization-induced phase separation (PIPS) (Yuhong 2013).

The increasing size of the growing polymer molecules and the resulting morphology (stabilized by gelation in crosslinking polymerization) are the main factors affecting PIPS. The PIPS processes are influenced by thermodynamic and kinetic factors that can determine: *stable systems* (where no phase separation is produced), *metastable systems* (when the phase separation may take place), or *unstable systems* (for which the phase separation does not take place).

The phase separation mechanism during the porous crosslinked polymer synthesis assumes

1. The sudden appearance of heterogeneity when the threshold concentrations of crosslinker and diluent are achieved
2. The formation of a porous structure with very large sizes compared to that of the primary nuclei
3. Important influences relative to the quantity and quality of diluent

The change in solvent quality from good to poor, from the increase in temperature, eventually leads to phase separation (Stieger et al. 2003).

When reaction takes place in both the dispersed and continuous phases, in spite of the low monomer solubility in water, the thermodynamic equilibrium between the continuous and dispersed phase is established before polymerization starts (Nzihou et al. 1997). A porogen can be used to increase the partition of a water soluble monomer in droplets when the continuous phase is water (Gokmen et al. 2012).

In order to obtain a heterogeneous system, phase separation must take place during the crosslinking process so that the dispersed structure can be fixed through the formation of new crosslinks.

A peculiarity of crosslinked (co)polymerization is the formation of networks at relatively low conversions at the moment when these structures (the networks) are strongly swollen in the mixture diluent-monomer, resulting in the phase separation starting close to the gel point.

The influence of the quantity and inert medium thermodynamic quality in the pore morphology formation depends on the moment in which phase separation takes place. As a function of the (co)polymerization reaction stage, phase separation, can take place at three different times:

- *Before the gel point*, when inferior coarse structures are obtained. This occurs because the difference between the amount of polymer phase and diluent is high (in this case, before the structure to be fixed by crosslinking, the coalescence phenomenon occurs).
- *Long after the gel point* (after the crosslinks formation), when the high crosslinking density from the polymer prevents the liquid microseparation and an exclusion occurs outside of the polymer (macrosinereza). A fine pore structure that determines the high specific surface appearance is created.
- *At the gel point.* The best moment to start phase separation is at the gel point, when the dispersal structure is fixed by the formation of new crosslinks.

During the formation of networks from the unstable systems, when the crosslinking density changes are very fast, the restoring balance takes place due to the crosslinked structure reorganization. Because of the rapid changes in composition, the phase separation is accompanied by a macrogranulation due to the coalescence of the increasing particles. At a high crosslinker concentration and a high dilution, the volume transition phenomenon appears together with the phase separation.

Macrosyneresis occurs after the gel point, when the crosslinks form a coherent network. The beginning of phase separation can be established by combining equilibrium effects that change in the system because of the conversion and the network swelling.

Microsyneresis is a consequence of the polymer–solvent interaction and occurs during the network formation (crosslinking). Microsyneresis is not stable in time; it is a nonequilibrium phase that passes in an equilibrium syneresis (macrosinereza). This transition from micro- to macrosyneresis is due to network deformation, which explains the volume change that takes place in the crosslinked structures.

The porous copolymer volume often is higher than the liquid phase volume (consisting of monomers and inert medium) because the network monomers adsorption on the swollen network determines the network's volume expansion.

In summary, the mechanism formation of the porous structures is due to the nonhomogeneous nature of the crosslinking (co)polymerization as well as to the phase separation process (determined by the thermodynamics of the system).

3.5 THE PARTICLE MORPHOLOGY AND CHARACTERIZATION METHODS

Particle morphology is the factor that determines the commercial applications for these polymers. In a dried state, the (co)polymer beads are characterized by their true (skeletal) and apparent density, total porosity, total volume of pores, pore sizes, distribution of pore size, and their specific surface area.

The shape and structure (morphology) of the beads depend on the polymerization conditions and on the physical properties of the system. Polymerization conditions that particularly influence particle morphology are as follows:

- The coalescing and dispersing drops before and during polymerization
- The density change between monomer phase and polymer phase
- The monomer's solubility in the polymer

In cases where the monomers are soluble during the copolymerization reaction, they are uniformly distributed in the reaction phase (monomer in the polymer droplet). But, in cases where the copolymer swells in monomers, two different compositions for the polymer droplet are obtained—one that is polymer-rich and another that is monomer-rich. If the compositions are sufficiently different, various structures can be obtained by the polymerization of the two phases.

The swelling capacity of various copolymers is a quantitative method for characterizing the crosslinked beads. The swelling and de-swelling volume depends on crosslinked density (Vlad and Mihailescu 2001).

The coalescence/dispersion factor is responsible for dispersing or particle agglomeration. Changes in the system's qualitative and quantitative components ratio during the crosslinking reaction can lead to particle collapse (the lower polymer density, when compared with that of the monomer, generates pearls with a lower volume than that of the initial droplet).

Control of the pore size distribution is of great importance in (macro) porous polymer design. For example, small pores and large specific surfaces are typical characteristics of the beads used in column chromatographic fillers, while beads with large pores are used in nucleic acid separation or in enzymes immobilization.

Polydispersity is characterized by the use of specific equipment for high precision optical measurements—atomic force microscopy (AFM), scanning electron microscopy (SEM), and transmission electron microscopy (TEM).

3.6 ADVANTAGES AND AREAS OF USE

Numerous suspension polymerization experiments have highlighted the technical and economic advantages:

- Low costs compared with the wide variety of particle properties.
- Excellent heat transfer and handling of materials during the monomers' transformation into the polymer.
- Ability to control the bead size.
- The use and separation of the resulting beads are simple procedures.
- There are a low number of components used in the polymerization system (compared to emulsion techniques).
- The purification procedure for the final product is relatively simple.
- May be stored in silos or tanks according to the manufacturer's instructions regarding storage temperature and bead shape.

Typical crosslinked polymers do not possess porosity other than when they are swollen in solvents. After solvation, the polymer chains are separated from each other, and they form pores that are places (spaces) filled with solvent. This limited swelling takes place for low crosslinked polymers or macroporous gels (products easily deformable under pressure) and generally precludes their use in bead form (in liquid chromatography, for example).

Macroporous polymer beads, synthesized by suspension polymerization, are a class of polymer beads containing a permanent, well-developed porous structure in a dry state and having a wide range of applications—such as support for catalysts, immobilization of enzymes, HPLC columns, liberation of active substances, or adsorbents.

The beads are obtained in the presence of porogen (inert substance to monomers) that remains embedded inside the polymer bead and, finally, is removed to create the grain pores. The internal structure of the macroporous beads consists of many interconnected cavities (pores) of different dimensions, with the beads' stiffness due to extensive crosslinks. Their porous structure improves the diffusion of different solutes through the polymeric network. Because these networks have particular properties, such as a high degree of crosslinking and a rigid structure both in swollen and dry state, the study of their synthesis conditions is very interesting (Gomez et al. 2012).

Macroporous rigid polar organic beads are commonly used for aqueous size exclusion chromatography (SEC), for hydrophobic interaction and affinity chromatography of biopolymers (Svec and Regnier 2002), or to investigate retention properties (Mingcheng et al. 2003).

Significant progress has been made in developing porous polymer designs used in fields that focus on improving life (cosmetics, eye implants, and bone) (Chalasani et al. 2007).

Recent greater interest in porous polymers is due to their controllable porous structure as well as their mass transfer and binding capacity. By using one-step swelling and the polymerization method, the monodispersed macroporous particles are used as a supporting matrix. This novel, macroporous core–shell-imprinted material may become a powerful tool for rapid and efficient enrichment and separation of target compounds from complicated samples (Yongliang et al. 2014).

Two types of morphology in crosslinked polymer resins are commercially available and have been investigated: a gel type and a macroporous resin. The relatively rigid macroporous resin is available as hard opaque beads with permanent pores (Van de Steene et al. 2014). By optimizing the porous polymer synthesis methods through the application of a freeze-drying technique, very efficient adsorbents for removing arsenic have been obtained (Taleb et al. 2015).

Storage is an essential condition in the development of hydrogen-fueled vehicles. In this regard, porous carbon materials have attracted considerable attention as attractive candidates for hydrogen storage due to their good adsorptive capacity, low density, low cost, and high surface area (Jianwei et al. 2012).

3.7 REMARKS AND FUTURE DIRECTIONS

This chapter discusses crosslinked polymers with permanent porosity and the factors affecting their synthesis, formation mechanism, and properties. A characterization of these polymers can be made based on network formation theory or on experimental results. Because of differences between experimental results and theoretical predictions, individual methods are used to indicate areas of uncertainty that need further exploration.

Until the last century, crosslinked polymers were studied only for their applicative properties. Today, characterization techniques allow the study of factors affecting the formation mechanism as well as the balance between reaction parameters and properties. Depending on the required properties, this suggests that a porous crosslinked polymer can be prepared for a specific purpose, considering the influences and mechanisms of phase separation that develop during synthesis.

ACKNOWLEDGMENTS

This work was supported by a grant from the Romanian National Authority for Scientific Research, CNCS-UEFISCDI, project no PN-II-ID-PCE-2011-3-0937, 2016 stage.

REFERENCES

Barbetta Andrea, Elena Barigelli, and Mariella Dentini. Porous alginate hydrogels: Synthetic methods for tailoring the porous texture. *Biomacromolecules*. 10 no. 8 (2009): 2328–2337.

Becher Paul, and Birkmeier R. L. The determination of hydrophile-lipophilc balance by gas-liquid chromatography. *The Journal of the American Oil Chemists' Society* (1964): 169–172.

Bradford E. B., and J. W. Vanderhoff. Electron Microscopy of Monodisperse Latexes. *J. Appl. Phys*. 26 (1955): 864–871.

Chalasani Rajeev, Poole-Warren Laura, Conway Max R., and Ben-Nissan Besim. Porous orbital implants in enucleation: A systematic review. *Surv. Ophthalmol.* 52 no. 2 (2007): 145–155.

Cheng Chieh-Min, Micale Fortunato Joseph, Vanderhoff John William, and El-Aasser Mohamed S. Synthesis and characterization of monodisperse porous polymer particles. *J. Polym. Sci. Part. A*. 30 no. 2 (1992): 235–244.

Degirmenci Isa, Avcı Duygu, and Aviyente Viktorya K. Density functional theory study of free-radical polymerization of acrylates and methacrylates: Structure-reactivity relationship. *Macromolecules*. 40 no. 26 (2007): 9590–9602.

Dragan Stela Ecaterina, and Vlad Cristina Doina. New developments in the synthesis of cross-linked (co)polymers as beads particles. In *New trends in nonionic (co)polymers and hybrids*, Ecaterina Stela DRAGAN (ed.), pp. 121–166. New York: Nova Science, 2006.

Dusek Karel. Special features of network formation by chain crosslinking copolymerization. *Collect. Czech. Chem. Commun*. 58 no. 10 (1993): 2245–2265.

Dusek Karel, and Duskova-Smrckova Miroslava. Network structure formation during cross-linking of organic coating systems. *Prog. Polym. Sci.* 25 no. 9 (2000): 1215–1260.

Dusek Karel, and MacKnight William J. Cross-linking and structure of networks. In *Cross-linked polymers*, Ray A. Dickie, S. S. Labana, Ronald S. Bauer (eds.), pp. 2–27. Washington, DC: ACS, 1988.

Elliott Jeannine E., and Bowman Christopher N. Effects of solvent quality during polymerization on network structure of cross-linked methacrylate copolymers. *J. Phys. Chem. B* 106 no. 11 (2002): 2843–2847.

Falk Benjamin, and Crivello James V. Synthesis of epoxy-functional microspheres by cationic suspension photopolymerization. *Chem. Mater.* 16 no. 24 (2004): 5033–5041.

Gokmen Talha M., and Du Prez Filip E. Porous polymer particles—A comprehensive guide to synthesis, characterization, functionalization and applications. *Prog. Polym. Sci.* 37 no. 3 (2012): 365–405.

Gomez Cesar G., Pastrana Gustavo, Zuzek Daniel Serrano Estanislao, Villar Marcelo A., and Strumia Miriam C. Macroporous poly (EGDMA-co-HEMA) networks: Morphological characterization from their behaviour in the swelling process. *Polymer.* 53 no. 14 (2012): 2949–2955.

Gridnev Alexei A., and Ittel Steven D. Catalytic chain transfer in free-radical polymerizations. *Chem. Rev.* 101 no. 12 (2001): 3611–3659.

Guerrero Sanchez Carlos, Erdmenger Tina, Sereda Petr, Wouters Daan, and Schubert Ulrich S. Water-soluble ionic liquids as novel stabilizers in suspension polymerization reactions: Engineering polymer beads. *Chem. Eur. J.* 12 no. 35 (2006): 9036–9045.

Hsiao Yu Ling, Maury Elise E., DeSimone Joseph M., Mawson Simon, and Johnston Keith P. Dispersion polymerization of methyl methacrylate stabilized with poly (1, 1–dihydroperfluorooctyl acrylate) in supercritical carbon dioxide. *Macromolecules.* 28 no. 24 (1995): 8159–8166.

Hulubei Camelia, Vlad Cristina Doina, Stoica Iuliana, et al. New polyimide-based porous crosslinked beads by suspension polymerization: Physical and chemical factors affecting their morphology. *J. Polym. Res.* 21 no. 9 (2014): 514–530.

Jacobeli H, M. Bartholin, and A. Guyot. Styrene-divinylbenzene copolymers II. Influence of the nature of the diluent on the texture of macroporous copolymers. *Anghew. Makromol. Chem.* 80 (1979): 31–51.

Jianwei Fu, Minghuan Wang, Chao Zhang, Peng Zhang, and Qun Xu. High hydrogen storage capacity of heteroatom-containing porous carbon nanospheres produced from cross-linked polyphosphazene nanospheres. *Mater. Lett.* 81 (2012): 215–218.

Kim Jeong F., Kim Ji Hoon, Lee Young Moo, and Drioli Enrico. Thermally induced phase separation and electrospinning methods for emerging membrane applications: A review. *AIChE. J.* 62 no. 2 (2016): 461–490. DOI: 10.1002/aic.15076.

Kotoulas Costas, and Kiparissides Costas. A generalized population balance model for the prediction of particle size distribution in suspension polymerization reactors. *Chem. Eng. Sci.* 61 no. 2 (2006): 332–346.

Krueger C. T, and J.E. Schwartz. Eur. Pat. Appl., 85116431 (December 21, 1985).

Kun Kenneth A., and Kunin Robert. Pore structure of some macroreticular ion exchange resins. *J. Polym. Sci. Part B.* 2 no. 8 (1964): 587–591.

Kun Kenneth A., and Kunin Robert. Macroreticular resins. III. Formation of macroreticular styrene-divinylbenzene copolymers. *J. Polym. Sci. Part A.* 6 no. 10 (1968): 2689–2701.

Leon Maria A., Roman Tschentscher, Nijhuis Alexander T., John van der Schaaf, and Schouten Jaap C. Rotating foam stirrer reactor: Effect of catalyst coating characteristics on reactor performance. *Ind. Eng. Chem. Res.* 50 no. 6 (2011): 3184–3193.

Mester Zoltan, Mohan Aruna, and Fredrickson Glenn H. Macro- and microphase separation in multifunctional supramolecular polymer networks. *Macromolecules.* 44 no. 23 (2011): 9411–9423.

Mingcheng Xu, Peterson Dominic S., Rohr Thomas, Svec Frantisek, and Frechet Jean M. J. Polar polymeric stationary phases for normal-phase HPLC based on monodisperse macroporous poly(2,3-dihydroxypropyl methacrylate-co-ethylene dimethacrylate) beads. *Anal. Chem.* 75 no. 4 (2003): 1011–1021.

Mohamed Mohamed H., and Wilson Lee D. Porous copolymer resins: Tuning pore structure and surface area with non-reactive porogens. *Nanomaterials.* 2 no. 2 (2012): 163–186.

Nzihou Ange, and Alain Ricard. Modeling of liquid-liquid equilibrium and mass transfer. Suspension copolymerization of acrylic acid and methyl methacrylate in water. *Ind. Eng. Chem. Res.* 36 no. (1997): 4815–4820.

Okay Oguz. Macroporous copolymer networks. *Prog. Polym. Sci.* 25 no. 6 (2000): 711–779.

Padilla Adora M., Chou Shin G., Luthra Sumit, and Pikal Michael J. The study of amorphous phase separation in a model polymer phase-separating system using Raman microscopy and a low-temperature stage: Effect of cooling rate and nucleation temperature. *J. Pharm. Sci.* 100 no. 4 (2011): 1362–1376.

Rabelo D., and Coutinho Fernanda Margarida Barbosa. Porous structure formation and swelling properties of styrene-divinylbenzene copolymers. *Eur. Polym. J.* 30 no. 6 (1994): 675–682.

Rouquerol Jean, Avnir David, Fairbridge Craig W., et al. Recommendations for the characterization of porous solids (Technical Report) *Pure. Appl. Chem.* 66 no. 8 (1994): 1739–1758.

Sederel Willem L., and De Jong Gerhardus J. Styrene–divinylbenzene copolymers. Construction of porosity in styrene divinylbenzene matrices. *J. Appl. Polym. Sci.* 17 no. 9 (1973): 2835–2846.

Silva Fabricio Machado, Lima Enrique Luis, and Pinto Jose Carlos. Acrylic Acid/Vinyl Acetate Suspension Copolymerizations. 2.Modeling and Experimental Results. *Ind. Eng. Chem. Res.* 43 no. 23 (2004): 7324–7342.

Stieger Markus, and Richtering Walter. Shear-induced phase separation in aqueous polymer solutions: Temperature-sensitive microgels and linear polymer chains. *Macromolecules.* 36 no. 23, (2003): 8811–8818.

Svec Frantisek, and Frechet Jean M. Temperature, a simple and efficient tool for the control of pore size distribution in macroporous polymers. *Macromolecules.* 28 no. 22 (1995): 7580–7582.

Svec Frantisek, and Regnier Fred E. Organic polymer support materials. In *HPLC of biological molecules*, Gooding K., Regnier F.E. (eds.), pp. 17–48. New York: Marcel Dekker, 2002.

Taleb Khaled, Markovski Jasmina, Milosavljevic Milutin, et al. Efficient arsenic removal by cross-linked macroporous polymer impregnated with hydrous iron oxide: Material performance. *Chem. Eng. J.* 279 (2015): 66–78.

Tang Yuan-hui, He Yan-dong, and Wang Xiao-lin. Investigation on the membrane formation process of polymer–diluent system via thermally induced phase separation accompanied with mass transfer across the interface: Dissipative particle dynamics simulation and its experimental verification. *J. Membr. Sci.* 474 (2015): 196–206.

Van de Steene Evelien, De Clercq Jeriffa, and Thybaut Joris W. Ion-exchange resin catalyzed transesterification of ethyl acetate with methanol: Gel versus macroporous resins. *Chem. Eng. J.* 242 (2014): 170–179.

Vivaldo-Lima Eduardo, Wood Philip E., and Hamielec Archie E. An updated review on suspension polymerization. *Ind. Eng. Chem. Res.* 36 no. 4 (1997): 939–965.

Vlad Cristina Doina. Suspension polymerization. In *Crosslinking (co)polymers obtained by suspension polymerization*, pp. 7–42. Iasi: PIM, 2008.

Vlad Cristina Doina, and Mihailescu Silvia. Cross-linked polymers based on diallylmaleic ester. *Eur. Polym. J.* 37 no. 1 (2001): 71–77.

Vlad Cristina Doina, Poinescu Ignat, and Barbu Maria. Behaviour of porous crosslinked poly(ethylene dimethacrylate esters). *Eur. Polym. J.* 30 no. 8 (1994): 863–868.

Vlad Cristina Doina, and Vasiliu Silvia. Crosslinking polymerization of polyfunctional monomers by free radical mechanism. *Polymery.* 55 no. 3 (2010): 181–185.

Weigang Lu, Yuan Daqiang, Zhao Dan, et al. Porous polymer networks: Synthesis, porosity, and applications in gas storage/separation. *Chem. Mater.* 22 no. 21 (2010): 5964–5972.

Yongliang Liu, He Yonghuan, Jin Yulong, Huang Yanyan, Liu Guoquan, and Zhao Rui. Preparation of monodispersed macroporous core–shell molecularly imprinted particles and their application in the determination of 2,4-dichlorophenoxyacetic acid. *J. Chromatogr. A.* 1323 (2014): 11–17.

Yuhong Liu. Polymerization-induced phase separation and resulting thermomechanical properties of thermosetting/reactive nonlinear polymer blends. *J. Appl. Polym. Sci.* 127 no. 5 (2013): 3279–3292.

4 Advanced Materials Based on Multicomponent Polymeric Systems

Mirela Teodorescu, Simona Morariu, and Maria Bercea

CONTENTS

4.1 OVERVIEW OF LITERATURE CONCERNING ADVANCED MATERIALS

About five decades ago, Wichterle and Lim developed a new kind of material called hydrogels (Wichterle and Lim 1960). Hydrogels are three-dimensional hydrophilic (co)polymeric networks that can swell in the presence of water or physiological fluids; they are highly applicable in a variety of fields—such as pharmaceutics, biomedicine, bioengineering, or the food industry. They can be obtained by physical junctions (such as crystalline structure formation, secondary forces, chain entanglements, or worsening the solvent quality) or through chemical crosslinks (covalent bonds) and can be designed to possess targeted characteristics. Among these characteristics, the most important are their biocompatibility with blood, body fluids, and natural tissue medium; their environmentally responsive properties (the ability to turn from a solution to a gel when a specific stimulus is applied); and their capacity to swell in large amounts of water without dissolving. Other characteristics of hydrogels—such as temperature and pH sensitivity as well as swelling degree—dictate their usage in drug delivery systems, wound dressings, tissue engineering, contact lenses, and many other areas.

At present, poly(vinyl alcohol) (PVA) is recognized as among the earliest and the most commonly used synthetic polymers for hydrogel preparation (Kamoun et al. 2015). PVA is a well-known water-soluble and biologically friendly polyhydroxy polymer that has been widely studied and utilized in various applications due to its very interesting chemical and physical properties. Among the most remarkable PVA properties are the following: its easily processable nature, well-established biocompatibility (Hassan and Peppas 2000; Curley et al. 2014), biodegradability (Abd El-Mohdy and Ghanem 2009; Jose et al. 2014), very low or nontoxicity, low cytotoxicity (DeMerlis and Schoneker 2003; Yang et al. 2010), excellent chemical resistance and thermostability, permeability, hydrophilic character, and the fact that it is noncarcinogenic (Hoffman 2012) and nonimmunogenic (Azuma et al. 2007). The high density of hydroxyl groups present in PVA gives it a pronounced hydrophilic character (Qiao et al. 2010) and also facilitates the formation of semicrystalline structures through intramolecular hydrogen bonds.

The solubility of PVA in water is mainly controlled by the degree of hydrolysis and tacticity. By increasing the degree of hydrolysis and syndiotacticity, a decrease in solubility in water was observed, resulting in a change of phase diagram. Choi investigated in detail the effect of stereoregularity on the physical gelation of PVA solution (Choi et al. 2001). With high syndiotacticity of PVA, the intermolecular hydrogen interactions among adjacent chains increase and gelation occurs without a serious crystallization. Solutions of PVA with syndiotactic content lower than 56% have shown gelation behavior similar to those of atactic PVA, and the liquid–liquid phase separation is involved in the gelation process for low polymer concentrations.

In order to reduce PVA's hydrophilic character, several methods can be used, such as blending (Abdelrazek et al. 2010; Bercea et al. 2011, 2014, 2015a; Nkhwa et al. 2014; Kamoun et al. 2015), crosslinking (Rakesh and Deshpande 2010; Jamnongkan et al. 2014), grafting (Sanli et al. 2007; Malikov et al. 2014), or copolymerization (Martens et al. 2003).

Moreover, PVA is highly capable of forming fibers, films, and gels, using various techniques (Chiellini et al. 2008; Li et al. 2013). Studies reported to date present three main categories for PVA hydrogel preparation methods: chemical (Kim et al. 2002; Tsai et al. 2010; Binetti et al. 2014; Bercea et al. 2015b), physical (Alves et al. 2011; Gupta et al. 2012; Bercea et al. 2013, 2015c; Nkhwa et al. 2014), and irradiative (Nho and Park 2002; Zheng et al. 2009). Chemical crosslinking can take place using a covalent crosslinking agent (e.g., glutaraldehyde, GA), *N,N′*-methylenebisacrylamide) or by ionizing radiation through the use of electron beams (Binetti et al. 2014), ultraviolet light (Nguyen and Liu 2013), or γ-radiation (Park and Nho 2003; Mahmoud et al. 2014). The major inconvenience in using crosslinking agents and reagents is their toxicity and how difficult it is to remove them from the resulting products. Moreover, if they are not properly removed, they can affect the nature of incorporated substances such as drugs, proteins, or cells. In contrast, radiation crosslinking is a very "clean" method, with easy processing control. It does not require the use of toxic initiators or crosslinking agents and also simultaneously provides sterilization (Nguyen and Liu 2013). One of the most commonly used methods of physically crosslinking PVA typically is carried out by freezing and thawing its aqueous solutions to obtain the well-known cryogels (Bercea et al. 2013; Gonzalez and Alvarez 2014). This method was pioneered by Peppas (Peppas 1975). The resulting PVA hydrogels are viscoelastic in nature and have excellent weight-bearing properties, a very good swelling capacity, and an ability to encapsulate a large amount of water (Gonzalez and Alvarez 2014). These aspects have been greatly explored in the design of biomaterials for a large variety of fields, among which the top applications are found in biomedical, pharmaceutical, and bioengineering areas.

PVA gained great importance in the biomedical and pharmaceutical fields, and it has been extensively studied, leading to biomaterials for tissue engineering (artificial pancreas, synthetic vitreous body [Feng et al. 2013], artificial skin, synthetic articular cartilage [Shi et al. 2014], and meniscus [Awadallah-F 2014]) and regenerative medicine (scaffolds for repairing and regenerating various tissues and organs [Ye et al. 2014]), as well as for drug delivery systems (topical applications as wound and burn dressings, transdermal systems, stimuli-responsive systems [Gao et al. 2010], controlled release applications [Jain et al. 2013]), and medical devices (cardiovascular device [Millon et al. 2007] and other vascular implanting [Alexandre et al. 2014]).

Other applications of PVA-based hydrogels include different coatings for metals (e.g., silver nanoparticles, or Ag NPs [Eisa et al. 2012]), carbon nanotubes (CNTs) (Malikov et al. 2014), or cellulose (Han et al. 2014), as well as paper coatings, water-soluble packaging films, textile sizing, and hydrophilic membranes for pervaporation (Zhang et al. 2012).

Previously, PVA has been blended with different synthetic polymers, or various other inorganic compounds, in order to change or improve its properties. Synergistic effects of PVA and the imbedded material (i.e., synthetic polymers or inorganic fillers) properties give rise to new and desirable characteristics, making the resulting multicomponent polymeric system very useful for a wide range of applications. Among the materials most frequently used in combination with PVA are the synthetic polymers and inorganic compounds that are presented in the following sections.

4.2 MIXTURES OF POLY(VINYL ALCOHOL) WITH SYNTHETIC POLYMERS

4.2.1 Poly(vinyl alcohol)/Poly(vinylpyrrolidone)

Poly(vinylpyrrolidone) (PVP) is one of the synthetic polymers that have drawn special attention in an attempt to improve the properties of poly(vinyl alcohol) (PVA) hydrogels.

It exhibits remarkable characteristics, such as biocompatibility, extremely low cytotoxicity, a complexing ability with numerous compounds and polymers, good solubility in water, easy processability, good environmental stability, and many others (Teodorescu and Bercea 2015). Either alone or in combination with another polymer, PVP has been used on a large scale for various applications in biomedical (Teodorescu and Bercea 2015), pharmaceutical (Folttmann and Quadir 2008), and cosmetic fields, as well as in the food industry (de Azeredo 2013) and in many technical applications.

The synergistic effects of associating PVA and PVP properties in PVA/PVP mixtures have led to a wide range of new biomaterials. Studies published so far confirm that PVA and PVP form a thermodynamically miscible pair (Lewandowska 2005; Abd El-Mohdy and Ghanem 2009) and because the proton-accepting carbonyl group is present in the PVP ring and the hydroxyl groups of PVA, their chains interact by hydrogen bonds. These types of interactions dictate the new properties in the resulting PVA/PVP mixtures in terms of solubility, improved lubrication (Shi and Xiong 2013; Shi et al. 2014), and better mechanical and tribological characteristics (Ma et al. 2009; Shi and Xiong 2013; Shi et al. 2014).

PVA/PVP hydrogels have been prepared using different procedures, including the following methods: chemical (Kim et al. 2002; Lu et al. 2003; Qiao et al. 2010), physical (freezing-thawing [Ma et al. 2009; Huang et al. 2011; Shi and Xiong 2013; Teodorescu et al. 2016]), solution casting (Ragab 2011), electrospinning, or radiation (γ-irradiation [Abd El-Mohdy and Ghanem 2009]), and ultraviolet (UV) light (Lu et al. 2003) that induce the formation of the crosslinking structure. It should be noted that, although many scientists have used the chemical crosslinking method, the trend now is to shift to more friendly methods such as physical (Teodorescu et al. 2016) or radiation crosslinking. Some authors use combinations of these techniques, such as a cyclic freezing-thawing method followed by γ-irradiation (Shi et al. 2014).

PVA/PVP blends processed as mentioned result in fibers, films (Ragab 2011; Eisa et al. 2012), gels (Abd El-Mohdy and Ghanem 2009; Huang et al. 2011), and interpenetrating polymer networks (IPNs) (Kim et al. 2002; Lu et al. 2003; Qiao et al. 2010), with a multitude of applications. Among the most explored applications we will review are wound dressings (Singh and Pal 2011; Gökmeșe et al. 2013), nucleus pulposus (Thomas et al. 2004; Joshi et al. 2006), synthetic articular cartilages (Katta et al. 2007; Ma et al. 2009; Shi and Xiong 2013; Shi et al. 2014), pervaporation membranes (Lu et al. 2003), and electrochemical applications (Qiao et al. 2010).

4.2.2 Poly(vinyl alcohol)/Poly(ethylene glycol)

Another synthetic polymer used in combination with PVA, in order to improve its properties, is poly(ethylene glycol) (PEG) (Martens et al. 2003; Mansur et al. 2004; Dutta 2012). This is a water-soluble, biocompatible, biodegradable, transparent, and viscous polymer that has been used in the industrial manufacturing of biomedical materials.

Studies focused on the applicability of PVA hydrogels for artificial articular cartilage concluded that because these hydrogels possess insufficient strength and toughness, they are not appropriate as a cartilage substitute (Stammen et al. 2001; Azuma et al. 2007). As a method of increasing the strength of PVA hydrogels, researchers tried to reduce the water content by thermal annealing (Bodugoz-Senturk et al. 2008). Although this led to stronger hydrogels, their lubricity decreased as a result of the reduction in equilibrium water content and the collapse of the pores (Bodugoz-Senturk et al. 2008). As a solution to this problem, Ruberti and Braithwaite proposed a new method (the theta-gel method), using a low molecular weight PEG, as gelling agent, to physically crosslink PVA hydrogels (Ruberti and Braithwaite 2003). The freezing of PVA aqueous solutions determines the formation of ice crystals in the amorphous region and growing polymer crystallites, which leads to a porous structure after thawing (Bercea et al. 2013; Morariu et al. 2015). By gradually thawing PVA/water systems, the polymer-enriched microphases and associations of polymer chains appear. During thawing, PVA chains inside these domains interact with hydrogen bonds, forming crystallites. As an alternative for the freezing-thawing method, adding PEG to the PVA aqueous solution contributes to phase separation and crystallization without freezing, thus forming a physically crosslinked hydrogel (see Figure 4.1a) (Ruberti and Braithwaite 2003; Bodugoz-Senturk et al. 2008). The inconvenience of the theta-method is that, although PEG fills and protects the pores during annealing, it diffuses out when the gel is rehydrated. Therefore, pores are permeated by water, which weakens the hydrogel strength (Bodugoz-Senturk et al. 2009).

To prevent this, Bodugoz-Senturk et al. proposed adding permanently crosslinked highly hydrophilic polymers such as poly(acrylamide) (PAAm) to the PVA network, which can protect it during annealing and also help maintain its strength and lubricity (see Figure 4.1b) (Bodugoz-Senturk et al. 2009).

A study published by Dutta presented PVA/PEG/calcium chloride ($CaCl_2$) hydrogels by γ-radiation crosslinking for applications as wound dressings (Dutta 2012). Both PEG and $CaCl_2$ contributed to the enhancement of hydrogel strength.

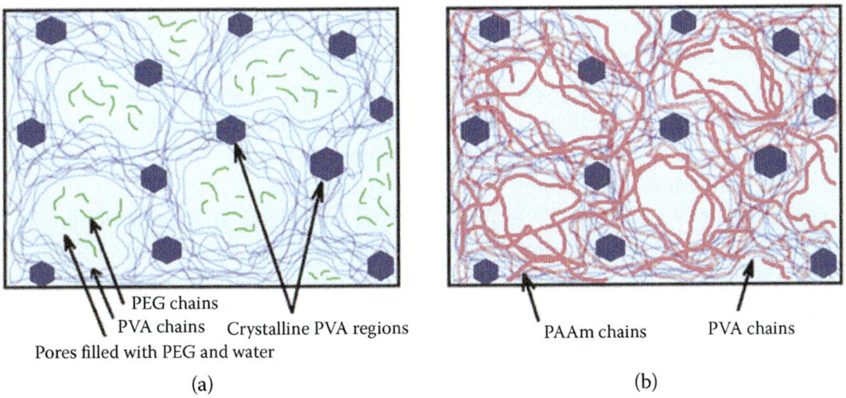

FIGURE 4.1 Schematic representation of a PVA hydrogel structure, when aqueous solutions of (a) PVA/PEG and (b) PVA/PAAm are used.

Moreover, the author performed antimicrobial and cytotoxicity tests, showing that the PVA/PEG/$CaCl_2$ hydrogels act as good barriers against microbes and that they also do not inhibit cell proliferation (Dutta 2012).

4.2.3 Poly(vinyl alcohol)/Poly(acrylamide)

Poly(acrylamide) (PAAm) is a polymer used for drug release systems due to its ease of processing, high water uptake capacity, and hydrophilic nature (Nagpal et al. 2013; Mahmoud et al. 2014). As mentioned earlier, it can be added to PVA solutions to obtain hydrogels with better physical and mechanical properties. The incorporation of an acrylamide (AAm) monomer into the PVA network can be performed by blending, grafting, or forming IPNs (Peppas et al. 2000; Bodugoz-Senturk et al. 2009). A study reported by Bodugoz-Senturk et al. presents the preparation of PVA-PAAm semi-IPNs by polymerization of AAm monomer in a physically crosslinked PVA network obtained by the freezing-thawing method (Bodugoz-Senturk et al. 2009).

The semi-IPNs obtained are further thermally annealed in order to increase their mechanical strength. Studies have found that the addition of AAm contributed to minimizing the loss of water content, thus maintaining the hydrogel's lubricity and preventing collapse of the pores (Bodugoz-Senturk et al. 2009). A more recent study reported on obtaining PVA/PAAm hydrogels by gamma radiation; tests for adsorption and release of Ketoprofen as a model drug were also performed (Mahmoud et al. 2014). The authors reported that hydrogels with a low thermal stability and decreased drug adsorption were obtained with a greater amount of AAm. Moreover, the in vitro Ketoprofen release study revealed a non-Fickian mechanism at $pH = 7$ (Mahmoud et al. 2014).

4.2.4 Poly(vinyl alcohol)/Poly(acrylic acid)

Poly(acrylic acid) (PAA, also known as carbomer) is a synthetic polymer that is used for obtaining super-absorbent materials that have the ability to absorb and retain a large amount of water (up to several times their original volume) (Jabbari and Karbasi 2004; Gao et al. 2013b). PAA is used in various applications as a thickening, dispersing, suspending, and emulsifying agent in bio/environmental-related, pharmaceutical, and cosmetics fields for products such as controlled release systems, membranes, ion exchange resins, disposable diapers, toothpastes, and so on (Gudeman and Peppas 1995; Hickey and Peppas 1997; Jose et al. 2014). In combination with PVA, PAA forms a miscible blend at the molecular level through hydrogen bond interactions (Zhang et al. 1992; Chen et al. 2002). The PVA/PAA blend characteristics are dictated by thermal stability and the excellent strength of PVA, while PAA contributes its ionic character and excellent swelling capacity. Also, PVA/PAA blends are biodegradable. A study concerning physical, mechanical, and thermal properties of PVA/PAA blends was published recently by Jose et al. (2014). The authors prepared the blends by mixing the individual solutions of PVA and PAA, using glycerol as a plasticizer, and cast them into films. Molecular level interactions were confirmed by Fourier transform infrared spectroscopy analysis for all blend compositions. It could be concluded that the incorporation of PAA determined the reduction in PVA storage modulus and its crystallinity, PVA becoming fully amorphous for mixtures with PAA content higher than 50 wt% in the blend (Jose et al. 2014). Other studies reported the swelling behavior of PVA/PAA hydrogels for different pH levels, for applications as pH sensitive bio-actuators/sensors (Quintero et al. 2010); drug delivery devices (Hickey and Peppas 1997); as materials for the dehydration process of alcohol, ethanol, or methanol (Rhim et al. 2002); as organic memory devices (Ayesh et al. 2013); and for the separation of organic solute molecules from water or electrolyte membranes for fuel cell applications (Rhim et al. 2002).

4.2.5 Poly(vinyl alcohol)/Poly(2-acrylamido-2-methyl-1-propanesulfonic acid)

Poly(2-acrylamido-2-methyl-1-propanesulfonic acid) (PAMPS) is a highly ionic conductive synthetic polymer that has been used in combination with PVA in order to produce an electroactive network membrane as an artificial muscle (Dai et al. 2009). The PVA/PAMPS blends were subjected to a heat treatment at high temperatures (above 60°C) to facilitate formation of physical crosslinks in the ionic network (see Figure 4.2).

Another study reports on the fabrication of an ionic polymer–metal composite membrane actuator in cantilever beam geometry that can perform an extensive bending movement (more than 100°) and that can remain curled up under constant voltage (Dai et al. 2009).

PVA-P(AA/AMPS) IPN hydrogels were synthesized by solution polymerization of AMPS and AA in the presence of PVA using *N*,*N*′-methylenebisacrylamide

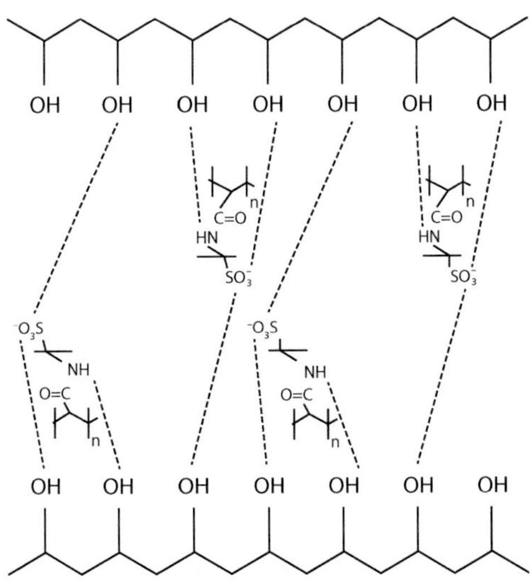

FIGURE 4.2 Physical interactions formed between PVA and PAMPS.

as a crosslinker and ammonium persulfate as an initiator (Lin et al. 2008). Investigations revealed that the copolymer and PVA form strong hydrogen interactions, and the crystallinity of PVA in IPN hydrogels decreased greatly. Moreover, it presented a swelling/deswelling behavior in aqueous Na_2SO_4 solution, in the sense that, when a water-swollen hydrogel was immersed in aqueous Na_2SO_4 solution, the hydrogel shrunk (Lin et al. 2008). The authors studied the influence of an applied electric field on the swelling/deswelling properties of the IPN hydrogels and observed that for low concentrations of Na_2SO_4 they swelled, while for high concentrations they shrunk. Furthermore, the electric stimuli response of the IPN hydrogels showed that they have a quick bending behavior dictated by the concentration of Na_2SO_4 solution, the applied voltage, and the hydrogel composition (Lin et al. 2008).

4.2.6 Poly(vinyl alcohol)/Poly(aniline)

Another component that has been blended with PVA is poly(aniline) (PANI), which is a conducting polymer (Gao et al. 2010; Ghaffari-Moghaddam and Eslahi 2014; Nath et al. 2014). In this regard, some studies have been published using PVA/PANI blends for application as electrochemical actuators (Ismail et al. 2009; Gao et al. 2010). Authors used a hydrogel mat template of electrospun nanofibers, and they concluded that, due to template's flexible structure and excellent electrical conductivity, the mechanical properties and actuation performance were significantly improved (Ismail et al. 2009; Gao et al. 2010). A major drawback of these kinds of materials is that their actuation performances primarily are confined to one- or two-dimensional extension/contraction movements. Gao et al. have proposed a three-dimensional monolithic PVA/PANI nanocomposite actuator with tunable stimuli-responsive properties for applications as micromechanical systems, on-demand devices, chemical sensors, or biomedical materials (Gao et al. 2010). Tested in an acid/base medium, the authors concluded that PVA/PANI nanocomposite actuators present superior bending/recovery responsive behavior and enhanced mechanical properties in contrast with conventional PANI actuators.

A recent study reported the synthesis, characterization, and antibacterial properties of a new nanocomposite material based on PVA/PANI and Ag NPs (Ghaffari-Moghaddam and Eslahi 2014). Using X-ray diffraction and scanning electron microscopy analysis, the authors discovered that the size of Ag NPs is in the range of 10–17 nm, with a favored triangle shape (which is a benign shape for antibacterial investigation). The investigation revealed that, although the PVA/PANI nanocomposite material did not exhibit good antibacterial activity, incorporation of Ag NPs led to a greater efficiency against gram-positive bacteria *Staphylococcus aureus* and gram-negative bacteria *Escherichia coli* (Ghaffari-Moghaddam and Eslahi 2014).

4.2.7 Poly(vinyl alcohol)/Polysiloxane

Semi-interpenetrated networks of PVA and polysiloxane (POS) have been synthesized by de Melo-Junior et al., through a sol-gel process, for application as a solid support in antibody immobilization (de Melo-Junior et al. 2008). This immobilization is achieved through covalently binding an anti-S100 protein antibody via GA; the protein-loading capacity of PVA/POS porous network spherical discs, with a uniform diameter size of 4.0 mm^2, was seen to be approximately 210 μg/disc. These authors also investigated the shelf-life of PVA/POS discs exposed to different types of environmental storage for six months, and they concluded that for discs incubated at 4°C and hydrated with phosphate buffer solution, a better protein-loading capacity than that of the dry discs kept at room temperature was observed (de Melo-Junior et al. 2008).

4.3 NANOCOMPOSITES OF POLY(VINYL ALCOHOL) WITH INORGANIC COMPOUNDS

4.3.1 Poly(vinyl alcohol)/Nanoparticles

Recently, the synthesis of nanoparticles (NPs) has attracted considerable research interest due to their increased surface area to volume ratio (Ghaffari-Moghaddam and Eslahi 2014) and the possible applications in a variety of fields, such as biomedicine, biosensors and biotechnologies, microelectronics and optics, and so on (Bryaskova et al. 2010). However, due to their tendency to easily aggregate because of their high surface energy, NPs must be incorporated into a dielectric polymer matrix. Because PVA has a high density of hydroxyl groups along the chain, it can stabilize the NPs by wrapping them and preventing agglomeration (Filippo et al. 2009). This leads to obtaining PVA/NP nanocomposites with combined/enhanced properties from both components.

4.3.1.1 Silver Nanoparticles

Silver (Ag) is among the metal NPs most commonly used in combination with PVA. Ag NPs are used on a large scale in a variety of fields (e.g., biomedical and cosmetic industries; food packaging; water, wastewater, and air treatment; textile and clothing; household applications; pesticides) due to their extraordinary chemical and physical properties (Nischala et al. 2011; de Azeredo 2013; Ghaffari-Moghaddam and Eslahi 2014). Their activity as an antibacterial, antivirus, and antifungal material is well known, and they are more sustainable, effective, and easy to process in comparison with other antibacterial agents (Yu et al. 2007; Wang et al. 2012; Gao et al. 2013a). Contact of Ag NPs with microorganisms is facilitated by the nanoparticles' well-developed surface and large surface area to volume ratio, which enables them to penetrate inside the bacteria and kill them (Bryaskova et al. 2010). For this purpose, Ag NPs need to be stable and to possess a specific size and shape. Their stabilization can be achieved by introducing them into suitable polymer substrates or matrices that can be obtained from PVA (as published research has proven). Among the various methods used so far, some have reported successful results for incorporation of Ag NPs into a PVA matrix—for example, in situ chemical reduction, solvent evaporation, ultraviolet (UV) light, electron radiation, or thermal annealing (Filippo et al. 2009; Bryaskova et al. 2010, 2013; Eisa et al. 2012).

Filippo et al. reported obtaining a promising colorimetric hydrogen peroxide sensor based on localized surface plasmon resonance of PVA-capped Ag NPs for applications in medical and environmental fields (Filippo et al. 2009). The Ag NPs were synthesized by reducing silver nitrate ($AgNO_3$) through thermal treatment in aqueous medium in the PVA matrix. The Ag NPs–PVA solution was observed to be highly sensitive and effective for the decomposition of hydrogen peroxide. This decomposition leads to the degradation of Ag NPs, which has been observed through the gradual changing of an Ag NPs–polymer solution from yellow to transparent (due to the change in the localized surface plasmon resonance absorbance strength). Thermal treatment of the transparent solution contributes to the re-obtaining of the Ag NPs, which can be confirmed by reappearance of the yellow color.

One inconvenience in using PVA for biomedical applications is its sensibility toward moisture, which reduces the strength of the resulted products. In order to remove this impediment, some researchers have proposed the incorporation of networked silica into the PVA matrix, using tetraethoxysilane (TEOS), γ-aminopropyltriethoxysilane (APTEOS), and 3-mercaptopropyltriethoxysilane (MPTEOS), via the sol-gel method (Bryaskova et al. 2010; Georgieva et al. 2012). The method proved to be successful, leading to PVA/silica hybrid materials with significant improvement in mechanical properties. Therefore, these hybrid matrices are more suitable for hosting Ag NPs as antibacterial coating materials. This was reported by Shanmugam, who incorporated Ag NPs into a PVA/TEOS composite film and demonstrated good antibacterial activity against *Escherichia coli* (Shanmugam et al. 2006). Bryaskova et al. reported obtaining a hybrid material with biological activity based on PVA/TEOS/Ag NPs synthetized using the sol-gel method (Bryaskova et al. 2010, 2013; Pencheva et al. 2012). First, they prepared the PVA/TEOS matrix, and then two different methods were applied for the synthesis of Ag NPs in the PVA/TEOS matrix—in situ reduction of silver ions by thermal annealing of the films and preliminary preparation of Ag NPs by chemical reduction using PVA as a reduction agent and stabilizer. The hybrid material proved to possess a strong antibacterial activity toward *Staphylococcus aureus*, *Escherichia coli*, and *Pseudomonas aeruginosa* and supposedly can be applicable as an antibacterial coating or as a catalytically active membrane for direct electron transfer in the electrochemical biosensor field (Pencheva et al. 2012). A later study reported the preparation of PVA/TEOS/Ag NPs matrices using the sol-gel method and their investigations in terms of physicochemical surface properties, biocompatibility, cytotoxicity, and cell adhesion—properties that make them possible candidates for applications in tissue engineering (Bryaskova et al. 2013).

4.3.1.2 Cadmium Sulfide Nanoparticles

Other interesting nanoparticles that have been explored for various applications are those of cadmium sulfide (CdS). These are semiconductor nanoparticles possessing remarkable photoluminescence properties that can be used in optoelectronics, bio-labeling, lasing, photovoltaics, and so on. For these reasons, they must be embedded into a polymeric matrix in order to obtain films or other structures with enhanced photoluminescence efficiency and electrical conductivity (Pattabi et al. 2007; Amma et al. 2014). In this regard, PVA is a promising candidate as a polymeric host for CdS NPs due to its good transparency, exceptional film forming capacity, and compatibility with additives (Qian et al. 2000).

Pattabi et al. reported obtaining self-standing flexible nanocomposite films based on PVP capped CdS NPs embedded in a PVA matrix with photoluminescence properties (Pattabi et al. 2007) and electrical conductivity (Amma et al. 2014). CdS NPs were obtained from cadmium nitrate and hydrogen sulfide by a nonaqueous method and then were dispersed in a PVA matrix. With an average particle size of approximately 3–5 nm, they contributed to the obtaining of photoluminescence emission spectra with two peaks, at 502 and 636 nm (Pattabi et al. 2007). The electrical conductivity of the (PVP-CdS) PVA films was seen to increase with temperature (Amma et al. 2014).

4.3.1.3 Titanium Dioxide Nanoparticles

Titanium dioxide (TiO_2) is another class of semiconductor NPs that has been studied as active photocatalysts for photodegradation of various organic contaminants of dye wastewater (Filippo et al. 2015). It has been reported that TiO_2 NPs represent a nontoxic, low-cost solution for clean advanced oxidation technology, and that they have long-term stability (Filippo et al. 2015). A study published by Filippo et al. outlines the preparation of TiO_2-PVA hybrid NPs with improved visible light photocatalytic activity through a two-step synthesis procedure (Filippo et al. 2015). First, the pure TiO_2 NPs are synthesized using a rapid microwave-assisted nonaqueous route, and then they are coated with PVA through a simple solution method followed by relatively low temperature treatment (Filippo et al. 2015). The authors affirm that the TiO_2/PVA core/shell hybrid NPs, as prepared, present promising photocatalytic activity and can be successfully used as low cost, environmentally friendly photocatalysts (Filippo et al. 2015).

4.3.1.4 Magnetic Nanoparticles

Magnetic nanoparticles are utilized in various biomedical applications in drug delivery systems, magnetic cell seeding, cell sheet construction, cancer hyperthermia treatment, or cell expansion (Ngadiman et al. 2015), and they are known to improve the properties of PVA when this is used as a carrier matrix (Idris et al. 2012; Shao et al. 2013). Some of these magnetic nanoparticles that have been successfully embedded into PVA matrix blends are maghemites (γ-Fe_2O_3). A recent study reported obtaining novel electrospun mats as promising tissue engineering scaffolds by using maghemite (γ-Fe_2O_3) NPs mixed into a PVA matrix (Ngadiman et al. 2015). Maghemite NPs are known to possess unique magnetic properties and also to enhance the cell growth rate. Compared with neat PVA nanofiber mats, PVA/γ-Fe_2O_3 mats were demonstrated to have higher cell proliferation rates when a 3-(4,5-dimethylthiazol-2-yl)-2,5-diphenyltetrazoliumbromide assay was done using human skin fibroblast cells (Ngadiman et al. 2015).

4.3.2 Poly(vinyl alcohol)/Carbon Nanotubes

There has been significant interest in carbon nanotubes (CNTs) in recent years owing to their unique electrical, mechanical, and thermal properties; their high aspect ratio and low density; as well as their potential for surface functionalization (Zhang et al. 2003; Shi et al. 2005; Huang et al. 2011). A major challenge in using CNTs is to overcome their well-known tendency to form bundles, entanglements, and agglomerates due to van der Waals forces among tubes, which makes their dispersion and alignment in a polymer matrix very difficult (Liu et al. 2005). The functionalization of CNTs is a successful method for addressing the problem of entanglements and efficiently embedding CNTs in a polymer matrix.

There have been several reports on obtaining different nanocomposites based on CNTs and PVA—in the form of fibers, films, or bulk materials—with a variety of applications, such as soft implants, implantable biosensors and devices, drug delivery systems, or tissue engineering scaffolds (Liu et al. 2005; Huang et al. 2011). Zhang reported the fabrication of composite films of PVA embedded with PVP and sodium dodecyl sulfate-covered single-wall carbon nanotubes (SWCNTs) (Zhang et al. 2003). Investigations on these composite films revealed that they present good nanotube dispersion and load transfer to the SWCNTs from the PVA matrix. A high-performance PVP-wrapped multi-wall carbon nanotube (MWCNT) reinforced PVA composite hydrogel for possible application in the biomedical field was obtained by the freezing-thawing process (Huang et al. 2011). PVP contributes to the uniform dispersion of nanotubes and improves the interface between them and the PVA hydrogel network, enhancing their extraordinary mechanical and physical properties. A new method for grafting PVA to MWCNTs was recently reported (Malikov et al. 2014). This method was performed via a Fisher esterification route, which permanently connects the MWCNTs to the PVA matrix via ester groups. Besides helping with the reduction of nanotube adhesion and entanglement during dispersion, PVA grafting with MWCNTs contributes to the orientation of the obtained grafted structures by exploiting the conformational changes of PVA chains (Malikov et al. 2014). These authors propose the use of this PVA/MWCNT composite as a precursor in manufacturing advanced plastic materials or as an additive used in construction materials for strengthening purposes.

4.4 CONCLUSIONS

This chapter presents a comprehensive overview of published research on multicomponent polymeric systems based on PVA and synthetic polymers or inorganic compounds (see Table 4.1).

Different fabrication techniques (e.g., chemical, physical, and radiation methods) for advanced materials in biomedical/pharmaceutical fields or other areas also have been outlined here. Moreover, the properties of PVA-synthetic polymer and/or inorganic filler have been discussed in detail, emphasizing the synergistic influence that the second (and the third) component has on the characteristics of the resulting PVA-based multicomponent system.

TABLE 4.1
Advanced Materials based on PVA and Synthetic Polymers and/or Inorganic Compounds and Their Applications in a Variety of Fields

	Second Component		Applications
PVA	Synthetic polymers	PVP	Wound and burn dressings, nucleus pulposus, intervertebral discs, synthetic articular cartilages, human lens substitution, magnetic drug targeting, pervaporation membranes, electrochemical cells
		PEG	Wound dressings, vehicles for nucleic acid delivery, cartilage tissue engineering, pH sensitive drug carriers, tissue engineering scaffolds, cardiovascular devices
		PAAm	Drug release systems, synthetic articular cartilages, ultrafiltration membranes
		PAA	pH-sensitive bio-actuators/sensors; drug delivery devices; materials for the dehydration process of alcohols, ethanol or methanol; organic memory devices; separation of organic solute molecules from water or electrolyte membranes for fuel cell applications
		PAMPS	Electroactive network membranes as artificial muscles, direct methanol fuel cells
		PANI	Electrochemical actuators, micromechanical systems, on-demand devices, chemical sensors, antibacterial nanocomposite materials, dye sensitized solar cells, microwave absorbing materials
		POS	Supports for antibody immobilization
		PVAc	pH-responsive luminescent sensors, direct methanol fuel cells, food packaging, drug delivery systems
	Inorganic compounds	Ag NPs	Biomedical and cosmetic industry; food packaging; water, wastewater, and air treatment; textile and clothing; household applications; pesticides
		CdS NPs	Optoelectronics, bio-labeling, lasing, photovoltaics
		TiO_2 NPs	Environmentally friendly active photocatalysts for photodegradation of various organic contaminants of dye wastewater, dielectric layers for electronic applications
		γ-Fe_2O_3 NPs	Drug delivery systems, magnetic cell seeding, cell sheet construction, cancer hyperthermia treatment, cell expansion, tissue engineering scaffolds
		CNTs	Dye-sensitized solar cells, soft implants, implantable biosensors and devices, drug delivery systems, tissue engineering scaffolds, advanced plastic materials

ACKNOWLEDGMENTS

This work was supported by a grant from the Romanian National Authority for Scientific Research, CNCS-UEFISCDI, project no PN-II-ID-PCE-2011-3-0199 (contract 300/2011).

REFERENCES

Abd El-Mohdy H. L., and Ghanem S. Biodegradability, antimicrobial activity and properties of PVA/PVP hydrogels prepared by γ-irradiation. *J. Polym. Res.* 16 no. 1 (2009): 1–10.

Abdelrazek E.M., Elashmawi I.S., El-khodary A., and Yassin A. Structural, optical, thermal and electrical studies on PVA/PVP blends filled with lithium bromide. *Curr. Appl. Phys.* 10 no. 2 (2010): 607–613.

Alexandre N., Ribeiro J., Gärtner A., et al. Biocompatibility and hemocompatibility of polyvinyl alcohol hydrogel used for vascular grafting—In vitro and in vivo studies. *J. Biomed. Mater. Res. A.* 102 no. 12 (2014): 4262–4275.

Alves Marie-Helene, Jensen Bettina E. B., Smith Anton A. A., and Zelikin Alexander N. Poly(vinyl alcohol) physical hydrogels: New vista on a long serving biomaterial. *Macromol. Biosci.* 11 no. 10 (2011): 1293–1313.

Amma B. Saraswarti, Pattabi Manjunatha, Parala Harish, and Shripathi T. Electrical conductivity of PVP capped Cds nanoparticles in PVA matrix. *J. Nanosci. Nanotechnol.* 2 no. 7 (2014): 817–820.

Awadallah-F Ahmed. Five years in vitro study of (poly vinyl alcohol/poly vinyl pyrrolidone/poly acrylic acid) hydrogel to mimic the knee joint meniscus. *Polym. Adv. Technol.* 25 no. 5 (2014): 581–587.

Ayesh A. I., Qadri S., Baboo V. J., Haik M. Y., and Haik Y. Nano-floating gate organic memory devices utilizing Ag-Cu nanoparticles embedded in PVA-PAA-glycerol polymer. *Synth. Met.* 183 (2013): 24–28.

Azuma Chinatsu, Yasuda Kazunori, Tanabe Yoshie, et al. Biodegradation of high-toughness double network hydrogels as potential materials for artificial cartilage. *J. Biomed. Mater. Res. A.* 81 no. 2 (2007): 373–380.

Bercea Maria, Bibire Elena-Livia, Morariu Simona, and Carja Gabriela. Chitosan/poly(vinyl alcohol)/LDH biocomposites with pH-sensitive properties. *Int. J. Polym. Mater.* 64 no. 12 (2015b): 628–636.

Bercea Maria, Bibire Elena-Livia, Morariu Simona, Teodorescu Mirela, and Carja Gabriela. pH influence on rheological and structural properties of chitosan/poly(vinyl alcohol)/layered double hydroxide composites. *Eur. Polym. J.* 70 (2015c): 147–156.

Bercea Maria, Darie Raluca Nicoleta, Nita Loredana Elena, and Morariu Simona. Temperature responsive gels based on Pluronic F127 and poly(vinyl alcohol). *Ind. Eng. Chem. Res.* 50 no. 7 (2011): 4199–4206.

Bercea Maria, Morariu Simona, Nita Loredana Elena, and Darie Raluca Nicoleta. Investigation of poly(vinyl alcohol)/Pluronic F127 physical gels. *Polym. Plast. Technol. Eng.* 53 no. 13 (2014): 1354–1361.

Bercea Maria, Morariu Simona, and Rusu Daniela. In situ gelation of aqueous solutions of entangled poly(vinyl alcohol). *Soft Matter.* 9 no. 4 (2013): 1244–1253.

Bercea Maria, Nita Loredana Elena, Morariu Simona, and Chiriac Aurica P. *In-situ* gelling system based on Pluronic F127 and poly(vinyl alcohol) for smart biomaterials. *Rev. Roum. Chim.* 60 no. 7–8 (2015a): 787–795.

Binetti Valerie R., Fussell Garland W., and Lowman Anthony M. Evaluation of two chemical crosslinking methods of poly(vinyl alcohol) hydrogels for injectable nucleus pulposus replacement. *J. Appl. Polym. Sci.* 131 no. 19 (2014): 40843-1–40843-8.

Bodugoz-Senturk Hatice, Choi Jeeyoung, Oral Ebru, et al. The effect of polyethylene glycol on the stability of pores in polyvinyl alcohol hydrogels during annealing. *Biomaterials.* 29 no. 2 (2008): 141–149.

Bodugoz-Senturk Hatice, Macias Celia E., Kung Jean H., and Muratoglu Orhun K. Poly(vinyl alcohol)–acrylamide hydrogels as load-bearing cartilage substitute. *Biomaterials.* 30 no. 4 (2009): 589–596.

Bryaskova Rayna, Georgieva Nelly, Andreeva Tonya, and Tzoneva Rumiana. Cell adhesive behavior of PVA-based hybrid materials with silver nanoparticles. *Surf. Coat. Technol.* 235 (2013): 186–191.

Bryaskova Rayna, Pencheva Daniela, Kale Girish M., Lad Umesh, and Kantardjiev T. Synthesis, characterisation and antibacterial activity of PVA/TEOS/Ag-Np hybrid thin films. *J. Colloid Interface Sci.* 349 no. 1 (2010): 77–85.

Chen Qun, Kurosu Hiromichi, Ma Lin, and Matsuo Masaru. Elongation-induced phase separation of poly(vinyl alcohol)/poly(acrylic-acid) blends as studied by ^{13}C CP/MAS NMR and wide-angle X-ray diffraction. *Polymer.* 43 no. 4 (2002): 1203–1206.

Chiellini Emo, Cinelli Patrizia, Ilieva Vassilka I., and Martera Martina. Biodegradable thermoplastic composites based on polyvinyl alcohol and algae. *Biomacromolecules.* 9 (2008): 1007–1013.

Choi Jin Hyun, Ko Sohk-Won, Kim Byoung Chul, Blackwell John, and Lyoo Won Seok. Phase behavior and physical gelation of high molecular weight syndiotactic poly(vinyl alcohol) solution. *Macromolecules.* 34 no. 9 (2001): 2964–2972.

Curley Colin, Hayes Jennifer C., Rowan Neil J., and Kennedy James E. An evaluation of the thermal and mechanical properties of a salt-modified polyvinyl alcohol hydrogel for a knee meniscus application. *J. Mech. Behav. Biomed. Mater.* 40 (2014): 13–22.

Dai Chi-An, Chang Chun-Jie, Kao An-Cheng, et al. Polymer actuator based on PVA/PAMPS ionic membrane: Optimization of ionic transport properties. *Sensor. Actuat. A: Physical.* 155 no. 1 (2009): 152–162.

de Azeredo Henriette M. C. Antimicrobial nanostructures in food packaging. *Trends Food Sci. Technol.* 30 no. 1 (2013): 56–69.

de Melo-Junior Mario Ribeiro, Alves Luiz Carlos, dos Santos Fábio Brayner, Beltrão Eduardo Isidoro Carneiro, and de Carvalho Luiz Bezerra. Polysiloxane–polyvinyl alcohol discs as support for antibody immobilization: Ultra-structural and physical–chemical characterization. *React. Funct. Polym.* 68 no. 1 (2008): 315–320.

DeMerlis C. C., and Schoneker D. R. Review of the oral toxicity of polyvinyl alcohol (PVA). *Food Chem. Toxicol.* 41 no. 3 (2003): 319–326.

Dutta Joydeep. Synthesis and characterization of γ-irradiated PVA/PEG/$CaCl_2$ hydrogel for wound dressing. *Am. J. Chem.* 2 no. 2 (2012): 6–11.

Eisa Wael H., Abdel-Moneam Yasser K., Shabaka A. A., and Hosam Abd Elhameed M. In situ approach induced growth of highly monodispersed Ag nanoparticles within free standing PVA/PVP films. *Spectrochim. Acta Mol. Biomol. Spectrosc.* 95 (2012): 341–346.

Feng Songfu, Chen Han, Liu Yaqin, et al. A novel vitreous substitute of using a foldable capsular vitreous body injected with polyvinylalcohol hydrogel. *Sci. Rep.* 3 (2013): 1–11.

Filippo Emanuela, Carlucci Claudia, Capodilupo Agostina Lina, et al. Facile preparation of TiO_2–polyvinyl alcohol hybrid nanoparticles with improved visible light photocatalytic activity. *Appl. Surf. Sci.* 331 (2015): 292–298.

Filippo E., Serra A., and Manno D. Poly(vinyl alcohol) capped silver nanoparticles as localized surface plasmon resonance-based hydrogen peroxide sensor. *Sensor. Actuat. B-Chem.* 138 no. 2 (2009): 625–630.

Folttmann Hubertus, and Quadir Anisul. Polyvinylpyrrolidone (PVP)—One of the most widely used excipients in pharmaceuticals: An overview. *Drug Deliv. Technol.* 8 no. 6 (2008): 22–27.

Gao Hainan, Zhang Junhu, Yu Weili, et al. Monolithic polyaniline/polyvinyl alcohol nanocomposite actuators with tunable stimuli-responsive properties. *Sensor. Actuat. B-Chem.* 145 no. 2 (2010): 839–846.

Gao Minjie, Sun Lei, Wang Zhiqiang, and Zhao Yanbao. Controlled synthesis of Ag nanoparticles with different morphologies and their antibacterial properties. *Mater. Sci. Eng. C.* 33 no. 1 (2013a): 397–404.

Gao Xiaoye, He Chaoliang, Xiao ChunSheng, Zhuang Xiuli, and Chen Xuesi. Biodegradable pH-responsive polyacrylic acid derivative hydrogels with tunable swelling behavior for oral delivery of insulin. *Polymer.* 54 no. 7 (2013b): 1786–1793.

Georgieva Nelly, Bryaskova Rayna, and Tzoneva Rumiana. New polyvinyl alcohol-based hybrid materials for biomedical application. *Mater. Lett.* 88 (2012): 19–22.

Ghaffari-Moghaddam Mansour, and Eslahi Hassan. Synthesis, characterization and antibacterial properties of a novel nanocomposite based on polyaniline/polyvinyl alcohol/Ag. *Arabian J. Chem.* 7 no. 5 (2014): 846–855.

Gökmeşe Faruk, Uslu Ibrahim, and Aytimur Arda. Preparation and characterization of PVA/PVP nanofibers as promising materials for wound dressing. *Polym. Plast. Technol. Eng.* 52 no. 12 (2013): 1259–1265.

Gonzalez Jimena S., and Alvarez Vera A. Mechanical properties of polyvinylalcohol/hydroxyapatite cryogel as potential artificial cartilage. *J. Mech. Behav. Biomed. Mater.* 34 (2014): 47–56.

Gudeman Linda F., and Peppas Nikolaos A. Preparation and characterization of pH-sensitive, interpenetrating networks of poly(vinyl alcohol) and poly(acrylic acid). *J. Appl. Polym. Sci.* 55 no. 6 (1995): 919–928.

Gupta Siddhi, Goswami Sudipta, and Sinha Arvind. A combined effect of freeze-thaw cycles and polymer concentration on the structure and mechanical properties of transparent PVA gels. *Biomed. Mater.* 7 no. 1 (2012): 1–8.

Han Jingquan, Lei Tingzhou, and Wu Qinglin. High-water-content mouldable polyvinyl alcohol-borax hydrogels reinforced by well-dispersed cellulose nanoparticles: Dynamic rheological properties and hydrogel formation mechanism. *Carbohyd Polym.* 102 (2014): 306–316.

Hassan Christie M., and Peppas Nikolaos A. Structure and morphology of freeze/thawed PVA hydrogels. *Macromolecules.* 33 no. 7 (2000): 2472–2479.

Hickey Alla S., and Peppas Nikolaos A. Solute diffusion in poly(vinyl alcohol)/poly(acrylic acid) composite membranes prepared by freezing/thawing techniques. *Polymer.* 38 no. 24 (1997): 5931–5936.

Hoffman Allan S. Hydrogels for biomedical applications. *Adv. Drug Deliv. Rev.* 64 (2012): 18–23.

Huang Yanyi, Zheng Yudong, Song Wenhui, et al. Poly(vinyl pyrrolidone) wrapped multi-walled carbon nanotube/poly(vinyl alcohol) composite hydrogels. *Compos. Part A Appl. S.* 42 no. 10 (2011): 1398–1405.

Idris Ani, Misran Effaliza, Hassan Nursia, Abd Jalil Aishah, and Seng Chan Eng. Modified PVA-alginate encapsulated photocatalyst ferro photo gels for Cr(VI) reduction. *J. Hazard. Mater.* 227–228 (2012): 309–316.

Ismail Yahya, Shin Min Kyoon, and Kim Seon Jeong. A nanofibrous hydrogel templated electrochemical actuator: From single mat to a rolled-up structure. *Sensor. Actuat. B-Chem.* 136 no. 2 (2009): 438–443.

Jabbari Esmaiel, and Karbasi Saeed. Swelling behavior and cell viability of dehydrothermally crosslinked poly(vinyl alcohol) hydrogel grafted with N-vinyl pyrrolidone or acrylic acid using γ-radiation. *J. Appl. Polym. Sci.* 91 (2004): 2862–2868.

Jain Neelam, Banik Arunabha, and Gupta Amit. Novel interpenetrating polymer network microspheres of lepidium sativum and poly(vinyl alcohol) for the controlled release of Simvastatin. *Int. J. Pharm. Pharm. Sci.* 5 Suppl. 1 (2013): 125–130.

Jamnongkan Tongsai, Wattanakornsiri Amnuay, Wachirawongsakorn Piyada, and Kaewpirom Supranee. Effects of crosslinking degree of poly(vinyl alcohol) hydrogel in aqueous solution: Kinetics and mechanism of copper(II) adsorption. *Polym. Bull.* 71 no. 5 (2014): 1081–1100.

Jose Jobin, Shehzad Farrukh, and Al-Harthi Mamdouh A. Preparation method and physical, mechanical, thermal characterization of poly(vinyl alcohol)/poly(acrylic acid) blends. *Polym. Bull.* 71 (2014): 2787–2802.

Joshi Abhijeet, Fussell Garland, Thomas Jonathan, et al. Functional compressive mechanics of a PVA/PVP nucleus pulposus replacement. *Biomaterials.* 27 no. 2 (2006): 176–184.

Kamoun Elbadawy A., Chen Xin, Eldin Mohamed S. Mohy, and Kenawy El-Refaie S. Crosslinked poly(vinyl alcohol) hydrogels for wound dressing applications: A review of remarkably blended polymers. *Arabian J. Chem.* 8 no. 1 (2015): 1–14.

Katta Jayanth K., Marcolongo Michele, Lowman Anthony, and Mansmann Kevin A. Friction and wear behavior of poly(vinyl alcohol)/poly(vinyl pyrrolidone) hydrogels for articular cartilage replacement. *J. Biomed. Mater. Res. A.* 83 no. 2 (2007): 471–479.

Kim Seon Jeong, Park Sang Jun, Kim In Young, Lee Yong Hee, and Kim Sun. I. Thermal characteristics of poly(vinyl alcohol) and poly(vinylpyrrolidone) IPNs. *J. Appl. Polym. Sci.* 86 no. 8 (2002): 1844–1847.

Lewandowska Katarzyna. The miscibility of poly(vinyl alcohol)/poly(N-vinylpyrrolidone) blends investigated in dilute solutions and solids. *Eur. Polym. J.* 41 no. 1 (2005): 55–64.

Lin Song-Bai, Yuan Cong-Hui, Ke Ai-Ru, and Quan Zhi-Long. Electrical response characterization of PVA–P(AA/AMPS) IPN hydrogels in aqueous Na_2SO_4 solution. *Sensor. Actuat. B-Chem.* 134 no. 1 (2008): 281–286.

Liu L., Barber A. H., Nuriel S., and Wagner H. D. Mechanical properties of functionalized single-walled carbon-nanotube/poly(vinyl alcohol) nanocomposites. *Adv. Funct. Mater.* 15 no. 6 (2005): 975–980.

Li Wei, Li Xueyong, Chen Yang, et al. Poly(vinyl alcohol)/sodium alginate/layered silicate based nanofibrous mats for bacterial inhibition. *Carbohyd. Polym.* 92 no. 2 (2013): 2232–2238.

Lu Jie, Nguyen Quangtrong, Zhou Jiqing, and Ping Zheng-Hua. Poly(vinyl alcohol)/poly(vinyl pyrrolidone) interpenetrating polymer network: Synthesis and pervaporation properties. *J. Appl. Polym. Sci.* 89 (2003): 2808–2814.

Mahmoud Ghada A., Hegazy Dalia E., and Kamal H. In-vitro release of ketoprofen behavior loaded in polyvinyl alcohol/acrylamide hydrogels prepared by gamma irradiation. *Arab J. Nuclear Sci. Appl.* 47 no. 1 (2014): 28–40.

Malikov E. Y., Muradov M. B., Akperov O. H., et al. Synthesis and characterization of polyvinyl alcohol based multiwalled carbon nanotube nanocomposites. *Physica E Low Dimens. Syst. Nanostruct.* 61 (2014): 129–134.

Mansur Herman S., Oréfice Rodrigo L., and Mansur Alexandra A. P. Characterization of poly(vinyl alcohol)/poly(ethylene glycol) hydrogels and PVA-derived hybrids by small-angle X-ray scattering and FTIR spectroscopy. *Polymer.* 45 no. 21 (2004): 7193–7202.

Martens Penny J., Bryant Stephanie J., and Anseth Kristi S. Tailoring the degradation of hydrogels formed from multivinyl poly(ethlene glycol) and poly(vinyl alcohol) macromers for cartilage tissue engineering. *Biomacromolecules.* 4 no. 2 (2003): 283–292.

Ma Ruyin, Xiong Dangsheng, Miao Feng, Zhang Jinfeng, and Peng Yan. Novel PVP/PVA hydrogels for articular cartilage replacement. *Mater. Sci. Eng. C.* 29 no. 6 (2009): 1979–1983.

Millon Leonardo E., Nieh Mu-Ping, Hutter Jeffrey L., and Wan Wankei. SANS characterization of an anisotropic poly(vinyl alcohol) hydrogel with vascular applications. *Macromolecules.* 40 no. 10 (2007): 3655–3662.

Morariu Simona, Bercea Maria, and Brunchi Cristina-Eliza. Effect of cryogenic treatment on the rheological properties of chitosan/poly(vinyl alcohol) hydrogels. *Ind. Eng. Chem. Res.* 54 no. 45 (2015): 11475–11482.

Nagpal Manju, Singh Shailendra Kumar, and Mishra Dinanath. Synthesis characterization and in vitro drug release from acrylamide and sodium alginate based superporous hydrogel devices. *Int. J. Pharm. Investig.* 3 no. 3 (2013): 131–140.

Nath Bikash Chandra, Gogoi Biswajit, Boruah Monalisha, et al. High performance polyvinyl alcohol/multi walled carbon nanotube/polyaniline hydrogel (PVA/MWCNT/PAni) based dye sensitized solar cells. *Electrochim. Acta.* 146 (2014): 106–111.

Ngadiman Nor Hasrul Akhmal, Idris Ani, Irfan Muhammad, et al. γ-Fe2O3 nanoparticles filled polyvinyl alcohol as potential biomaterial for tissue engineering scaffold. *J. Mech. Behav. Biomed. Mater.* 49 (2015): 90–104.

Nguyen Ngoc-Thang, and Liu Jui-Hsiang. Fabrication and characterization of poly(vinyl alcohol)/chitosan hydrogel thin films via UV irradiation. *Eur. Polym. J.* 49 no. 12 (2013): 4201–4211.

Nho Y. C., and Park K. R. Preparation and properties of PVA/PVP hydrogels containing chitosan by radiation. *J. Appl. Polym. Sci.* 85 no. 8 (2002): 1787–1794.

Nischala K., Rao Tata N., and Hebalkar Neha. Silica-silver core-shell particles for antibacterial textile application. *Colloids Surf. B Biointerfaces.* 82 no. 1 (2011): 203–208.

Nkhwa Shathani, Lauriaga Kristo Fernando, Kemal Evren, and Deb Sanjukta. Poly(vinyl alcohol): Physical approaches to designing biomaterials for biomedical applications. *Conf. Paper. Sci.* (2014): 1–7.

Park Kyoung Ran, and Nho Young Chang. Synthesis of PVA/PVP hydrogels having two-layer by radiation and their physical properties. *Radiat. Phys. Chem.* 67 no. 3–4 (2003): 361–365.

Pattabi Manjunatha, Amma B. Saraswathi, and Manzoor K. Photoluminescence study of PVP capped CdS nanoparticles embedded in PVA matrix. *Mater. Res. Bull.* 42 no. 5 (2007): 828–835.

Pencheva Daniela, Bryaskova Rayna, and Kantardjiev Todor. Polyvinyl alcohol/silver nanoparticles (PVA/AgNps) as a model for testing the biological activity of hybrid materials with included silver nanoparticles. *Mater. Sci. Eng. C.* 32 no. 7 (2012): 2048–2051.

Peppas Nikolaos A. Turbidimetric studies of aqueous poly(vinyl alcohol) solutions. *Die Makromolekulare Chemie.* 176 no. 11 (1975): 3433–3440.

Peppas N. A., Bures P., Leobandung W., and Ichikawa H. Hydrogels in pharmaceutical formulations. *Eur. J. Pharm. Biopharm.* 50 no. 1 (2000): 27–46.

Qian X. F., Yin J., Guo X. X., Yang Y. F., Zhu Z. K., and Lu J. Polymer-inorganic nanocomposites prepared by hydrothermal method: PVA/ZnS, PVA/CdS, preparation and characterization. *J. Mater. Sci. Lett.* 19 no. 24 (2000): 2235–2237.

Qiao Jinli, Fu Jing, Lin Rui, Ma Jianxin, and Liu Jianshe. Alkaline solid polymer electrolyte membranes based on structurally modified PVA/PVP with improved alkali stability. *Polymer.* 51 no. 21 (2010): 4850–4859.

Quintero S. M. M., Ponce F. R. V., Cremona M., et al. Swelling and morphological properties of poly(vinyl alcohol) (PVA) and poly(acrylic acid) (PAA) hydrogels in solution with high salt concentration. *Polymer.* 51 no. 4 (2010): 953–958.

Ragab H.M. Spectroscopic investigations and electrical properties of PVA/PVP blend filled with different concentrations of nickel chloride. *Physica B: Condensed Matter.* 406 no. 20 (2011): 3759–3767.

Rakesh G., and Deshpande Abhijit P. Rheology of crosslinking poly vinyl alcohol systems during film formation and gelation. *Rheol. Acta.* 49 no. 10 (2010): 1029–1039.

Rhim Ji-Won, Lee Sun-Wha, and Kim Youn-Kook. Pervaporation separation of water–ethanol mixtures using metal-ion-exchanged poly(vinyl alcohol) (PVA)/sulfosuccinic acid (SSA) membranes. *J. Appl. Polym. Sci.* 85 no. 9 (2002): 1867–1873.

Ruberti Jeffrey W., and Braithwaite Gavin J. C. Systems and methods for controlling and forming polymer gels. US7745532 B2 (2003).

Sanli Oya, Ay Nuran, and Işiklan Nuran. Release characteristics of diclofenac sodium from poly(vinyl alcohol)/sodium alginate and poly(vinyl alcohol)-grafted-poly(acrylamide)/sodium alginate blend beads. *Eur. J. Pharm. Biopharm.* 65 no. 2 (2007): 204–214.

Shanmugam Sangaraju, Viswanathan Balasubramanian, and Varadarajan Thirukkallam Kanthadai. A novel single step chemical route for noble metal nanoparticles embedded organic-inorganic composite films. *Mater. Chem. Phys.* 95 no. 1 (2006): 51–55.

Shao Dali, Qin Liqiao, and Sawyer Shayla. Optical properties of polyvinyl alcohol (PVA) coated In_2O_3 nanoparticles. *Opt. Mater.* 35 no. 3 (2013): 563–566.

Shi Jiahua, Guo Zhi-Xin, Zhan Bohan, et al. Actuator based on MWNT/PVA hydrogels. *J. Phys. Chem. B Lett.* 109 (2005): 14789–14791.

Shi Yan, and Xiong Dangsheng. Microstructure and friction properties of PVA/PVP hydrogels for articular cartilage repair as function of polymerization degree and polymer concentration. *Wear.* 305 no. 1–2 (2013): 280–285.

Shi Yan, Xiong Dangsheng, and Zhang Jinfeng. Effect of irradiation dose on mechanical and biotribological properties of PVA/PVP hydrogels as articular cartilage. *Tribol. Int.* 78 (2014): 60–67.

Singh Baljit and Pal Lok. Radiation crosslinking polymerization of sterculia polysaccharide-PVA-PVP for making hydrogel wound dressings. *Int. J. Biol. Macromol.* 48 no. 3 (2011): 501–510.

Stammen J. A., Williams S., Ku D. N., and Guldberg R. E. Mechanical properties of a novel PVA hydrogel in shear and unconfined compression. *Biomaterials.* 22 no. 8 (2001): 799–806.

Teodorescu Mirela, and Bercea Maria. Poly(vinylpyrrolidone)—A versatile polymer for biomedical and beyond medical applications. *Polym. Plast. Technol. Eng.* 54 no. 9 (2015): 923–943.

Teodorescu Mirela, Morariu Simona, Bercea Maria, and Sacarescu Liviu. Viscoelastic and structural properties of poly(vinyl alcohol)/poly(vinylpyrrolidone) hydrogels. *RSC Adv.* 6 no. 46 (2016): 39718–39727.

Thomas Jonathan, Gomes Kate, Lowman Anthony, and Marcolongo Michele. The effect of dehydration history on PVA/PVP hydrogels for nucleus pulposus replacement. *J. Biomed. Mater. Res. Part B, Appl. Biomater.* 69 no. 2 (2004): 135–140.

Tsai Chun-En, Lin Chi-Wen, and Hwang Bing-Joe. A novel crosslinking strategy for preparing poly(vinyl alcohol)-based proton-conducting membranes with high sulfonation. *J. Power Sources.* 195 no. 8 (2010): 2166–2173.

Wang Bai-Liang, Liu Xiang-Sheng, Ji Ying, Ren Ke-Feng, and Ji Jian. Fast and long-acting antibacterial properties of chitosan-Ag/polyvinylpyrrolidone nanocomposite films. *Carbohyd. Polym.* 90 no. 1 (2012): 8–15.

Wichterle Otto, and Lim Drahoslav. Hydrophilic gels for biological usage. *Nature.* 185 (1960): 117–118.

Yang Xiaomin, Yang Kang, Wu Shengwei, et al. Cytotoxicity and wound healing properties of PVA/ws-chitosan/glycerol hydrogels made by irradiation followed by freeze-thawing. *Radiat. Phys. Chem.* 79 no. 5 (2010): 606–611.

Ye Mao, Mohanty Pravansu, and Ghosh Gargi. Morphology and properties of poly vinyl alcohol (PVA) scaffolds: Impact of process variables. *Mater. Sci. Eng. C.* 42 (2014): 289–294.

Yu Haijun, Xu Xiaoyi, Chen Xuesi, et al. Preparation and antibacterial effects of PVA-PVP hydrogels containing silver nanoparticles. *J. Appl. Polym. Sci.* 103 no. 1 (2007): 125–133.

Zhang Wei, Zhao Xinzhen, Zhang Zhennan, Xu Yunxia, and Wang Xinping. Preparation of poly(vinyl alcohol)-based membranes with controllable surface composition and bulk structures and their pervaporation performance. *J. Membr. Sci.* 415–416 (2012): 504–512.

Zhang Xiaoqing, Takegoshi K., and Hikichi Kunio. Phase separation and thermal degradation of poly(vinyl alcohol)/poly(methacrylic acid) and poly(vinyl alcohol)/poly(acrylic acid) systems by ^{13}C C.p./m.a.s. N.m.r. *Polymer.* 33 no. 4 (1992): 718–724.

Zhang Xiefei, Liu Tao, Sreekumar T. V., et al. Poly(vinyl alcohol)/SWNT composite film. *Nano Lett.* 3 no. 9 (2003): 1285–1288.

Zheng Yudong, Huang Xiaoshan, Wang Yingjun, Xu Hong, and Chen Xiaofeng. Performance and characterization of irradiated poly(vinyl alcohol)/polyvinylpyrrolidone composite hydrogels used as cartilages replacement. *J. Appl. Polym. Sci.* 113 no. 2 (2009): 736–741.

5 Flow Domains of Polymer Solutions with Liquid Crystal Phase

Andreea Irina Barzic

CONTENTS

5.1 INTRODUCTION

The liquid crystalline phase, or mesophase, is a particular state of matter characterized on a macroscopic scale by long-range order; it is similar to fluid in that matter has the ability to flow (Percec and Tomazos 1992). Generally, molecules must have asymmetric shapes to form liquid crystals (LCs). For polymers, the mesogenic groups (basic structural units that form in liquid crystalline phases) are present either in the main chain with flexible segments or in the side chain connected to the flexible backbone. The mesogenic groups are known to give polymers their LC character. However, there are some macromolecular compounds that do not contain mesogenic groups but that are also capable of forming LC phases (Collyer 1992; Acierno and Brostow 1996; Brostow 1998). This aspect can be ascribed to their rigid or semi-rigid backbone, which creates steric effects or intramolecular hydrogen bonding that restricts the flexibility of the chain. Application of a shear deformation leads to complex flow behavior, and a specific texture results after relaxation of the chains that can be further induced to other materials (Cosutchi et al. 2010).

This chapter presents some fundamental aspects concerning liquid crystal polymers, highlighting the complex rheological properties of liquid crystal polymer solutions. If isotropic components are added in the multiphase system, the flow domains are affected by the interactions occurring among the counterparts. This chapter also describes the flow behavior of some polymer mixture solutions that have a lyotropic phase as a function of system composition, shear intensity, and temperature. The correlation between rheological properties and the changes in surface morphology induced by the system's anisotropic component is reviewed with regard to practical uses in electro-optics and tissue engineering.

5.2 FUNDAMENTALS OF LIQUID CRYSTAL POLYMERS

5.2.1 Definition and Main Types of Liquid Crystal Polymers

Polymers with rigid and semi-rigid backbones exhibit the ability to self-organize depending on various factors. The main categories of liquid crystal polymers are depicted in Figure 5.1. Factors such as concentration, temperature, and (in some cases) pressure determine the self-organization of such materials. Thus, lyotropic liquid crystal polymers are organized as a function of their concentration in an adequate solvent, while thermotropic compounds are organized in an appropriate temperature domain. In a few cases, liquid crystalline phases appear in polymers at elevated pressures.

5.2.2 Mesophases of Liquid Crystal Polymers

Differences among liquid crystal phases arise from the parameters that describe the organization inside the material. In a liquid crystal polymer, the macromolecules have a tendency to point in a preferential direction. This common axis is

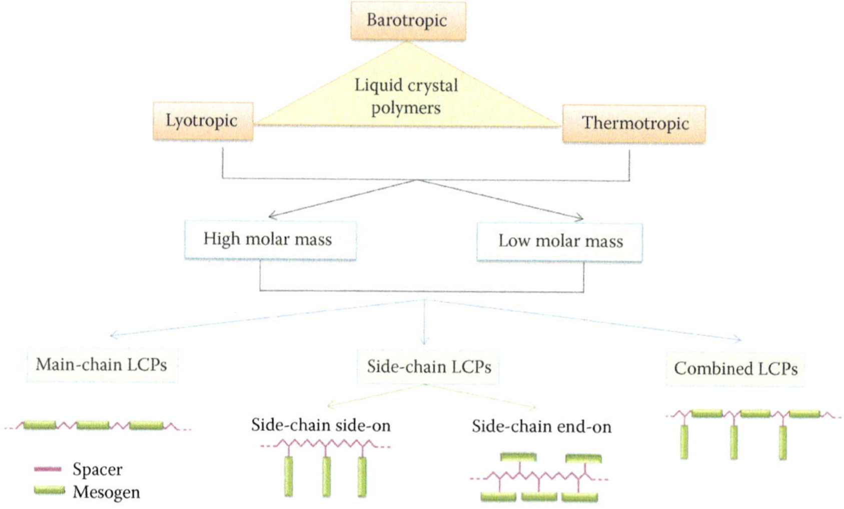

FIGURE 5.1 The schematic representation of the main types of liquid crystal polymers.

called the LC director. The inherent order in LC compounds is a combination of the long tridimensional order of solids and the disorganization present in isotropic liquids. Therefore, the phases of liquid crystal polymers may be described as having the following aspects (Acierno and Brostow 1996; Brostow 1998):

- Orientational order—A measure of the tendency of the molecules to align along the director on a long-range basis
- Positional order—The extent to which the position of an average molecule or group of molecules shows translational symmetry

The main phases of liquid crystal polymers are *nematic, cholesteric, smectic*, and *discotic*. Their ordering and corresponding textures as observed by optical microscopy are presented in Figure 5.2. In some cases, liquid crystal polymers display polymorphism, namely more than one phase is distinguished in the LC state. The *mesophases* in polymers consist in "subphases," in which molecular order ranges as a result of a possible imposed alignment or of a high degree of translational motion.

In the *nematic phase*, the molecules are able to move in all directions. This means that there is no positional order of the centers of mass; but, on the average, they keep their long axes locally parallel (the LC director tends to point in the same direction). Liquid crystal polymers are anisotropic materials so their physical characteristics are strongly dependent on the average alignment with the director (Acierno and Brostow 1996).

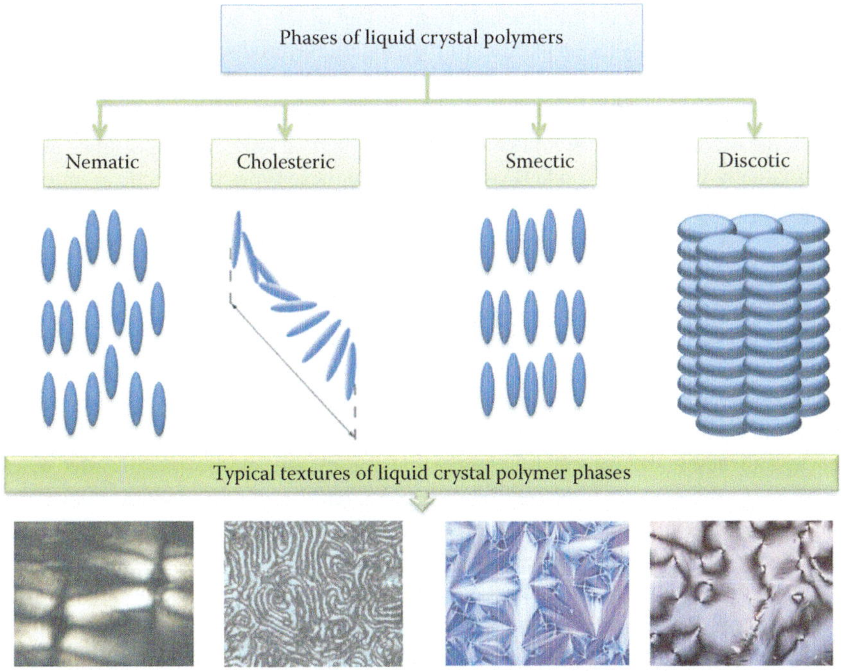

FIGURE 5.2 The main phases of liquid crystal polymers and the corresponding textures.

The *cholesteric phase* is a special case within the nematic phase where the molecules are chiral. This concept induces important alterations of the macroscopic arrangement of the LC molecules, thus modifying the optical behavior of the substance. Molecules follow a helical arrangement in the cholesteric phase. The helix axis, around which the molecules twist, lies perpendicular to the local director. The periodic structure along the helix axis is formed because the twist has a constant strength throughout the sample (a Liquid Crystal Organic Compound Engineering Essay, https://www.ukessays.com/essays/engineering/a-liquid-crystal-organic-compound-engineering-essay.php). It is not a positional periodicity but a directional one. In every half turn of the helix, the molecules are organized in the same way. If the full helical turn (generally called the pitch of the cholesteric) is of the same order of magnitude as the wavelength of electromagnetic radiations in a visible range, spectacular optical consequences are noticed. For instance, when a light wavelength corresponds to the pitch and passes through the LC sample along the helix axis, the circularly polarized component with the same handedness as the helix is totally reflected. Because the pitch changes with temperature, the color of the substance will also vary. In addition, the color also changes with the viewing angle because the light entering at an angle has a longer optical periodicity.

There are a great number of important structural variations in the *smectic* state. The main feature that characterizes this phase is the positional order along one dimension. However, in some smectic phases, the sample presents positional order in more than one dimension. Mention must be made that translational order of molecules is present only in the smectic state not in the nematic. In other words, in the smectic phase, the molecules keep the general orientational order of nematics and also tend to align themselves in layers or planes. In the smectic layered structure, the molecules are positioned parallel or tilted relative to the normal layer. The motion is restricted within these planes, and separate planes appear to flow past each other. Two smectic phases, denoted as smectic A and smectic C, have gained a special importance and are now relatively well understood (Brostow 1998). In these types of smectic compounds, the layers lack positional order. The molecules have some freedom to move within the layers but, as in all smectic phases, are unable to move between layers. Smectic compounds can therefore be viewed as stacks of two-dimensional fluids that behave like crystals across the layers. The absence of in-layer order contributes to their high potential for future electro-optic applications. There are several smectic phases that differ from one another in the director's tilt angle with respect to the normal layer and also in the arrangement of molecules within each layer. So far, eleven types of smectic phases have been reported. The simplest is the smectic A phase, which is characterized by a director parallel to the layer normal and by random positional order within the plane. In the smectic C phase, the molecules have the same random order within the layer but tilt relative to the normal layer (Scharf 2007). The tilt angle usually increases with decreasing temperature. Substances in the smectic A phase can also exhibit the smectic C phase when the temperature is lowered (Acierno and Brostow 1996; Brostow 1998). The other smectic phases are even more crystalline in that they also feature some positional order within the layers, and they may exhibit hexagonal packing of the molecules. For instance, in the smectic B phase, the LC director is normal to the plane containing the molecules arranged in a hexagonal structure.

When the smectic substance exhibits chirality (denoted by a star after the phase label), it can have a defect-free macroscopic structure. However, no twist can be allowed within the layers due to the layered arrangement of the smectic. Without breaking the layers, the orientation of the director is modified continuously but only between layers. Hence, the helix axis must lie perpendicular to the layer planes. No visible difference is observed when comparing the untilted A* phase with the achiral A phase. However, the chirality of A* generates new physical properties that are different from those of the A phase. When the director starts to tilt, such as in the C* phase, a helicoidal macroscopic structure is formed, with the azimuthal angle of the director varying with a constant value from layer to layer. In some cases, when the intrinsic twisting power of the compound is extremely high, the continuous layer structure of the smectic A* may break down, allowing a twist with its axis perpendicular to the director and similar to that of cholesterics (Collyer 1992). This means that the smectic structure tears apart into periodic stacks of layers, with a finite twist in between, that is mediated by a regular array of screw dislocations. This type of defect structure is known as the twist-grain-boundary (TGB) phase and corresponds to the blue phase that can appear in a cholesteric due to the very high twisting power. A corresponding TGB phase can be also obtained from a smectic C* phase (a Liquid Crystal Organic Compound Engineering Essay, https://www.ukessays.com/essays/engineering/a-liquid-crystal-organic-compound-engineering-essay.php). In both cases, the twist axis is along the smectic layers. The TGB phase is noticed only in a very narrow temperature range.

The *discotic* phase is distinctive from the aforementioned phases because the shapes appear like disks instead of rods. This mesophase consists of stacked columns of molecules. These columns are packed together to form a bidimensional crystalline array. The characteristics of this type of mesophase depend not only on the molecules' disposal in the columns but also on the columns' arrangement. The latter aspect can lead to new mesophases.

5.2.3 Phase Transitions in Liquid Crystal Polymers

In the LC phase, polymers exhibit a large variety of phase transitions. These are usually of a weak first order (enthalpy change is very small) or second order (critical) (http://grpfm.upc.edu/research/liquid-crystals-1). When analyzing the phase transitions, one must take into consideration the mechanisms that drive the self-organization of liquid crystal polymer materials. Thermotropic transitions are noticed in most LCs with certain changes of temperature. Thus, the LC state can be reached by increasing the temperature of a solid or by lowering the temperature of a liquid. Considering these aspects, one may identify two types of LC thermotropic behavior (Collyer 1992):

- *Enantiotropic* LCs—These can be changed into an LC state either from lowering the temperature of a liquid or raising the temperature of a solid.
- *Monotropic* LCs—These can only be changed into an LC from either increasing the temperature of a solid or decreasing the temperature of a liquid but not from both.

In most thermotropic transitions, mesophases occur due to anisotropic dispersion forces among the molecules and because of packing interactions. In contrast to thermotropic mesophases, lyotropic mesophases take place due to the influence of solvents and not temperature. The solvent-induced aggregation of the constituent mesogens into micelles is the main factor that determines the formation of the lyotropic mesophase. The mesogens are typically amphiphilic, meaning that they are formed from both lyophilic (solvent-attracting) and lyophobic (solvent-repelling) parts. This favors their reorganization into micellar structures in the presence of a solvent. Thus, the lyophobic ends will stay together, but the lyophilic parts will extend outward toward the solvent molecules. When increasing the solution concentration (and cooling the solution), the micelles will present higher dimensions and eventually will coalesce. This separates the newly formed LC state from the solvent.

The basic demanding aspects for LC formation, regardless of molecular size or shape, are as follows (Percec 1992):

- A first-order transition must occur between the true crystalline state at the lower temperature bound leading to the liquid crystalline state and another first-order transition leading to the isotropic liquid state (or another liquid crystal state) at the upper temperature bound of the liquid crystalline state.
- A liquid crystal must display one- or two-dimensional order only. True crystals have three-dimensional order, and the isotropic liquid is completely disordered.
- A liquid crystalline compound must exhibit a certain degree of fluidity, although for polymers the viscosity may be high.
- In most cases, the conditions for liquid crystal formation are best met when the molecules have at least some portion of their structure in the form of rods or disks.

Most investigations deal primarily with the nematic-isotropic (N-I) and the smectic A-nematic (SmA-N) phase transitions. The theory Maier–Saupe (Chen 2011) shows that the N-I transition results from competition between thermally excited forces that tend to disrupt the orientational order and molecular-field forces tending to orient the molecular axes. This phase transition will be first order in nature but, due to fluctuations of the nematic order parameter, the N-I transition could become second-order (http://grpfm.upc.edu/research/liquid-crystals-1). SmA-N phase transition is affected by both the nematic and the smectic order parameters. Thus, it must be treated differently. The order of this transition is influenced by the nematic range. When the latter is small, a relevant coupling between both order parameters occurs, and the phase transition is first-order in nature. However, as the nematic range becomes larger, the coupling changes and leads to a second-order transition. In contrast, other theoretical approaches (Halperin et al. 1974) show that the SmA-N phase transition is always weakly first order in nature.

Scientists have investigated uniaxial–biaxial phase transition for nematic liquid crystal polymers and have tried to describe it through the order parameters and also by considering the terms that account for the energy of elastic deformation and the energy of interactions within the field (if an external field is present). The existence of biaxial nematic phases was first theoretically predicted and subsequently proven experimentally (Yu and Saupe 1980). Optical microscopy and nuclear magnetic resonance (NMR) spectroscopy data revealed the presence of uniaxial and biaxial phases over a large range of concentrations and temperatures for LC in solvent mixtures. These first experimental observations led to a set of theories that attempted to describe the order of the uniaxial–biaxial phase transition. It was established that this transition can be second order in mean-field approximation, a property that is maintained when fluctuation effects are taken into consideration (Longa et al. 2005). Evidence of biaxial phases in thermotropic LCs of low molecular weight has not been found, though they were noticed in side-chain liquid crystal polymers. Modification of the dynamic Landau theory (Landau and Lifschitz 1984) for nonconserved order parameter around the uniaxial–biaxial nematic phase transition has shown that coupling of the order parameter with the strain field in liquid crystal polymers impedes the occurrence of second-order phase transitions and that, instead, first-order transitions or no transition at all may be observed. The reported investigation (Onsager 1949) demonstrated that the density of free energy contains a cubic term. This indicates that the transition will be either of the first order (where minimization of the free energy results in a physically meaningful order parameter) or will not exist at all (such as in case of a nonphysical order parameter being a solution—for example, if one defines order parameter as positive and solution as negative). This is in opposition to that of a monomeric liquid crystal, where the transition is of the second order because of the lack of coupling between strain and order parameters.

The phase behavior of a lyotropic rigid-chain liquid crystal polymer was analyzed using dissipative particle dynamics (DPD) by changing two factors—the solution concentration and the temperature (Zhao and Wang 2011). Each mesogenic polymer backbone is represented by a chain of fused DPD particles surrounded with the strongly interacted solvent molecules. The free solvent molecules are modeled as independent DPD particles, where each particle includes a lump of solvent molecules with the volume roughly equal to the solvated polymer segment (Zhao and Wang 2011). The simulation indicates that smectic-B (S(B)), smectic-A (S(A)), nematic (N), and isotropic (I) phases occur within certain regions in the temperature and concentration parameter space. The temperature-dependent S(B)/S(A), S(A)/N, and N/I phase transitions are observed in the high concentration range. The simulation indicates the coexistence of the anisotropic phases and the isotropic phase, at intermediate concentrations, where the anisotropic phases can be the S(B), S(A), or N phases. It was observed that the concentration of rigid rods in the isotropic phase increases as the temperature increases. Mole fraction and compositions of the coexisted phases are determined from the simulation. The coexistence of the two phases was outlined by fitting the orientational distribution function of the analyzed system. The distribution of the rigid rods in the two coexistence phases is quantitatively estimated from the parameter α (which resulted from the simulation). Introduction of the solvent particles in the DPD simulation is critical for predicting the phase coexistence and for obtaining the phase diagrams of the lyotropic rigid-chain polymer liquid crystal (Zhao and Wang 2011).

Phase transition of lyotropic rigid-chain liquid crystal polymers is dependent on the solvent used. This was indicated in the DPD simulation (Zhao and Wang 2013). The effect of solvent quality is reflected through the modifications of the repulsion parameter between the rods and solvent particles. The simulation data reveal that the type of solvent determines the phase transition behavior of the system, particularly when varying the concentrations before the solvent becomes extremely poor. The influence of the solvent quality is attributed to the interplay between the immiscibility-induced phase separation and nematic-isotropic phase transition, which can be described by the binodal lines and nematic-isotropic transition lines in the phase diagrams (Zhao and Wang 2013). When the system is located in one phase region, it undergoes a typical nematic to isotropic phase transition as the temperature increases. Conversely, when the system is located in the biphasic region, two different types of nematic-isotropic phase transitions may be noticed. They depend on whether the transition temperature from the biphasic region to the isotropic phase region is lower or higher than the nematic-isotropic transition temperature of the concentrated phase. The first type is ascribed to the transition from the biphasic region to the isotropic one-phase region, and the second type is attributed to the nematic-isotropic phase transition that takes place in the concentrated phase.

The phases of liquid crystal polymers determine their textures and implicitly impact the optical properties of these materials. The application range of such systems depends on the ease with which the textural changes (and therefore, the modification of optical characteristics) can be brought about by external applied forces. The macroscopic orientation of the macromolecules in the sample determines the texture. For instance, in homeotropic texture, the particles are arranged with their long axes parallel to the film normal throughout the macroscopic sample. However, in homogeneous texture, the long axes are oriented parallel to the film surface. The variety of liquid crystal polymer textures, observed through microscopic means, are directly related to the molecular structure of the material. In some cases, it is possible derive the molecular structure of the LC modifications from the observation of the texture. The main morphological features of the most important liquid crystal polymer phases are presented in Figure 5.2. Specific textures and textural changes may be generated by a variety of external factors. The methods include the use of shear, electric, and magnetic fields or changing the sample's temperature.

5.3 FLOW BEHAVIOR OF LIQUID CRYSTAL POLYMER SOLUTIONS

Rheology is one of the most common methods used to analyze the flow characteristics of various materials, including liquid crystal polymers (Mezzenga et al. 2005). There are several types of flow behaviors, which are broadly divided into two main categories (Shaw 2012; Osswald and Rudolph 2014):

- *Newtonian:* The shear stress and shear rate are always in direct proportion to each other. The viscosity of a Newtonian fluid will always be the same, regardless of the shear stress or shear rate (the shear stress versus shear rate graph will always be a straight line that will intercept the origin).

- *Non-Newtonian:* These are characterized by large dissolved or solvated molecules, with a tendency to re-associate and to exhibit a strong interaction with the solvent. They are further classified into the following types:
  - *Pseudoplastic:* When viscosity decreases with increasing shear rates, this flow is encountered with large swollen or dissolved particles.
  - *Bingham plastic:* A minimum amount of shear stress must be applied before initiating the flow.
  - *Ellis plastic:* This is similar to pseudoplastic flow except that a minimum amount of shear stress is needed to start the flow.
  - *Dilatant:* This type is encountered in suspensions that contain high concentrations of closely packed solids.

In most cases, the flow properties of polymers in solution or in a molten state are Newtonian, pseudoplastic, or a combination of both. In the case of liquid crystal polymer solutions, the flow behavior is more complex. The profound difference in the rheological behavior of ordinary and liquid crystalline polymers is due to the fact that, for the first ones, the molecular orientation is entirely determined by the flow process. The second ones are anisotropic materials already at equilibrium (Acierno and Brostow 1996). The spontaneous molecular orientation is already in existence before the flow and is switched on, varying in space, over distances of several microns or less (polydomain). If one ignores the latter, one can discuss the linear case (slow flow) as long as the rate of deformation due to flow (the magnitude of the symmetric part of the velocity gradient) is lower than the rate at which molecules rearrange their orientational spread by thermal motions.

The solution behavior of liquid crystal polymers was first described by Flory (1984) using the lattice model. His theory indicates that linear rod-like polymers would undergo a change from an isotropic phase to an ordered (anisotropic) phase in concentrated solutions. The phase separation arises from the favorable entropy of forming an ordered phase at high polymer concentrations. Both entropy and enthalpy determine the stability of mesophases, although for liquid crystal polymers, the dominant factor is the entropy term, which in turn is affected by the size and shape of the macromolecules. The assumption that polymer chains are completely rigid and rod-like is not entirely true because even the most rigid polymers have some degree of flexibility.

Special attention has been given to lyotropic liquid crystal polymer solutions that are characterized by a wide range of unusual rheological phenomena (Peuvrel and Navard 1990, 1991; Marrucci and Greco 1993). The viscosity of such solutions presents a particular variation with concentration. Isotropic solutions exhibit a monotonic increase in shear viscosity with their increasing concentration. The viscosity reaches a maximum value, which indicates the transition from isotropic to anisotropic state. Upon formation of the anisotropic phase, the viscosity starts to decrease. From then on, the viscosity begins to increase exponentially as the concentration continues to increase (Figure 5.3).

The hydrodynamic volume in the isotropic state is large due to the random orientation of the macromolecules. This limits the polymer diffusivity and causes an increase in viscosity. Upon anisotropic phase formation, the aligned chains

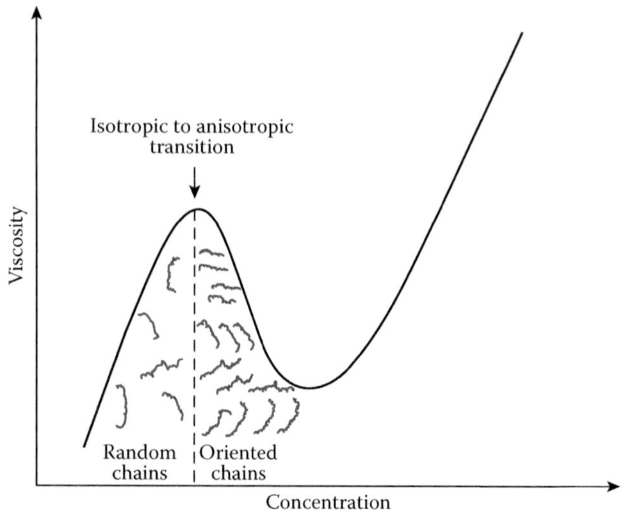

FIGURE 5.3 Viscosity dependence on concentration for a liquid crystal polymer solution. (Adapted from Appaw Collins. *Rheology and microstructure of cellulose acetate in mixed solvent systems.* Dissertation at Graduate Faculty of North Carolina State University, 2004, http://repository.lib.ncsu.edu/ir/bitstream/1840.16/3487/1/etd.pdf.)

determine a small hydrodynamic volume. The viscosity values are lowered because rotational diffusion is much easier with a net orientation. However, it has been proven that the viscosity maximum can be related to the occurrence of the anisotropic phase only when the shear rate is close to zero (Zugenmaier 2004). At a higher shear rate, the concentration at which the maximum viscosity occurs shifts to a smaller value. When polymer solution is subjected to higher shear rates, the entanglements of the randomly distributed chains are loosened, and shear-induced orientation (the pseudo-nematic phase) of macromolecules occurs. The viscosity maximum disappears, and a monotonic increase of viscosity as a function of concentration is then noticed. Thus, it can be stated that the viscosity of the nematic or pseudo-nematic mesophase is less sensitive to concentration than that of the chiral nematic phase. Anisotropic solutions of cellulose derivatives present a similar viscosity–concentration relationship.

Steady flow behavior is the most thoroughly investigated rheological property. Onogi and Asada (1980) assumed that the viscosity of liquid crystal polymer solutions can be described by the universal existence of three shear flow domains:

- Regime I: A shear thinning regime at low shear rates
- Regime II: A Newtonian plateau at intermediate shear rates
- Regime III: Another shear thinning regime at high shear rates

The first flow regime appears at low shear rates, typical for shear thinning, but also exhibits a yield stress similar to some plastic materials. This flow domain is characterized by distortional elasticity associated with spatial variation in the director field (average local molecular orientation). The second flow regime is a Newtonian plateau, reflecting a "dispersed polydomain" structure. This is followed by the last regime, which is a shear thinning zone, describing viscoelastic behavior. In Regimes I and II, the flow is not strong enough to affect molecular orientation. However, in Regime III, the flow field is very strong and generates molecular orientation during shearing. In the case of some liquid crystalline polymer solutions, such as those of cellulose derivatives, not all regimes appear because not every regime lies in the accessible shear rate range.

Polymeric liquids behave nonlinearly if the velocity gradient of a flow is large enough so that the rate of molecular relaxation is taken over by the deformation rate due to flow (Acierno and Brostow 1996). The most interesting phenomenon in the steady flow of liquid crystalline polymers is the behavior of the normal stress differences, N1 and N2, in simple shear flow. The curves show complex nonmonotonic features, and both N1 and N2 change sign; although in ordinary polymer solutions, the normal stress differences vary monotonically with shear rate. In particular, negative values of first normal stress difference (N1)—that is, two sign changes in N1 as a function of shear rate—were reported (Grizzuti et al. 1990). In contrast, isotropic solutions only show positive N1 at all shear rates. This peculiar behavior of polymer liquid crystal solutions results from a coupling of molecular tumbling with the local molecular-orientation distribution (Navard 1986; Ernst and Navard 1989). At low shear rates, the director tumbles with the flow, and N1 will be positive. At intermediate shear rates, nonlinear viscoelastic effects are reflected in the director tumbling that competes with the steady director alignment along the flow. In this flow domain, the director oscillates around a steady value, and N1 becomes negative. A further increase in shear rate causes the director to align along the flow, and N1 is positive again. Generally, it is supposed that an interruption of the director tumbling takes place at shear rate at which N1 is minimum, thus orienting the macromolecules. (Baek et al. 1993). However, for polymer solutions in the concentrated domain, the negative N1 behavior is no longer observed due to polymer–polymer friction interactions (Ernst and Navard 1989).

Negative values of N1 have been noted in many lyotropic polymer solutions (Ernst and Navard 1989; Baek et al. 1994). For monodomain samples, optical evidence of tumbling at the start-up of a shear flow has also been reported (Acierno and Brostow 1996). It was noticed that tumbling at low shear rates is more typical for lyotropic polymers because negative normal stresses are rarely observed for thermotropic main-chain liquid crystal polymers. This might arise from the fact the macromolecules of thermotropics are less rigid than lyotropic ones and also, as yet, there is no good theory of the dynamics of semi-rigid chains in a nematic phase.

5.4 FLOW-INDUCED PHENOMENA OF POLYMER LIQUID CRYSTALS

The complex aspects of polydomain dynamics might serve as a qualitative guidance for describing flow-induced molecular orientation, which is mainly observed for main chain liquid crystal polymers (Brostow 1998). This arises from the cooperative nature of a nematic phase where, at least locally, the molecules are already mutually oriented in an almost parallel manner. To produce a global orientation, the focus of attention must change from thermal agitation to the weak forces that are mandatory for reorienting the director.

A good orientation is more readily obtained in the case of elongational flow, particularly when preexisting defects are missing or when they find the time or the topological conditions to disappear under the influence of flow. For shear-dominated flows, a possible misunderstanding could be that tumbling completely destroys the orientation, effectively producing an isotropic situation at the mesoscopic level. The mesoscopic orientation, which is lower than that of the monodomain, roughly occurs along the shear direction in spite of director tumbling (Acierno and Brostow 1996). Even if the latter is continuously rotating, most of the time it stays close to the shear direction, sometimes flipping over and taking a half turn. The directors in the polydomain are not randomly distributed, and they are predicted to be placed in the shear plane. A more complete, and complex, analysis of the polydomain situation can be made if one considers the following aspects: the initial orientation in the quiescent sample domain interactions are due to Frank elasticity defects—specifically core defects—that are believed to confine actual tumbling of the surrounding nematic through an anchoring effect (Marrucci and Greco 1993).

Another important aspect that should be mentioned is that shear flow of tumbling liquid crystal polymers plays a major role in orienting the molecules along the shear direction. This takes place if the shear rate is high enough to reach the nonlinear range, where the material undergoes a transition from tumbling to flow-aligning behavior. For instance, at high shearing, a phenomenon that consisted of the formation of transverse striations perpendicular to the direction of shear was reported (Navard 1986; Cosutchi et al. 2010). These generally are known as "bands" as distinguished from "stripes" (which are oriented parallel to flow). The formation of bands was reported for both lyotropic and thermotropic polymers (Navard 1986; Kwiatkowski and Hinrichsen 1990). Initially, it was thought that steady-state bands are formed during shear. Most of the investigations focused on band structures attempted to elucidate the kinetics and morphology of band structures formed after cessation of shear. The negative N1 and band structures respond to shear in different ways. The observation of negative N1 was noticed for few lyotropic systems (Baek et al. 1993), although the formation of band structures essentially is universal in both lyotropes and thermotropes (Brostow 1998). Even more highly complex variations on band structure have been studied, such as the "helical coarse rope" structure in PBC solutions (Kisst and Porter 1980), a "torsad" structure in HPC solutions (Ding et al. 1995), and a "tractor" or "herringbone" texture in thermotropic polyesters (Kwiatkowski and Hinrichsen 1990).

The process of band formation is characterized by several parameters: the period of time during which the solution is sheared, the shear rate, the time necessary for band occurrence, and the length of time the bands persist before disappearing. Starting with these aspects, a literature survey on this subject revealed the following (Navard 1986; Brostow 1998):

- The bands do not appear until a critical shear is reached; the latter is inversely proportional to molecular weight.
- There is an upper critical shear rate after which bands no longer occur.

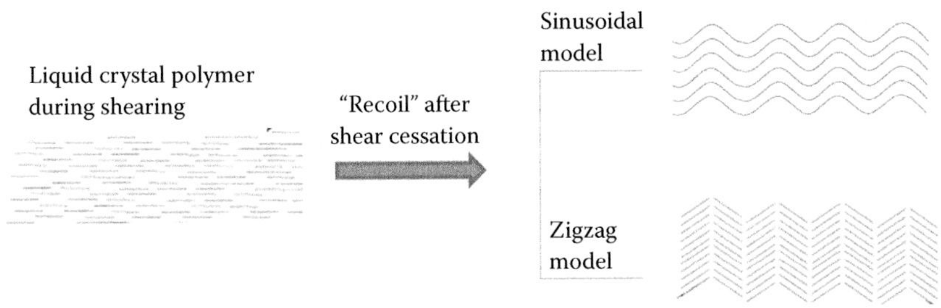

FIGURE 5.4 A schematic representation of the models for band texture formation.

- The shearing time must be long enough, before removing the deformation field, to allow the shearing to be applied depending on the force magnitude. If prior shear of a sufficiently high rate is imposed for a long enough time, band formation occurs after a waiting period ranging from zero to several minutes.
- After the bands are formed, they gradually begin to lose their parallelism and start to merge with adjacent bands. After several minutes, they disappear, leaving a texture similar to that of the undeformed state.
- The fineness of the band texture is enhanced as the rate of prior shear is increased. However, since band formation occurs more rapidly and is then followed by broadening, it could be stated that the band spacing as a function of time is independent of the rate of prior shear.

In order to determine the disposition of the molecules within the bands, several studies have focused on dried films of sheared lyotropic solutions (Wang and Labes 1992; Patnaik et al. 1995; Appaw 2004; Cosutchi et al. 2011). The disposition of the molecules has been described as similar to a sinusoid or a zigzag, and indeed, a "recoil" phenomenon was revealed upon cessation of shear (Ding et al. 1995). These aspects are schematically represented in Figure 5.4.

5.5 FLOW DOMAINS OF POLYMER SYSTEMS WITH LYOTROPIC PHASE

The specific band texture of lyotropic polymer systems can be explored in electronic and biomedical applications, where control of surface morphology is mandatory. Moreover, the pattern of such materials can be further induced to an isotropic polymer (Cosutchi et al. 2010, 2011). Flow behavior of multiphase systems based on hydroxyproyl cellulose (HPC) in lyotropic form combined with a poly(amic acid) or a polyimide reveals the mutual influence of the components. Both polyimide and its precursor exhibit a constant viscosity region over the entire shear rates range, namely a Newtonian behavior. On the contrary, the cellulose derivative in dimethylacetamide (60 wt%) presents typical rheological properties for a liquid crystal in that its flow curve is composed from Regime II (Newtonian plateau) and Regime III (shear thinning). The viscosity versus concentration at very low shear rates (0.7 1/s) denotes a maximum value at about 47 wt% (Cosutchi et al. 2010). This designates anisotropic phase formation and supports the fact that the HPC solution is found in liquid crystal form at 60 wt%. Analysis of the flow domains of the multiphase solutions containing the lyotropic phase reveals a complex behavior.

Introduction of a higher amount of HPC limits the length of the Newtonian regime. At the working concentrations, the Newtonian regime is more similar in character to the isotropic polymer than to HPC. Similarly, in the case of individual components, enhancing the temperature parameter determines a decrease in shear viscosity and a slight increase in the Newtonian domain. The flow activation energy values reflect the interactions occurring among the polymers but also those taking place between the individual counterparts and the dimethylacetamide solvent. The poly(amic acid) and HPC liquid crystal may interact with the solvent through hydrogen bonds. On the contrary, the polyimide lacks functional groups and cannot interact in such a manner with the solvent. Thus, its flow activation energy is higher when compared with its precursor—the poly(amic acid). In addition, the aliphatic and flexible sequences contained in the polyimide and its precursor structure determine lower energetic barriers comparatively with the lyotropic HPC compound. The shapes of curves representing log shear stress versus log shear rate plots denote two flow regimes, reflecting the combined behavior of the liquid crystal and the isotropic counterpart (Cosutchi et al. 2010).

The morphology of such blends can be explained based on the typical texture of HPC. Optical micrographs of the lyotropic sample indicate alternating bright and dark lines perpendicular to the shear direction, which result from the reorientation of the director, caused by the relaxation of the elastic energy stored in the material during flow (Cosutchi et al. 2011). When the deformation is removed, relaxation leads to the formation of the bands perpendicular to the flow direction, while the flow or deformation orthogonal to bulk orientation gives rise to bands parallel to the second deformation direction. Atomic force microscopy investigations support the idea of two types of periodicities at the HPC film surface after shear cessation (Cosutchi et al. 2011). Blending HPC with the semi-aliphatic polyimide or its precursor changes the surface pattern. However, the bands are still observed—particularly for the HPC/poly(amic acid)—because the hydrogen bond interactions formed between polyimide precursor and the liquid crystalline component stabilized the resulted morphology. In case of HPC/polyimide, the bands are discernible but are more pronounced for high amounts of liquid crystal. However, the lack of interactions among the polymers from the multiphase system is reflected in a less-defined band texture, and the small bands along the shear direction are not observed (Cosutchi et al. 2010). The bands' thickness and the distance between them depend significantly on the composition of the system. For 50/50 HPC/poly(amic acid), the surface organization is extended to a third direction (Cosutchi et al. 2011). If this composition is immersed in a good solvent for the liquid crystal but a nonsolvent for the poly(amic acid) for removal of the HPC matrix, drying and surface analyses reveal that the band pattern remains on the surface of the anisotropic polymer.

Polymer surfaces with such a tunable pattern are of great importance in electronics where they can be used as mixed alignment layers. Nematics brought in contact with this type of polymer support can orient themselves, opening new perspectives for obtaining high performance alignment layers used in display devices (Cosutchi et al. 2011; Barzic et al. 2015).

Pure and applied research related to cell growth has shown that a polymer substrate with specific topography can guide the cells to grow in a specific manner (Hiruma et al. 2011). These aspects are essential in tissue engineering, where tridimensional cell spheroids can be obtained on patterned polymers, leading to new insights in tissue reconstruction.

5.6 GENERAL REMARKS AND FUTURE DIRECTIONS

The self-organizing nature of liquid crystal polymers is reflected in their complex flow behavior. The relaxation phenomenon of lyotropic polymer solution after shear cessation leads to band texture morphology that can be further induced to isotropic materials. Future research should focus on solvents' influence on band size implicitly related to the induced pattern in polymers with different structures. Another aspect that could be explored is the imidization of patterned polyimide precursors and those conditions in which the texture is still maintained.

ACKNOWLEDGMENTS

This work was supported by a grant from the Romanian National Authority for Scientific Research and Innovation, CNCS–UEFISCDI, project PN-II-RU-TE-2014-4-2976, no. 256/1.10.2015. Special thanks are addressed to Dr. Silvia Ioan and Dr. Camelia Hulubei from "Petru Poni" Institute of Macromolecular Chemistry–Romania for supervising the rheological experiments of liquid crystal polymer solutions and for synthesis of some imidic polymers containing aliphatic units, respectively.

REFERENCES

Acierno D., and Brostow Witold. *Rheology and processing of liquid crystal polymers.* London: Springer, 1996.

Appaw Collins. *Rheology and microstructure of cellulose acetate in mixed solvent systems.* Dissertation at Graduate Faculty of North Carolina State University, 2004, http://repository.lib.ncsu.edu/ir/bitstream/1840.16/3487/1/etd.pdf

Baek S.G., Magda J.J., and Cementwala S. Normal stress difference in liquid-crystalline hydroxypropylcellulose solutions. *J. Rheol.* 37 no. 5 (1993): 935–945.

Baek S.G., Magda J.J., Larson R.G., and Hudson S.D. Rheological differences among liquid-crystalline polymers. II. Disappearance of negative N1 in densely packed lyotropes and thermotropes. *J. Rheol.* 38 no. 5 (1994): 1473–1503.

Barzic Andreea Irina., Hulubei Camelia., Avadanei Mihaela Iuliana., Stoica Iuliana., and Popovici Dumitru. Polyimide precursor pattern induced by banded liquid crystal matrix: Effect of dianhydride moieties flexibility. *J. Mater. Sci.* 50 no. 3 (2015): 1358–1369.

Brostow Witold. *Mechanical and thermophysical properties of polymer liquid crystals.* London: Chapman & Hall, 1998.

Chen Robert H. Maier–Saupe theory. In *Liquid crystal displays: Fundamental physics and technology*, ed. Lowe Anthony C., pp. 147–159. London: Wiley, 2011.

Collyer A.A. *Liquid crystal polymers: From structures to applications*. New York: Elsevier, 1992.

Cosutchi Andreea Irina, Hulubei Camelia, Stoica Iuliana, and Ioan Silvia. Morphological and structural-rheological relationship in epiclon-based polyimide/hydroxypropylcellulose blend systems. *J. Polym. Res.* 17 no. 4 (2010): 541–550.

Cosutchi Andreea Irina, Hulubei Camelia, Stoica Iuliana, and Ioan Silvia. A new approach for patterning epiclon-based polyimide precursor films using a lyotropic liquid crystal template. *J. Polym. Res.* 18 no. 6 (2011): 2389–2402.

Ding Jiadong, Feng Jun, and Yang Yuliang. Sinusoidal supermolecular structure of band textures in a presheared hydroxypropyl cellulose film. *Polym. J.* 27 no. 11 (1995): 1132–1138.

Ernst Benoit, and Navard Patrick. Band textures in mesomorphic (hydroxypropyl)cellulose solutions. *Macromolecules.* 22 no. 3 (1989): 1419–1422.

Flory P.J. Molecular theory of liquid crystals. *Adv. Polym. Sci.* 59 (1984): 1–35.

Grizzuti Nino, Cavella Silvana, and Cicarelli Paolo. Transient and steady-state rheology of liquid crystalline hydroxypropylcellulose solution. *J. Rheol.* 34 no. 8 (1990): 1293–1310.

Halperin B.I., Lubensky T.C., and Ma Shang-Keng. First-order phase transitions in superconductors and smectic-a liquid crystals. *Phys. Rev. Lett.* 32 no. 6 (1974): 292–295.

Hiruma Hitomi, Asayama Shoichiro, and Kawakami Hiroyoshi. Control of cell morphology on the polyimide surface patterned by rubbing and ion-irradiation. *Polym. Adv. Technol.* 22 no. 8 (2011): 1311–1314.

Kisst Gabor, and Porter Roger S. Rheo-optical studies of liquid crystalline solutions of helical polypeptides. *Mol. Cryst. Liq. Cryst.* 60 no. 4 (1980): 267–280.

Kwiatkowski Miroslaw, and Hinrichsen Grete. A new banded texture in sheared thermotropic liquid crystalline polymers. *J. Mater. Sci.* 25 no. 3 (1990): 1548–1550.

Landau Lev D., and Lifschitz Evgeny M. *Electrodynamics of continuous media. Course of theoretical physics*. Oxford: Butterworth-Heinemann, 1984.

Longa Lech, Grzybowski Piotr, Romano Silvano, and Virga Epifanio. Minimal coupling model of the biaxial nematic phase. *Phys. Rev. E.* 71 no. 5 (2005): 051714.

Marrucci Giuseppe, and Greco Francesco. Flow behavior of liquid crystalline polymers. In *Advances in Chemical Physics*, eds. Prigogine I., Rice Stuart A., 331–404. Hoboken, NJ: Wiley, 1993.

Mezzenga Raffaele, Meyer Cedric, Servais Colin, Romoscanu Alexandre I., Sagalowicz Laurent, and Hayward Ryan C. Shear rheology of lyotropic liquid crystals: A case study. *Langmuir.* 21 no. 8 (2005): 3322–3333.

Navard Patrick. Formation of band textures in hydroxypropylcellulose liquid crystals. *J. Polym. Sci. B.* 24 no. 2 (1986): 435–442.

Onogi Shigharu, and Asada Tadahiro. Rheology and rheo-optics of polymer liquid crystals. In *Rheology*, volume 1, eds. Astarita Giovanni Marrucci Giuseppe, Nicolais Luigi, pp. 127–147. New York: Plenum, 1980.

Onsager Lars. The effects of shape on the interaction of colloidal particles. *Ann. N. Y. Acad. Sci.* 51 (1949): 627–659.

Osswald Tim A., and Rudolph Natalie. *Polymer rheology: Fundamentals and applications*. Munich: Carl Hanser Verlag GmbH & Co. KG, 2014.

Patnaik Soumya S., Bunning Timothy J., and Adams W. Wade. Atomic force microscopy and high-resolution scanning electron microscopy study of the banded surface morphology of hydroxypropylcellulose thin films. *Macromolecules.* 28 no. 1 (1995): 393–395.

Percec Virgil, and Tomazos Dimitris. Molecular engineering of liquid crystalline polymers. In *Comprehensive polymer science*, eds. Allen G, Bevington J.C., pp. 1–53. Oxford: Pergamon Press, 1992.

Peuvrel Edith, and Navard Patrick. Shear velocity profiles in hydroxypropyl cellulose solutions. *Macromolecules.* 23 no. 22 (1990): 4874–4875.

Peuvrel Edith, and Navard Patrick. Band textures of liquid crystalline polymers in elongational flows. *Macromolecules.* 24 no. 20 (1991): 5683–5686.

Scharf Toralf. *Organic optical materials. In Polarized Light in Liquid Crystals and Polymers*, pp. 103-141. Hoboken, NJ: :Wiley, 2007.

Wang J., and Labes M.M. Control of the anisotropic mechanical properties of liquid crystal polymer films by variations in their banded texture. *Macromolecules.* 25 no. 21 (1992): 5790–5793.

Yu L.J., and Saupe A. Observation of a biaxial nematic phase in potassium laurate-1-decanol-water mixtures. *Phys. Rev. Lett.* 45 no. 12 (1980): 1000–1003.

Zhao Tongyang, and Wang Xiaogong. Phase behavior of lyotropic rigid-chain polymer liquid crystal studied by dissipative particle dynamics. *J. Chem. Phys.* 135 (2011): 244901.

Zhao Tongyang, and Wang Xiaogong. Solvent effect on phase transition of lyotropic rigid-chain liquid crystal polymer studied by dissipative particle dynamics. *J. Chem. Phys.* 138 (2013): 024910.

Zugenmaier Peter. Characterization and physical properties of cellulose acetates. *Macromol. Symp.* 208 no. 1 (2004): 81–166.

6 Microstructure Evolution during Reinforcement of Polymer Solutions

Silvia Ioan

CONTENTS

6.1 INTRODUCTION

Determination of the macroscopic properties of reinforced composite polymers from the corresponding local properties of the different phases constitutes the fundamental challenge in obtaining new materials with reinforced components. This aspect is motivated by the essential physical observation that the constitutive properties of these materials appear to be homogeneous at the length scale of practical applications, despite the sharp variation of their local properties at the length scale of heterogeneity, and also by the fairly complex spatial distribution of their phases. In this context, the realization of high-performance materials has stimulated considerable research on reinforced polymeric systems.

6.2 REINFORCED POLYMERS BASED ON BLENDS OF POLYMERS WITH THERMOTROPIC LIQUID CRYSTALLINE POLYMERS

Generally, a thermotropic liquid crystalline polymer (LCP) can function as a processing aid and a reinforcing filler, when added to an immiscible polymer (e.g., polystyrene) (Weiss et al. 1987). In this case, the thermal, rheological, and mechanical properties of the pure components and blends exhibit immiscibility, and the LCP forms an elongated fibrous phase oriented in the flow direction when an extensional component of flow is present during processing. Thus, the viscosity at melt temperature decreases as the LCP concentration increases as a result of the high extensional forces present at the entrance of the capillary that deform and orient the LCP domains in the flow direction. Because of the unusually high relaxation times of LC polymers, the aligned domains formed in the entrance region remain deformed and aligned in the capillary. The LCP polymer has a much lower viscosity than the polymer, although, in the blend, the oriented LCP phase tends to lubricate the melt after cooling—the dispersed fibrous LCP phase being preserved in the solidified material as microfibers (which behave as a reinforcing component and improve the mechanical properties of the blend).

Some studies (Kim et al. 2009) have described the composite materials obtained from poly(ethylene 2,6-naphthalate) (PEN) resin (in dimethylterephthalate of 0.1 wt% to enhance the processability of the polymer–polymer composite resins) as a matrix polymer and a flexible thermotropic liquid crystal copolyester (TLCP; produced by synthesizing poly(p-hydroxybenzoate) and poly(ethyleneterephthalate) with a molar ratio of 80:20) used as a filler. Thermotropic liquid crystal polymers possess a high degree of orientation under shear and elongational flow fields. This property contributes to deformation of liquid crystal domains and formation of fibrils (which act as reinforcing fillers in the blend, reduce melt viscosity, and enhance processability because of their low melt viscosity) without sacrificing the mechanical and thermal properties of the pure poly(ethylene 2,6-naphthalate). Authors have shown that these reinforced composites can be used for the production of high-performance fibers in various industrial applications, such as tire cords, air bags, and others (Kim et al. 2009). Rheological parameters of TLCP/PEN composite materials with a weight ratio of 0.5/99.5, 1/99, 3/97, and 5/95 (prepared for obtaining optimal spinnability) are presented in Figure 6.1a and b.

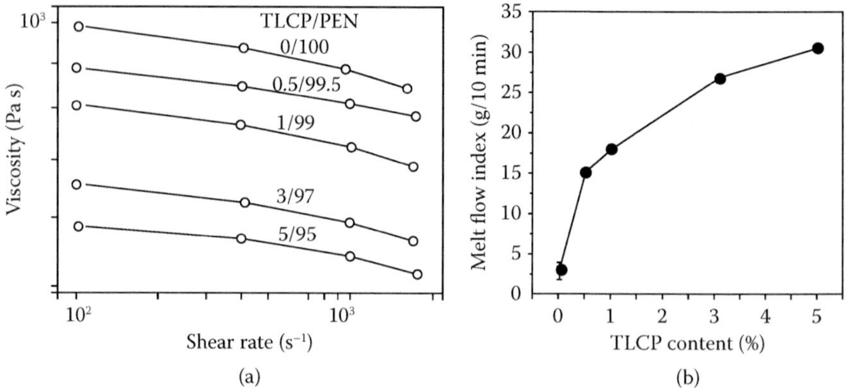

FIGURE 6.1 Effect of shear rate on viscosity at different TLCP/PEN compositions (a) and the effect of TLCP/PEN compositions on the melt flow index (b). (Adapted from Kim Yun Seong et al. *Composites: Part A.*, 40, no. 5, 607–612, 2009.)

A typical shear thinning behavior as well as a decrease in viscosity and an increase in the melt flow index while increasing the content of thermotropic liquid crystal copolyester was recorded. In addition, the oriented and ordered fibril structure of thermotropic liquid crystal copolyesters generated in the poly(ethylene 2,6-naphthalate) matrix is illustrated in Figure 6.2.

A TLCP, having a trimethylene terephthalate (TT) unit and a terephthaloyl mesogenic triad unit, was also synthesized, and its blends with poly(trimethylene terephthalate) (PTT) were prepared for TLCP-reinforced fiber spinning (Choi et al. 2015). The thermal, mechanical, and morphological properties show that the thermotropic liquid crystalline copolymer acted as a reinforcing material in the poly(trimethylene terephthalate) matrix, leading to an increased initial modulus and tensile strength of the blend fibers, with increasing the thermotropic liquid crystalline copolymer content. This chapter, summarizing literature data, points out that TLCPs have received much attention with regards to their processability due to their shear-thinning effect, low melt viscosity, slow relaxation time, and their orientation in flow direction. This fibrillar morphology of TLCP can induce self-reinforcement in the matrix polymer.

A liquid crystal polymer (LCP) used to improve the physical properties of glass fiber (GF) reinforced polypropylene (GFRPP) was also studied (Li et al. 2015). The mechanical and heat resistance properties of the GFRPP/LCP composite were improved by the presence of the liquid crystal polymer. For these composites, the yield strength increases by 62.7% and the impact strength increases by 18.1 wt% compared with the GFRPP with 30 wt% glass fiber. Also, the crystallinity of the polypropylene matrix in GFRPP/LCP composites increases for 5% LCP and then decreases as the LCP content increases. The γ-phase crystals for the polypropylene matrix occurred in the shear layer of the injection molded GFRPP/LCP samples. The improved adhesion interface between the glass fiber and the polypropylene matrix contributed to reinforce and toughen the GFRPP/LCP composites with a low addition of liquid crystal polymer.

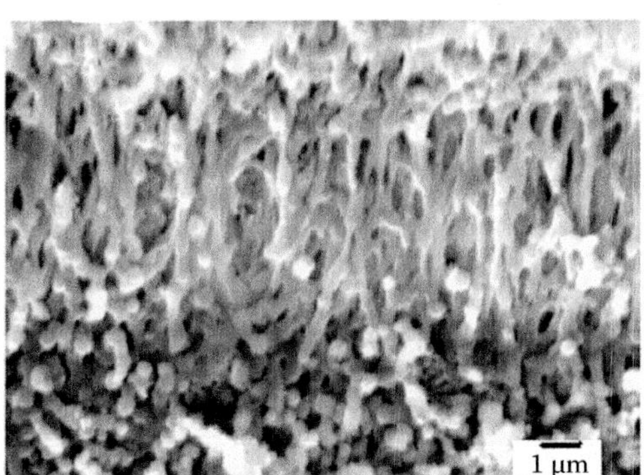

FIGURE 6.2 Scanning electron microscopy image of TLCP/PEN composite with weight ratio of 5/95. (Adapted from Kim Yun Seong et al. *Composites: Part A.*, 40, no. 5, 607–612, 2009.)

6.3 PROCESSING OF POLYMER NANOCOMPOSITES REINFORCED WITH POLYSACCHARIDE NANOCRYSTALS

The mechanical properties and reinforcing potential as well as the biodegradability of polysaccharides make them candidates for the processing of polymer nanocomposites (Dufresne 2010; Thakur 2013, 2015).

The homogeneous dispersion of cellulose nanoparticles in a polymer matrix in order to obtain nanomaterials is due to their size, which allows penetration in hydrosoluble or at least hydrodispersible structures (as latex-form polymers) as well as dispersion of polysaccharide nanocrystals in nonaqueous media especially using surfactants and chemical grafting. Thus, one of the processing techniques of polymer nanocomposites reinforced with polysaccharide nanocrystals was carried out using hydrosoluble or hydrodispersible polymers. In this respect, the literature has reported preparation of polysaccharide particles with reinforced starch (Svagan et al. 2009), silk fibroin (Noishiki et al. 2002), poly(oxyethylene) (POE) (Samir et al. 2006), polyvinyl alcohol (PVA) (Zimmermann et al. 2005), hydroxypropyl cellulose (HPC) (Zimmermann et al. 2005), carboxymethyl cellulose (CMC) (Choi and Simonsen 2006), or soy protein isolate (SPI) (Zheng et al. 2009).

The preparation procedures of polysaccharide nanocrystal reinforced polymer nanocomposites using latexes consist of mixing and casting the two aqueous suspensions, followed by

- Water evaporation performed above the glass transition temperature of the polymer. During water evaporation, the solid content in the medium increases and the latex particles get closer, adopting a polyhedral form. In this way, a continuous polymeric film is created, containing the dispersed polysaccharide nanoparticles.
- Freeze-drying and hot-pressing of the mixture—or its freeze-drying, extrusion, and hot-pressing—where the dispersion of its nanoparticles in the nanocomposite film strongly depends on the processing technique and conditions.

The polymeric particle size seems to play a predominant role (Dubief et al. 1999), whereas the latex large or small particles provide superior or poor mechanical properties, respectively. These procedures were presented in the literature (Dufresne 2008, 2010) as follows:

- Copolymerization of styrene and butyl acrylate [poly(S-*co*-BuA)] in association with tunicin whiskers (the cellulose extracted from a tunicate—a marine animal).
- Copolymerization of styrene and butyl acrylate in association with wheat straw, sugar beet cellulose nanocrystals, potato starch nanocrystals, squid pen, and Riftia tube chitin whiskers.
- Poly(β-hydroxyoctanoate), polyvinylchloride, waterborne epoxy, natural rubber (NR), and polyvinyl acetate (PVAc) were also used as matrices.

- Stable, aqueous nanocomposite dispersions containing cellulose whiskers and a poly(styrene-*co*-hexyl-acrylate) matrix were prepared via miniemulsion polymerization.

When using nonaqueous systems, the processes of nonpolar polymer nanocomposites reinforced with polysaccharide nanoparticles involve dispersion in the organic medium of nanoparticles, coating with a surfactant, or surface chemical modification of the nanoparticles to reduce their surface energy and to increase dispersibility in nonpolar media. The decrease of surface energy ensures dispersion of nanoparticles in organic liquids with low polarity.

Long chain surface chemical modification of polysaccharide nanoparticles, consisting of grafting agents bearing a reactive end group and a long "compatibilizing" tail, was also reported in the literature for increasing the apolar character of the nanoparticle. Scheme 6.1 describes two ways of surface chemical modification of polysaccharide nanoparticles selected from the literature—grafting "onto" and grafting "from" procedures (Dufresne 2010).

Surface modifications can act as binding sites for active agents in drug delivery systems or for toxins in purifying and treatment systems. These surface modifications may inter-diffuse upon heating to form the polymer matrix phase. The covalent linkage between reinforcement and matrix will result in a near-perfect stress transfer at the interface, resulting in exceptional mechanical properties of the composite.

As an example of a graft ("onto") procedure, the literature shows the chemical modification of starch-derived nanocrystals as follows:

- Poly(ethylene glycol) methyl ether (PEGME) and stearic acid chloride (Thielemans et al. 2006). Crystallization of the stearate moieties grafted onto the starch nanoparticles forms a crystalline hydrophobic shell around the hydrophilic starch nanocrystal, reducing hydrogen bonding and the polar interactions among individual particles.
- Poly(tetrahydrofuran), poly(caprolactone), and poly(propylene glycol) monobutyl ether (PPGBE) chains using toluene 2,4-diisocyanate as a linking agent (Labet et al. 2007)
- Poly(caprolactone)-grafted ramie cellulose whiskers and PCL-grafted waxy maize starch nanocrystals by directly grafting the polymer on the surface of the nanoparticles with a coupling agent.

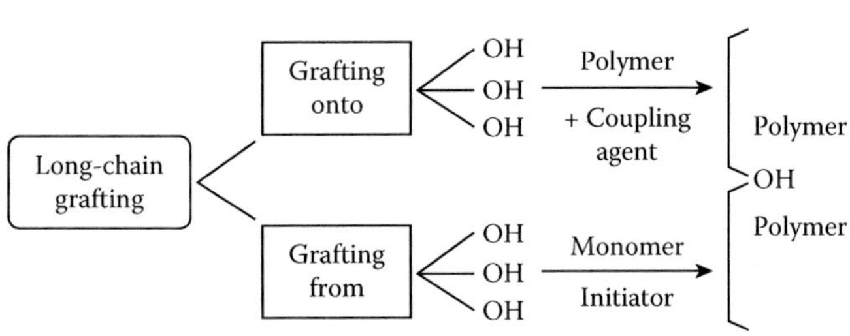

SCHEME 6.1 Chemical modification of polysaccharide nanoparticles.

The literature notes another graft version, poly(caprolactone)-grafted cellulose whiskers, which consist of mixing the nanoparticles with the monomer and an initiator, and with the polymer growing directly on the surface of the nanoparticles (Habibi et al. 2008). This compound was thermoformed to produce sheets with good mechanical properties (Chen 2009) and, finally, a high hydrophobicity of the composite.

The literature presents similar results for injection-molded poly(caprolactone)–grafted crab shell chitin whiskers (Feng et al. 2009) or poly(caprolactone) grafted onto the surface of starch nanocrystals via microwave-assisted ring-opening polymerization (ROP) (Yu et al. 2008; Chang et al. 2009; Lin et al. 2009). In another system (Lin et al. 2009), cellulose whiskers were surface-grafted with poly(caprolactone) via microwave-assisted ring-opening polymerization and yielding foliaceus cellulose whisker-graft-poly(caprolactone) nanocrystals incorporated into the polylactic acid as a matrix. The graft-forming procedure has been applied in other studies on chemical surface modifications of microfibrillated cellulose (Stenstad et al. 2008), extrusion and characterization of functionalized cellulose whisker reinforced polyethylene nanocomposites (de Menezes et al. 2009), surface grafting and processing of waterborne polyurethane-cellulose nanocrystal nanocomposites (Cao et al. 2009), chemical grafting of cellulose nanoparticles with a long chain isocyanate (Siqueira et al. 2010), and nanocomposites prepared by in situ enzymatic polymerization of phenol with TEMPO-oxidized nanocellulose (Li et al. 2010).

Extrusion and impregnation, electrospinning, and multilayer films are other processes applicable for reinforced nanocomposites. Thus, preparation of cellulose whiskers reinforced with polylactic acid nanocomposites (by melt extrusion carried out by pumping the suspension of nanocrystals into the polymer melt during the extrusion process), using polyvinyl alcohol as a compatibilizer for the dispersion of cellulose whiskers within the polylactic acid matrix, was reported (de Menezes et al. 2009); bacterial cellulose whiskers incorporated into poly(oxyethylene) nanofibers by electrospinn to enhance the mechanical properties of electrospun fibers (Peresin et al. 2010) or the use of the layer-by-layer assembly technique, which maximizes the interaction between cellulose whiskers and a polar polymeric matrix (Bruno et al. 2009; Aulin et al. 2010), are some examples of nanocomposites reinforced by the last three methods.

6.4 REINFORCED POLYMERS BASED ON POLYMER BLENDS WITH INORGANIC PARTICLES

Blending of polymers with inorganic particles represents a method for producing new reinforced materials. Polymer nanocomposites are two-phase materials in which the polymers are reinforced. Generally, the filler material belongs to the smectite class of aluminum silicate clays, of which the most representative one is montmorillonite. The layered silicates, uniformly dispersed and exfoliated in a polymer matrix, ensure improved properties such as strength, higher modulus, thermal stability, barrier properties, and lower flammability (Nguyen and Baird 2006; Banc et al. 2015). Generally, aggregates of nanoparticles are often formed, and the inclusion of a polymer inside them may increase the concentration of hard matter and may introduce volume spanning and, on occasion, fractal, percolating structures.

The literature outlines the considerable efforts dedicated to the measurement of chain conformation in polymer nanocomposites, sometimes yielding contradicting results depending on the chemical system, size ratio between polymer and filler, and filler concentration (Jouault et al. 2010a, 2010b; Crawford et al. 2013; Tung et al. 2013).

As a function of microscopic nanoparticle dispersion, polymer nanocomposites combine the viscoelastic properties of polymer matrices with the strength of embedded nanoparticles, leading to highly different rheological or mechanical behavior (Banc et al. 2014, 2015). The authors studied the filler structure and conformation of poly(methyl methacrylate) chains in silica–latex nanocomposites using different procedures. First, they analyzed the presence and distribution of aggregates for nanocomposites made with matrix chains of different masses. Therefore, the effect of silica content has been studied in four different systems made of small and big filler nanoparticles and short and long chains. Typical transmission electron microscopy images are shown in Figure 6.3 (Banc et al. 2015) for a fixed silica volume fraction of 3 vol. %, where the best dispersion appears to be realized for big nanoparticles and long chains (see Figure 6.3c; the particle shapes are better defined in Figure 6.3c and d, for bigger nanoparticles).

On the contrary, it has been found that the rheological properties of silica–latex nanocomposites are a result of matrix properties and of their modification in the presence of silica filler. The filler structure of the nanocomposites is governed by the

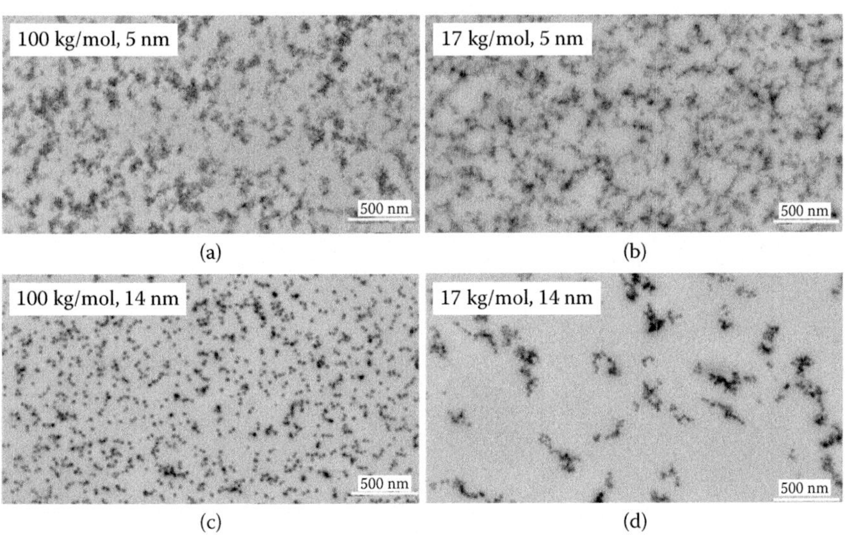

FIGURE 6.3 TEM pictures of 3 vol% silica–latex nanocomposites for two polymer molecular weights and silica particles sizes: (a) 100 kg/mol, 5 nm; (b) 17 kg/mol, 5 nm; (c) 100 kg/mol, 14 nm; and (d) 17 kg/mol, 14 nm. (Adapted from Banc Amelie et al. *Macromolecules.*, 48, no. 18, 6596–6605, 2015.)

mass of the polymer matrix chains and also by the rheology of the system during film formation and annealing. Linear rheology, investigated with oscillatory shear experiments, shows a feature related to the silica structure at low frequencies, the amplitude of which can be described by two power laws separated by the percolation threshold of aggregates. Rheological properties of nanocomposites realized on the different poly(ethyl methacrylate) matrices and increasing amounts of silica (up to 10 vol.%) have been measured over the same frequency domain. All three plots from Figure 6.4 (Banc et al. 2014) show the same dependence of storage modulus, G′, on the silica volume fraction at low frequencies. At high frequencies, the curves increase with the content of silica volume fraction and tend to display a common plateau.

These rheological results, coupled with other experimental techniques, offer a control parameter of the silica contribution, and of the evolution of the corresponding phenomena, with volume fractions. Also, these studies contribute to a better understanding of the reinforcement effect in nanocomposites.

Recently, Jouault et al. formulated nanocomposites from solutions of polystyrene/poly(2-vinylpyridine) diblocks with silica nanoparticles (diameter ≈ 14 nm) (Jouault et al. 2014a, 2014b) and investigated the influence of casting solvents on the spatial dispersion of nanoparticles in polymer nanocomposites. They found out that poly(2-vinylpyridine) adsorbs on silica with a nominal grafting density of ≈0.01 chains/nm^2 for a saturated concentration of the adsorbed layer corresponding to ≈6 chains/nanoparticle, in close agreement with the estimate of the number of latex beads surrounding a nanoparticle. The small number of adsorbed chains is a feature of small nanoparticles. In this study, nanocomposites were prepared from bare silica nanoparticles and poly(2-vinylpyridine) by casting from two different solvents, methyl ethyl ketone (theta solvent) and pyridine (good solvent), for both polymer and nanoparticles:

- In methyl ethyl ketone, poly(2-vinylpyridine) strongly adsorbs onto the silica surface to create a temporally stable bound polymer layer. The resulting hairy particles are sterically stabilized against agglomeration,

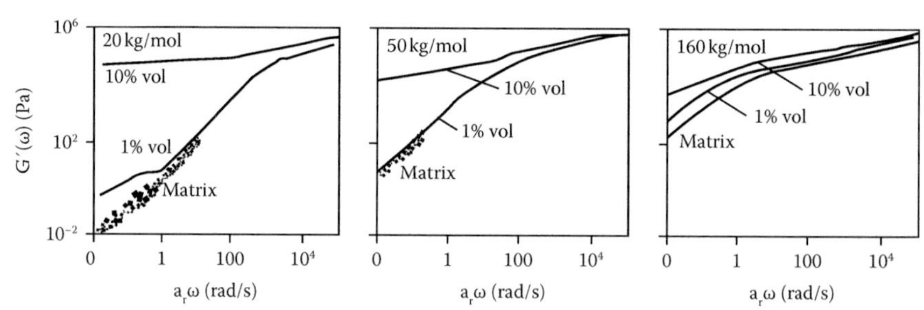

FIGURE 6.4 Storage modulus, G′, as a function of angular frequency, ω, at a temperature of 180°C for master curves of matrix and of nanocomposites with poly(ethyl methacrylate) having a mass of 20, 50, and 160 kg/mol and silica volume fractions of 1 and 10 vol%. (Adapted from Banc Amelie et al. *Macromolecules*., 47, no. 9, 3219–3230, 2014.)

and thus good nanoparticle dispersion in polymer nanocomposites is always achieved independent of poly(2-vinylpyridine) molecular weight, concentration, or nanoparticle loading.
- In pyridine, poly(2-vinylpyridine) does not adsorb on silica nanoparticles, and broadening and apparent increase of size distribution are due to the coexistence of bare nanoparticles and aggregates.

Figure 6.5 shows that good dispersion in poly(2-vinylpyridine) is due to the equilibrium among electrostatic repulsion, polymer-induced depletion attraction, and kinetic slowdown of nanoparticles (Jouault et al. 2014b). Thus, in contrast to the good dispersion obtained with methyl ethyl ketone, nanoparticles form large aggregates when pyridine is used as a solvent.

Smectite clays and other layered inorganic materials that can be broken down into nanoscale building blocks are good alternatives to the sol–gel process for the preparation of organic–inorganic nanocomposites.

Polymer–clay composites can be divided into three categories:

1. Conventional composites (where clay tactoids exist in their original, aggregated state with no intercalation of the polymer matrix into the clay).
2. Intercalated nanocomposites (where insertion of a polymer into the clay structure occurs in a crystallographically regular fashion, regardless of the clay-to-polymer ratio; an intercalated nanocomposite is normally interlayered only within a few molecular layers of polymer, and the composite properties typically resemble those of the ceramic host) and exfoliated nanocomposites (where the individual thick clay layers of 10 Å are separated in a continuous polymer matrix at average distances that depend on loading).
3. An exfoliated nanocomposite with a monolithic structure having properties related to those of the starting polymer (Nguyen and Baird 2006).

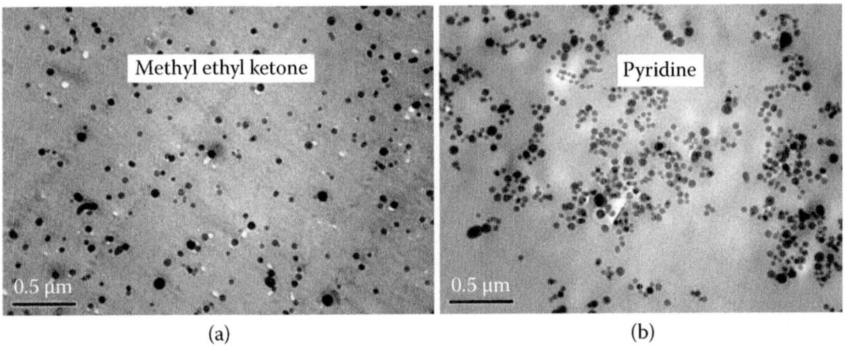

FIGURE 6.5 Transmission electron microscopy images of as-cast 10 wt% 50-nm bare silica nanoparticles in a 14.7 K P2VP matrix prepared by the solvent casting method using (a) methyl ethyl ketone and (b) pyridine as solvents. (Adapted from Jouault Nicolas et al. *Macromolecules*., 47, no. 15, 5246–5255, 2014b.)

Polymer layered silicate nanocomposites are obtained by:

- Intercalation of a suitable monomer and subsequent in situ polymerization when the monomer is used directly as a solubilization agent—in a suitable solvent—for swelling the layered silicate. After combining the silicate layers and the monomer, the polymerization is initiated, thus allowing formation of polymer chains between the intercalated sheets.
- Intercalation of the polymer from solution (involving a solvent that dissolves the polymer) and swelling of the silicate layers. The layered silicate is dispersed into the polymer solution, after which the solvent is removed by polymer precipitation or evaporation.
- Polymer melt intercalation (involving a polymer and a layered silicate mixture heated under discontinuous or continuous shear above the softening point of the polymer; thus, the polymer chains diffuse from the molten polymer into the silicate galleries and form intercalated or exfoliated sheets).

The degree of exfoliation of a particular nanocomposite and the establishment of the nanocomposite's structure are investigated by wide-angle X-ray diffraction analysis (WAXD), to determine the interlayer spacing of the silicate layers in the original layered silicates and in intercalated nanocomposites, and by transmission electron microscopy (TEM), for a qualitative understanding of the internal structure, spatial distribution of the various phases, and direct visualization of the defect structure. The intercalated or exfoliated nanocomposite structures are studied by observing the position, shape, and intensity of the basal reflections from the distributed silicate layers. The small-angle X-ray scattering (SAXS) method can also be used to characterize the structure of nanocomposites, when layer spacings exceed 6–7 nm in intercalated nanocomposites or when the layers become relatively disordered in exfoliated nanocomposites.

Polymer mixtures with a low content of layered smectite clays may increase mechanical, dimensional, permeable, and flame-retardant properties relative to the pure polymer. Thus, the literature reports that (Solomon et al. 2001)

- Mixtures of nylon-6 and smectite clay montmorillonite exhibit heat distortion temperatures higher than that of pure nylon for a less than 5 wt% clay loading (Kojima et al. 1993).
- Epoxy/clay hybrids at 10 wt% concentration have a higher tensile modulus than pure epoxy (Lan and Pinnavaia 1994).
- Improvements in the fire resistance of nylon-6/clay hybrids (Nguyen and Baird 2006) occur.
- Polymer/clay hybrid nanocomposites, used in automotive, packaging, and aerospace applications (Dagani 1999), may be prepared.

Among the known methods for obtaining nanocomposite fibers, electrospinning ensures a low cost and high speed of preparation. The method involves application of a high voltage between the tip of a needle and a collecting electrode.

A pendent drop of solution is charged under surface tension, inducing charges that are distributed all over the surface. The Columbic repulsive force overcomes the solution surface tension and thus forces the ejection of a liquid jet, which results in the formation of a Taylor cone. The electrified jet of the viscoelastic solution then undergoes a stretching process and forms thin fibers on the collector. The literature presents polyacrylonitrile (PAN) nanocomposite fibers with different amounts of iron oxide (Fe_3O_4) nanoparticle loadings, fabricated by a high voltage electrospinning process (Zhang et al. 2009). Thus, these nanocomposites are prepared from a PAN/*N*,*N*-dimethylformamide solution of different concentrations, in which different amounts of Fe_3O_4 have been added by a high voltage electrospinning process. Dispersion of Fe_3O_4 particles in solution was performed through ultrasonication, after which the solution was used for fiber electrospinning fabrication and subsequent rheological investigation (Figure 6.6) (Zhang et al. 2009). The studies presented here demonstrate the capacity of rheological behavior for obtaining reinforced polymers based on blends of polymer with inorganic particles. Solution viscosity and solution surface tension play important roles in determining the morphology of the electrospun nanofibers.

A slight decrease in viscosity is observed at a concentration of 6.5 wt%, after which this decrease is intensified. It is even more pronounced starting from 10 wt%, in 50–1200 s^{-1} shear rate domain. Thus, orientation of the macromolecular chains at high concentration causes a pseudoplastic behavior and thinning phenomena. For a composite polymer solution, the strong interaction between the Fe_3O_4 nanoparticles and PAN increases solution inertia when shear stress is applied, creating an alignment of PAN molecules and a more pronounced shear thinning behavior.

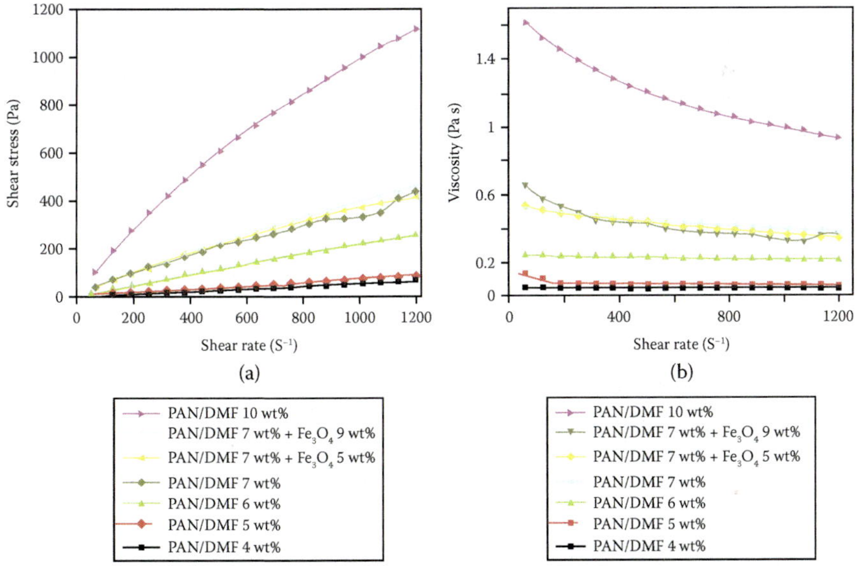

FIGURE 6.6 (a) Shear stress and (b) viscosity versus shear rate of PAN and PAN/Fe_3O_4 solution systems. (Adapted from Zhang Di et al. *Polymer.*, 50, no. 17, 4189–4198, 2009.)

The authors conclude that the optimum viscosity for producing uniform fibers by the electrospinning method falls within a large range of values, which is consistent with other observations related to obtaining uniform PAN nanofibers with different polymer concentrations. The nanoparticles from the composite nanofibers become magnetically harder, with a much larger coercivity than that of the dried nanoparticles.

6.5 FIBER-REINFORCED POLYMER COMPOSITES

A fiber-reinforced polymer represents a composite material consisting of a polymer matrix (usually an epoxy, vinylester, or polyester thermosetting plastic, and phenol formaldehyde resins) reinforced with fibers (especially from glass, carbon, and aramid, and sometimes from wood or asbestos), used in almost every type of advanced engineering structure, ranging from aircraft, helicopters, and spacecraft, to boats, ships, and offshore platforms, and on to automobiles, sporting goods, chemical processing equipment, and civil infrastructure (such as bridges and buildings) (Masuelli 2013). Microstructure evolution during polymer reinforcement refers to a material functionalized via a structuring process. A promising way to obtain new materials based on core-shell fibers is represented by electrospinning of a polymer solution or melt. Following this procedure, a variety of polymer fibers are prepared for different applications, including templates for reinforced materials. The literature describes the processing of core-shell nano-/mesofibers by co-electrospinning of two materials. Compound capillary liquid–liquid jets of Newtonian and non-Newtonian fluids are introduced in ink-jet printing. Both liquids outflowing from the core- and the surrounding concentric annular nozzles are polymer solutions or a combination of polymer solution and melt. A compound droplet sustained at the edge of such compound nozzles should undergo transformation into a Taylor cone compound with a compound jet co-electrospin from its tip. Then, as in the ordinary electrospinning process, the jet will be pulled by the electric field and stretched through bending instability far enough from the droplet. The solvent will evaporate and the compound jet will solidify, resulting in compound core-shell nanofibers. Research has observed that this process can be successfully used for the formation of coaxial fibers made of pairs of different materials: poly(ethylene oxide)–poly(dodecylthiophene), poly(L-lactide)-Pd or poly(ethylene oxide)–polysulfone, as well as pairs of identical polymers poly(ethylene oxide)–poly(ethylene oxide) contrasted by dyeing agents, such as bromophenol (Sun et al. 2003). Studies on fiber-reinforced polymers are currently increasing due to the vast range of traditional and special applications related to their characteristics and specific properties.

Shaohua et al. used short electrospun fibers for the first time for the preparation of homogeneous polymer nanofiber/polymer composites, improving the mechanical properties by an order of magnitude comparative with electrospun long and continuous nanofibers (Shaohua et al. 2013). Thus, the formation of electrospun fiber nonwovens was carried out by dispersion casting of short electrospun polyimide (PI) nanofibers and the preparation of homogeneous polymer/polymer nanofiber composite films. The self-reinforced PI/PI-short nanofiber composite films were obtained from mixtures of short electrospun PI nanofibers and polyamic acid after imidization. Significant improvement in mechanical properties such as tensile strength and modulus in comparison to the neat PI film was observed by using 2 wt% short fibers.

Electrospinning has been recognized as an efficient technique for the fabrication of polymer nanofibers. The basic criteria for a polymer solution or melt to be electrospinnable are its ability to carry an electric charge and having sufficient viscosity to be stretched without breaking up into droplets (Mohammadzadehmoghadam et al. 2015). Generally, three components initiate the electrospining process: a high voltage supplier, a capillary tube with a small diameter needle, and a grounded collecting screen. The morphology, density, and diameter of electrospun nanofibers depend on the following parameters: solution parameters (solution viscosity, solution concentration, molecular weight, surface tension, electrical conductivity, dipole moment, dielectric strength), processing parameters (flow rate, electric field strength, tip-to-collector distance, needle shape, collector composition, geometry), and ambient parameters (temperature, humidity, air flow). Various polymers have been successfully electrospun into ultrafine fibers, mostly in solvent solution and some in melt form. Potential applications based on such fibers, specifically their use as reinforcements in nanocomposite development, have been realized. A comprehensive review has been presented on the research and developments related to electrospun polymer nanofibers including processing, structure and property characterization, applications, and modeling and simulations (Huang et al. 2003). Information on these polymers, together with processing conditions for the electrospinning of ultrafine fibers, has been summarized in the literature.

Agoras, in his dissertation, showed that determination of the effective or macroscopic properties of composite materials from the corresponding local properties of the constituent phases constitutes the fundamental issue in the mechanics of composites (Agoras et al. 2009a, 2009b; Agoras 2010). This aspect is motivated by the constitutive properties of these materials that may be homogeneous or uniform at the length scale of practical applications, despite the sharp variation of their local properties at the length scale of heterogeneity, and also by the fairly complex spatial distribution of their phases. The author discusses the macroscopic mechanical properties of "multi-scale" viscoplastic composites, that is, composite systems with viscoplastic constituents exhibiting heterogeneity at more than one, well-separated length-scale, exemplifying the case of semicrystalline polymers such as polyethylene, polypropylene, nylon-6, and so on that constitute the largest class of polymers used in technological applications.

Artificial blood vessels consisting of natural polymer nanofibers grafted into human bodies to facilitate the recovery of damaged blood vessels have been studied in recent years. It is known that the electrospun nanofibers (ENs) of biocompatible materials, such as chitosan, have poor mechanical properties. For this reason, studies have been developed that describe the design and fabrication of artificial blood vessels composed of a blend of chitosan and poly(ε-caprolactone) electrospun nanofiber and coated with poly(ε-caprolactone) strands, using rapid prototyping technology (Lee et al. 2015). The resulting tubular vessels exhibit excellent mechanical properties and show that this process may be useful for vascular reconstruction.

6.6 CARBON NANOTUBE-BASED FIBERS AND COMPOSITES

Significant progress has been recorded on carbon nanotubes, thus enabling significant improvements in the technology of carbon nanotube-based composites. In this context, the literature (Chou et al. 2010) summarizes the main approaches for the production of carbon nanotubes fibers:

- Spinning from a lyotropic liquid crystalline suspension of nanotubes, in a wet-spinning process similar to that used for polymeric fibers, such as aramids
- Spinning from multi-walled carbon nanotubes previously grown on a substrate as "semi-aligned" carpets
- Spinning directly from an aerogel of single-walled and multi-walled carbon nanotubes during their formation in a chemical vapor deposition reactor

Also, carbon nanotubes act as nucleating agents for polymer crystallization and as templates for polymer orientation. Due to its considerable length, single-walled carbon can act as a nucleating agent. It is suggested that the next generation of carbon fibers likely will be processed from their composites with carbon nanotubes. According to the literature (Chae and Kumar 2008; Xu et al. 2013), such carbon nanotube fibers could have ten times the specific strength of the strongest commercial fiber available today.

On the contrary, Kevlar (a para-aramid synthetic fiber) and carbon fibers possess ultrahigh strength and stiffness, owing to the regularly aligned linear polymer chains and planar graphitic layers, and now are an integral part of modern industry. High-performance fibers have also been fabricated from highly aligned carbon nanotubes. Graphene, a newly discovered two-dimensional carbon allotrope consisting of carbon atoms arranged in a honeycomb-like network (to the thickness of an atom), called the material of the future, can rival carbon nanotube fibers in advanced materials and electronic devices because of its mechanical and electrical properties (Bai et al. 2011; Wang et al. 2012).

Several studies have tried to incorporate carbon nanotubes into polymer nanofibers produced through electrospinning. However, a combination of dielectrophoretic forces, caused by dielectric or conductivity mismatch between carbon nanotubes and the polymer solution and the high shear forces induced by spinning, determines the alignment of carbon nanotubes or of their bundles along the fiber direction. The resulting nanocomposite does exhibit the mechanical properties as much as one would expect (Huang et al. 2003). Works into which carbon single wall nanotubes are dispersed include

- A polyacrylonitrile solution that was electrospun into ultrafine fibers. The obtained nanocomposite fibrils were characterized by the structure, composition, and physical properties of the resulting nanocomposite fibrils (Ko et al. 2002).
- A polyimide solution for electrospinning (Park et al. 2002).
- Ultrafine fibers from polycaprolactone/gold or ZnO, polyacrylnitrile/TiO_2, PVA/silica, nylon-6/montmorillonite (Fong et al. 2002).

The applications of reinforced composite based on electrospun polymer nanofibers generally refer to domains listed in Table 6.1 (Huang et al. 2003).

Carbon nanotubes are regarded as ideal filler materials for polymeric fiber reinforcement due to their exceptional mechanical properties and cylindrical geometry (nanometer-size diameter). Polymer chains in the vicinity of carbon nanotubes (interphase) have been observed to have a more compact packing, higher orientation, and better mechanical properties than bulk polymers due to the carbon nanotube polymer interaction. The existence of interphase polymers in composite fibers, their structural characterization, and fiber properties are summarized and discussed in the literature (Liu and Satish 2014). Besides improvements in tensile properties, the presence of carbon nanotubes in polymeric fibers also influences other factors (thermal stability, thermal transition temperature, fiber thermal shrinkage, chemical resistance, electrical conductivity, and thermal conductivity).

TABLE 6.1
Application Areas of Composite Reinforcements Based on Electrospun Polymer Nanofibers

Application of Composite Reinforcements Based on Electrospun Polymer Nanofibers	Areas of Applicability
Applications in life science	Drug delivery carrier
	Hemostatic devices
	Wound dressing
Tissue engineering scaffolding	Porous membrane for skin
	Tubular shapes for blood vessels and nerve regenerations
	Three-dimensional scaffolds for bone and cartilage regenerations
Filter media	Liquid filtration
	Gas filtration
	Molecular filtration
Nanosensor	Thermal sensor
	Piezoelectric sensor
	Biochemical sensor
	Fluorescence optical chemical sensor
Cosmetic skin mask	Skin cleansing
	Skin healing
	Skin therapy with medicine
Military protective clothing	Minimal impedance to air
	Efficiency in trapping aerosol particles
	Anti-biochemical gases
Other industrial applications	Micro/nano electronic devices
	Electrostatic dissipation
	Electromagnetic interference shielding
	Photovoltaic devices—nano–solar cell
	LCD devices
	Ultra-lightweight spacecraft materials
	Higher efficient and functional catalysts

Another study describes a process for obtaining electroconductive fibers from cellulose and carbon nanotubes spun from an aqueous alkaline/urea solution (Qi et al. 2015). The microstructure and physical properties of the resulting fibers were investigated by scanning electron microscopy, Raman microscopy, wide-angle X-ray diffraction, tensile tests, and electrical resistance measurements. These flexible composite fibers have sufficient mechanical properties and good electrical conductivity, with volume resistivities in the range of about 230–1 Ohm cm, for 2–8 wt% carbon nanotube loading. Also, it has been established that the multifunctional sensing behavior of these nanotube/cellulose composite fibers with regard to tensile strain, temperature, and environmental humidity are promising properties for their use in electronics and for the design of various smart materials. Also, carbon nanotubes have a great potential for use as electrical wires because of their outstanding electrical and mechanical properties (Guo et al. 2015). Lightweight carbon nanotube fibers have been fabricated with electrical conductivity higher than that of the stainless steel from macroscopic carbon nanotube films; these properties are obtained by their being drawn through diamond wire-drawing dies. The entangled carbon nanotube bundles are reinforced by suffering tension, which improves the alignment of fibers.

6.7 REMARKS AND FUTURE DIRECTIONS

This review is designed as a source for polymer nanocomposite research, including fundamental structure properties, manufacturing techniques, and applications of reinforced nanocomposite materials. In addition to presenting the scientific framework for the advances recorded in polymer nanocomposite research, this review focuses on the scientific principles and mechanisms related to the evolutive processes in solutions for obtaining reinforced nanocomposites, with a discussion on commercial applications. This chapter offers information on the technology, modeling, characterization, processing, manufacturing, and applications of such composites. Reinforced composite materials have gained major importance as structural elements for space engineering; aeronautical, naval, automotive, and civil engineering; and for sporting goods and other consumer products. Reinforcement represents an effective manufacturing strategy for enhancing the mechanical strength and other properties of composite materials. Among the methods for obtaining reinforced materials, electrospinning has gained widespread interest because of its ability to fabricate continuous ultrafine nanofibers with advanced features. The impact of reinforced materials on composite polymer synthesis and processing, characterization, and applications in different domains such as electronics, tissue scaffolding, drug delivery, and nanofiltration are extensively studied in the literature, with interesting perspectives in the near future.

ACKNOWLEDGMENTS

This work was supported by grant from the Romanian National Authority for Scientific Research, CNCS-UEFISCDI, project no PN-II-ID-PCE-2011-3-0937, 2016 stage.

REFERENCES

Agoras Michalis. *On macroscopic constitutive relations and microstructure evolution in multi-scale viscoplastic composites*. Dissertation at University of Pennsylvania ScholarlyCommons, 2010, http://repository.upenn.edu/edissertations/193

Agoras Michalis, Lopez-Pamies Oscar, and Ponte Castañeda Pedro. A general hyperelastic model for incompressible fiber-reinforced elastomers. *J. Mech. Phys. Solids*. 57 no. 2 (2009a): 268–286.

Agoras Michalis, Lopez-Pamies Oscar, and Ponte Castañeda Pedro. Onset of macroscopic instabilities in fiber-reinforced elastomers at finite strain. *J. Mech. Phys. Solids*. 57 no. 11 (2009b): 1828–1850.

Aulin Christian, Johansson Erik, Wågberg Lars, and Lindström Tom. Self-organized films from cellulose I nanofibrils using the layer-by-layer technique. *Biomacromolecules*. 11 no. 4 (2010): 872–882.

Bai Hua, Li Chun, and Shi Gaoquan. Functional composite materials based on chemically converted grapheme. *Adv. Mater.* 23 no. 9 (2011): 1089–1113.

Banc Amelie, Genix Anne-Caroline, Chirat Mathieue, et al. Tuning structure and rheology of silica-latex nanocomposites with the molecular weight of matrix chains: A coupled SAXS-TEM-Simulation Approach. *Macromolecules*. 47 no. 9 (2014): 3219–3230.

Banc Amelie, Genix Anne-Caroline, Dupas Christelle, et al. Origin of small-angle scattering from contrast-matched nanoparticles: A study of chain and filler structure in polymer nanocomposites. *Macromolecules*. 48 no. 18 (2015): 6596–6605.

Bruno Jean, Heux Laurent, Dubreuil Frederic, Chambat Gerard, and Cousin Fabrice. Non-electrostatic building of biomimetic cellulose-xyloglucan multilayers. *Langmuir*. 25 no. 7 (2009): 3920–3923.

Cao Xiaodong, Habibi Youssef, and Lucian Lucia A. One-pot polymerization, surface grafting, and processing of waterborne polyurethane-cellulose nanocrystal nanocomposites. *J. Mater. Chem.* 19 no. 38 (2009): 7137–7145.

Chae Gi Han, and Kumar Satish. Materials science. Making strong fibers. *Science*. 319 (2008): 908–909.

Chang Peter R., Ai Fujin, Chen Yun, Dufresne Alain, and Huang Jin. Effects of starch nanocrystal-graft-polycaprolactone on mechanical properties of waterborne polyurethane-based nanocomposites. *J. Appl. Polym. Sci.* 111 no. 2 (2009): 619–627.

Chen Guangjun, Dufresne Alain, Huang Jin, and Chang R. Peter. A novel thermoformable bionanocomposite based on cellulose nanocrystal-graft-poly(ε-caprolactone). *Macromol. Mater. Eng.* 294 no. 1 (2009): 59–67.

Choi Jae-Kon, Lee Bong-Woo, Choi Yoo-Sung, Jo Byung-Wook, and Choi Soo-Kyung. Reinforcing properties of poly(trimethyleneterephthalate) by a thermotropic liquid crystal polymer. *J. Appl. Polym. Sci.* 132 no. 5 (2015): 41408(1–9).

Choi Yong Jae, and Simonsen John. Cellulose nanocrystal-filled carboxymethyl cellulose nanocomposites. *J. Nanosci. Nanotechnol.* 6 no. 3 (2006): 633–639.

Chou Tsu-Wei, Gao Limin, Thostenson T. Erik, Zhang Zuoguang, and Byun Joon-Hyung. An assessment of the science and technology of carbon nanotube-based fibers and composites. *Compos. Sci. Technol.* 70 no. 1 (2010): 1–19.

Crawford M. K., Smalley R. J., Cohen G., et al. Chain conformation in polymer nanocomposites with uniformly dispersed nanoparticles. *Phys. Rev. Lett.* 110 no. 19 (2013): 196001-1–196001-5.

Dagani R. Nanocomposites. *Chem. Eng. News*. 77 no. 23 (1999): 25–37.

de Menezes Aparecido J., Siqueira Gilberto, Curvelo Antonio A.S., and Dufresne Alain. Extrusion and characterization of functionalized cellulose whisker reinforced polyethylene Nanocomposites. *Polymer*. 50 no. 19 (2009): 4552–4563.

Dubief David, Samain Eric, and Dufresne Alain. Polysaccharide microcrystals reinforced amorphous poly(β-hydroxyoctanoate) nanocomposite materials. *Macromolecules.* 32 no. 18 (1999): 5765–5771.

Dufresne Alain. Polysaccharide nanocrystal reinforced nanocomposites. *Can. J. Chem.* 86 no. 6 (2008): 484–494.

Dufresne Alain. Processing of polymer nanocomposites reinforced with polysaccharide nanocrystals. *Molecules.* 15 no. 6 (2010): 4111–4128.

Feng Liangdong, Zhou Ziyan, Dufresne Alain, Huang Jin, Wei Ming, and An Lijia. Structure and properties of new thermoforming bionanocomposites based on chitin whisker-graft-polycaprolactone. *J. Appl. Polym. Sci.* 112 no. 5 (2009): 2830–2837.

Fong Hao, Liu Weidong, Wang Chyi-Shan, and Vaia A. Richard. Generation of electrospun fibers of nylon 6 and nylon 6-montmorillonite nanocomposite. *Polymer.* 43 no. 3 (2002): 775–780.

Guo Fengmei, Li Can, Wei Jinquan, et al. Fabrication of highly conductive carbon nanotube fibers for electrical application. *Mater. Res. Express.* 2 (2015): 095604.

Habibi Youssef, Goffin Anne L., Schiltz Nancu, Duquesne Emmanue, Dubois Philippe, and Dufresne Alain. Bionanocomposites based on poly(ε-caprolactone)-grafted cellulose nanocrystals by ring opening polymerization. *J. Mater. Chem.* 18 no. 9 (2008): 5002–5010.

Huang M. Zheng, Zhang Y.Z., Kotaki M., and Ramakrishna S. A review on polymer nanofibers by electrospinning and their applications in nanocomposites. *Compos. Sci. Technol.* 63 no. 15 (2003): 2223–2253.

Jouault Nicolas, Dalmas Florent, Said Sylvère, et al. Direct measurement of polymer chain conformation in well-controlled model nanocomposites by combining SANS and SAXS. *Macromolecules.* 43 no. 23 (2010a): 9881–9891.

Jouault Nicolas, Dalmas Florent, Said Sylvère, et al. Direct small-angle neutrons scattering observations of stretched chain conformation in nanocomposites: More insight on polymer contributions in mechanical reinforcement. *Phys. Rev. E.* 82 no. 3 (2010b): 031801.

Jouault Nicolas, Lee Danmi, Zhao Dan, and Kumar K. Sanat. Block-copolymer-mediated nanoparticle dispersion and assembly in polymer nanocomposites. *Adv. Mater.* 26 no. 24 (2014a): 4031–4036.

Jouault Nicolas, Zhao Danmi, and Kumar K. Sanat. Role of casting solvent on nanoparticle dispersion in polymer nanocomposites. *Macromolecules.* 47 no. 15 (2014b): 5246–5255.

Kim Yun Seong, Kim Hun Seong, Lee Hwan Seung, and Youn Ryoun Jae. Internal structure and physical properties of thermotropic liquid crystal polymer/poly(ethylene 2,6-naphthalate) composite fibers. *Composites: Part A.* 40 no. 5 (2009): 607–612.

Ko Frank K., Khan Sakina, Ali Ashraf, et al. Structure and properties of carbon nanotube reinforced nanocomposites. Collection of Technical Papers -AIAA/ASME/ASCE/AHS/ASC Structures. *Struct. Dynam. Mater. Conf.* 3 (2002): 1779–1787.

Kojima Yoshitsugu, Usuki Arimitsu, Kawasumi Masaya, et al. Mechanical properties of nylon 6-clay hybrid. *J. Mater. Res.* 8 no. 5 (1993): 1185–1189.

Labet Marianne, Thielemans Wim, and Dufresne Alain. Polymer grafting onto starch nanocrystals. *Biomacromolecules.* 8 no. 9 (2007): 2916–2927.

Lan Tie, and Pinnavaia J. Thomas. Clay-reinforced epoxy nanocomposites. *J. Chem. Mater.* 6 no. 12 (1994): 2216–2219.

Lee Jin Sang, Heo Nyoung Dong, Park Sun Ji, et al. Characterization and preparation of bio-tubular scaffolds for fabricating artificial vascular grafts by combining electrospinning and a 3D printing system. *Phys. Chem. Chem. Phys.* 17 no. 5 (2015): 2996–2999.

Li Youbing, Shi Wen, Li Jingyuan, et al. Structure and properties of glass fiber reinforced polypropylene/liquid crystal polymer blends. *J. Macromol. Sci. Part B.* 54 no. 9 (2015): 1144–1152.

Li Zhuo, Renneckar Scott, and Barone Justin R. Nanocomposites prepared by in situ enzymatic polymerization of phenol with TEMPO-oxidized nanocellulose. *Cellulose.* 17 no. 1 (2010): 57–68.

Lin Ning, Chen Guangjun, Huang Jin, Dufresne Alain, and Chang R. Peter. Effects of polymer-grafted natural nanocrystals on the structure and mechanical properties of poly(lactic acid): A case of cellulose whisker-graft-polycaprolactone. *J. Appl. Polym. Sci.* 113 no. 5 (2009): 3417–3425.

Liu Yaodong, and Satish Kumar. Polymer/carbon nanotube nano composite fibers—A review. *ACS Appl. Mater. Interf.* 6 no. 9 (2014): 6069–6087.

Masuelli A. Martin. *Fiber reinforced polymers—The technology applied for concrete repair.* Croatia: In Tech, 2013.

Mohammadzadehmoghadam Soheila, Dong You, and Davies Jeffery Ian. Recent progress in electrospun nanofibers: Reinforcement effect and mechanical performance. *J. Polym. Sci. Part B.* 53 no. 17 (2015): 1171–1212.

Nguyen T. Quang, and Baird G. Donald. Preparation of polymer–clay nanocomposites and their properties. *Adv. Polym. Technol.* 25 no. 4 (2006): 270–285.

Noishiki Yasutomo, Nishiyama Yoshiharu, Wada Masahisa, Kuga Shigenori, and Magoshi Jun. Mechanical properties of silk fibroin-microcrystalline cellulose composite films. *J. Appl. Polym. Sci.* 86 no. 13 (2002): 3425–3429.

Park C., Ounaies Z., Watson K.A., et al. Polymer-single wall carbon nanotube composites for potential spacecraft applications. *Making Functional Materials with Nanotubes Symposium (Materials Research Society Symposium Proceedings).* 706 (2002): 91–96.

Peresin S. Maria, Habibi Youssef, Zoppe O. Justin, Pawlak J. Joel, and Rojas J. Orlando. Nanofiber composites of polyvinyl alcohol and cellulose nanocrystals: Manufacture and characterization. *Biomacromolecules.* 11 no. 3 (2010): 674–681.

Qi Haisong, Schulz Björn, Vad Thomas, et al. Novel carbon nanotube/cellulose composite fibers as multifunctional materials. *ACS Appl. Mater. Interfaces.* 7 no. 40 (2015): 22404–22412.

Samir My Ahmed Saïd Azizi, Alloin Fannie, and Dufresne Alain. High performance nanocomposite polymer electrolytes. *Compos. Interf.* 13 no. 406 (2006): 545–559.

Shaohua Jiang, Duan Gaigai, Schöbel Judith, Agarwal Seema, and Greiner Andreas. Short electrospun polymeric nanofibers reinforced polyimide nanocomposites. *Compos. Sci. Technol.* 88 no. 14 (2013): 57–61.

Siqueira Gilberto, Bras Julien, and Dufresne Alain. New process of chemical grafting of cellulose nanoparticles with a long chain isocyanate. *Langmuir.* 26 no. 1 (2010): 402–411.

Solomon J. Michael, Almusallam S. Abdulwahab, Seefeldt F. Kurt, Somwangthanaroj Anongnat, and Varadan Priya. Rheology of polypropylene/clay hybrid materials. *Macromolecules.* 34 no. 6 (2001): 1864–1872.

Stenstad Per, Andresen Martin, Tanem Bjorn Steinar, and Stenius Per. Chemical surface modifications of microfibrillated cellulose. *Cellulose.* 15 no. 1 (2008): 35–45.

Sun Zaicheng, Eyal Zussman, and Alexander L. Yarin. Compound core-shell polymer nanofibers by co-electrospining. *Adv. Mater.* 15 no. 22 (2003): 1929–1932.

Svagan J. Anna, Hedenqvist S. Mikael, and Berglund Lars. Reduced water vapour sorption in cellulose nanocomposites with starch matrix. *Compos. Sci. Technol.* 69 no. 3–4 (2009): 500–506.

Thakur Kumar Vijay. *Green composites from natural resource.* Boca Raton, FL: CRC Press, 2013.

Thakur Kumar Vijay. *Cellulose-based graft copolymers.* Boca Raton, FL: CRC Press, 2015.

Thielemans Wim, Belgacem Mohamed N., and Dufresne Alain. Starch nanocrystals with large chain surface modifications. *Langmuir.* 22 no. 10 (2006): 4804–4810.

Tung Wei-Shao, Bird Vikki, Composto J. Russell, Clarke Nigel, and Winey I. Karen. Polymer chain conformations in CNT/PS nanocomposites from small angle neutron scattering. *Macromolecules*. 46 no. 13 (2013): 5345–5354.

Wang Zhijuan, Wu Shixin, Zhang Juan, et al. Comparative studies on single-layer reduced graphene oxide films obtained by electrochemical reduction and hydrazine vapor reduction. *Nanoscale Res. Lett.* 7 no. 1 (2012): 161–167.

Weiss A. Robert, Huh Wansoo, and Nicolais L. Novel reinforced polymers based on blends of polystyrene and a thermotropic liquid crystalline polymer. *Polym. Eng. Sci.* 27 no. 9 (1987): 684–691.

Xu Zhen, Sun Haiyan, Zhao Xiaoli, and Gao Chao. Ultrastrong fibers assembled from giant graphene oxide sheets. *Adv. Mater.* 25 no. 2 (2013): 188–193.

Yu Jiahui, Ai Fujin, Dufresne Alain, Gao Shanjun, Huang Jin, and Chang R. Peter. Structure and mechanical properties of poly(lactic acid) filled with (starch nanocrystal)-graft-poly(ε-caprolactone). *Macromol. Mater. Eng.* 293 no. 9 (2008): 763–770.

Zhang Di, Karki B. Amar, Rutman Dan, et al. Electrospun polyacrylonitrile nanocomposite fibers reinforced with Fe_3O_4. *Polymer.* 50 no. 17 (2009): 4189–4198.

Zheng Hua, Ai Fujin, Chang R. Peter, Huang Jin, and Dufresne Alain. Structure and properties of starch nanocrystal-reinforced soy protein plastics. *Polym. Compos.* 30 no. 4 (2009): 474–480.

Zimmermann Tanja, Pöhler Evelyn, and Schwaller Patrick. Mechanical and morphological properties of cellulose fibril reinforced nanocomposites. *Adv. Eng. Mater.* 7 no. 12 (2005): 1156–1161.

Section II

Multiphase Polymer Systems in Solid Phase

7 Polymer Dispersed Liquid Crystal Design

New Challenges and Applications

Luminita Marin and Ridvan Karapinar

CONTENTS

7.1 INTRODUCTION

Polymer dispersed liquid crystals (PDLCs) are a relatively new class of composite materials consisting of micrometric liquid crystalline droplets dispersed into a polymer matrix (Bronnikov et al. 2013). Thus, they combine the unique optical properties of liquid crystals (LCs) with the film-forming ability and mechanical properties of a polymer matrix—resulting in appropriate materials for a large variety of flexible opto-electronic applications (Bouteiller et al. 1996; Bronnikov et al. 2013; Dierking 2000; Drzaic 2006; Kitzerow 1994; Smith 1993). As evidenced by

Fergason in 1984 (Fergason 1985), the development of PDLC composites started once their potential as active substrate in displays had been determined (Drzaic 2006). Applications as smart windows, holographic systems, microlenses, or lasers kept researchers interest in this challenging domain that grew constantly (Bouteiller et al. 1996; Bronnikov et al. 2013; Dierking 2000; Drzaic 2006; Kitzerow 1994; Smith 1993). Recent years have brought new achievements in this area, such as the use of PDLCs for bio-applications—for example, smart food packaging (Perju et al. 2011, 2015), tunable artificial iris modulating light intensity through human eyes for assisting patients of aniridia (Hsu et al. 2011; Lapointe et al. 2010), bio-sensors for biologically active matter (Sidiq et al. 2014), or other envisaged bio-applications (Woltman et al. 2007). Such innovative bio-applications use biocompatible materials. To apply PDLC systems as the next generation of biomaterials addressed in biotechnology and medicine, friendly soft composites based on biocompatible polymers need to be developed. This is the reason why the ability of biopolymers or biocompatible polymers to act as polymer matrices for PDLC composites is currently of interest.

7.2 OBTAINING POLYMER DISPERSED LIQUID CRYSTALS: MATERIALS, METHODS, ADVANTAGES, AND DISADVANTAGES

PDLC properties and their ability to be used for intelligent opto-electronic applications dramatically depend on the morphology of the PDLCs—the intrinsic properties of the liquid crystal and polymer components and the interfacial phenomena that determine the anchoring of the LC into the carrying matrix. The dispersion of micrometric droplets into the polymeric matrix leads to a high surface-to-volume ratio. For instance, a liquid crystal cell of 1 cm^2 in area and 5 μm in thicknesses results in a total surface of 2 cm^2. The same dimensions of a PDLC film containing droplets of 1 μm with uniform distribution will result in a total surface of 30 cm^2—an increment 15 times greater (Drzaic 1995). Therefore, it is understandable why the anchoring conditions are important. Previous reviews (Bronnikov et al. 2013; Jeon et al. 2007; Liu and Sun 2008; Mucha 2003) underlined the importance of all these aspects in obtaining performance PDLC systems, especially those designed for LC displays and holographic systems. Also, these reviews stressed the importance of the polymer carrying matrix for application of the resulting PDLCs. The polymer matrix favors the formation of the LC droplets assuring them mechanical support and also determining their size and distribution, as well as the orientation of the director field via anchoring forces (features that ultimately influence the optical properties of the PDLC composites). Moreover, the chemical structure of the polymers affects the optical properties of the PDLCs due to their intrinsic optical properties, which guide the selection of the application field. Thus, the proper choice of the polymer matrix is very important and will determine the application direction.

The most used liquid crystals are those that are cylindrical in shape, with a single axis in front of the incident beam, which makes them able to rotate the polarization direction of the incident light (Drzaic 1995). The liquid crystals are electrically controllable and the optical modes of their droplets are different from the ones in isotropic compounds due to LC birefringence and the specific configuration of the director field within the droplet, which is a consequence of the physical interaction between the liquid crystal and matrix (Musevic and Humar 2011). Those based on cyano-biphenyl units are especially preferred due to their good chemical stability, long relaxation time, and dielectrical anisotropy (Mucha et al. 2001). Primarily, obtaining the PDLC is based on microencapsulation and phase separation processes.

The *microencapsulation* method developed out of microemulsion technology was invented by Fergason (1985) as the first attempt to produce PDLCs. In this method, a liquid crystal is mixed with a polymer dissolved in water; under fast stirring, an emulsion is formed. Because the organic liquid crystals are insoluble in aqueous media, the phase separation is naturally achieved. The PDLC material is obtained as the water evaporates, when tiny "capsules" of liquid crystal are surrounded by a layer of polymer and distributed throughout the newly formed material. Due to the immiscibility of the two phases, the plasticization effect of the polymer matrix is actually absent—an important advantage of this method. A disadvantage is the size polydispersity of the produced droplets which can be interconnected with one another. Water-soluble polyvinyl alcohol (PVA) was the first matrix polymer and one of the most widely used (in combination with many mesogenic materials) in obtaining PDLCs (Mucha 2003). PVA's great properties (e.g., nontoxicity, water solubility—and thus avoidance of unpleasant organic solvents, completely biodegradable, good chemical stability, and film forming ability) recommend it for many human and environmentally friendly applications. PVA promotes a strong planar anchoring of the liquid crystal into the PDLC composite (Dolgov et al. 2012). Studies dedicated to obtaining homeotropic alignment of liquid crystals into PVA matrices concluded that a third component of the PDLC composite, a surfactant, is needed to achieve this goal (Teperek et al. 1995). The surfactant physically interacts with liquid crystal molecules and determines their perpendicular alignment on the polymer surface and, therefore, the change of the director configuration as the capability of planar anchoring of the PVA is turned into a homeotropic one. In addition, the interactions that develop between LC molecules and surfactants provide surface stabilization and thus small droplet size. But the use of an additional composite component brought about new problems regarding composite formulation and extra cost due to the complexity of the processes utilized. Another bio-friendly matrix recently discovered as promising for encapsulation of droplets is gelatin in combination with gum arabic and color pigments—in order to create color PDLC devices (Kim et al. 2015).

The *phase-separation method* was developed at Kent State University in the late 1980s. It consists of in situ segregation of liquid crystalline microdroplets from a homogeneous mixture of liquid crystal and polymer/prepolymer, induced by a factor that gave the technique its name, as follows: (a) *polymerization-induced phase separation* (PIPS); (b) *solvent-induced phase separation* (SIPS); and (c) *thermally induced phase separation* (TIPS).

Polymerization-induced phase separation consists of segregation of the liquid crystal from a homogeneous mixture with a prepolymer; as the polymerization proceeds and polymerization medium hardens, the liquid crystal droplets are forced out by decreasing solubility and diffusion rate. This method proved to be the most versatile and useful in producing PDLC films because it allows easy control of the dimension and distribution of the droplets by controlling the speed of polymerization. The main condition required is the miscibility of the liquid crystal compound with the prepolymer material. Obtaining PDLC via the PIPS method is dominated by the use of polyacrylates as polymer matrix, forming in situ from different acrylate pre-polymers (Bronnikov et al. 2013; Mucha 2003). The polyacrylates meet the main requirements of the carrying matrix polymers for PDLCs. They are transparent; resistant to breakage; elastic; highly stable with ultraviolet (UV) light, oxygen, or ozone; and offer serviceability within a relatively wide temperature range (Acton 2013). Another important advantage is the high purity of the acrylic monomers and their availability via a variety of polymerization methods. The problems arise from waste management because they are not as easily biodegradable (Andrews and Clary 1986). The PIPS method also works well for the monomers that produced epoxy resins from polycondensation (Zhang et al. 2012) or for other vinyl monomers such as styrene (Srivastava et al. 2012), thiol-ene (White et al. 2007), and various curable monomers (Ellahi et al. 2014).

Thermally induced phase separation consists of mixing the liquid crystal with the melted polymer; the microdroplet segregation occurs as the mixture is slowly cooled down at a specific rate and the polymer hardens. In this case, the main factor affecting the droplet size is the cooling rate of the molten liquid crystal/polymer; experimental evidence notes that droplet size decreases with an increased rate of cooling. In this method, the melting point of the polymer matrix is lower than its initial decomposition temperature and higher than the volatile point of the liquid crystal. The TIPS method is typically used in obtaining PDLCs based on a thermoplastic polymeric matrix. Frequently, the TIPS method is used in combination with the SIPS method (Bronnikov et al. 2013). Among the methods for obtaining PDLC, TIPS has been used to obtain the greatest number of PDLCs by using a large variety of thermoplastic polymers: polystyrene (Tercjak et al. 2008), poly(vinyl chloride) (Chen and Shanks 2007), polymethacrylates (Bauer et al. 2010; Srivastava et al. 2011), poly(bisphenol A carbonate)s, and polysiloxanes (Tercjak et al. 2008), to underline only a few. Hence, an important advantage of this method is the larger possibility of choosing appropriate polymers for the targeted application. However, the morphological features of the PDLCs obtained by the TIPS method appear to be less favorable for good opto-electronic properties compared with those obtained via the PIPS method. However, the stability of the internal structure remains a problem in both cases because of the component miscibility. To further improve the morphological properties, some additives (e.g., ionic liquids) were used that facilitated a more uniform distribution of the LC droplets and a uniform alignment of the LC director inside the droplets in a similar manner as those with surfactants in the microencapsulation method (Cupelli et al. 2011). On the contrary, the use of additives introduced the disadvantage of a difficult formulation and more complicated preparation procedure.

Solvent-induced phase separation occurs when both liquid crystal and polymer are combined in a homogeneous solution; the liquid crystal grows (as microdroplets) as the solvent evaporates. In such a segregation mechanism, the rate of solvent removal plays the crucial role in droplet size and in their dimensional polydispersity; droplet size decreases with an increase in the rate of solvent evaporation (Bronnikov et al. 2013).

SIPS is the proper method when the polymer matrix has a melting temperature higher than that of the liquid crystal or when the TIPS method cannot be applied. An important parameter in controlling the size and uniform dispersion of the LC droplets is the difference between the solubility parameter of the low molecular weight liquid crystal and polymer matrix (Chen and Shanks 2007). For small solubility differences in the two components, there is a more uniform distribution of the LC droplets into the polymer matrix.

7.3 SIGNIFICANCE OF EXPERIMENTAL METHODS IN POLYMER DISPERSED LIQUID CRYSTAL CHARACTERIZATION

Polarized light microscopy (POM) is the visual method used to display the segregation of LC droplets dispersed into the polymeric matrix, by observing their birefringence (Demus and Richter 1978). Additionally, POM supplies bidimensional data related to the shape, localization, size, and distribution of the droplets in the observed plane. When the microscope is coupled with a heating stage, the thermal dynamic of the droplets also can be visualized. The droplet texture provides important clues regarding the anchoring type, while the reproducibility of the thermotropic behavior during multiple heating/cooling cycles indicates the thermal stability of the PDLCs.

Scanning electron microscopy (SEM) clearly shows tridimensional geometry (the size and distribution) of the liquid crystalline droplets in the carrying matrix; these parameters drastically influence the optical properties of the PDLCs. Systematic studies of the relationship between morphology and photophysical properties have demonstrated that optical performance properties were obtained with uniform distribution of droplets having a micrometric diameter smaller than 10 μm; sizes in the submicrometric domain (greater than 10 μm) proved to be inefficient in light scattering (Drzaic 1995). Moreover, it was demonstrated that the higher the concentration of liquid crystals in PDLCs, the greater the increase in device stability; the micrometric droplets are able to preserve the intrinsic properties of the pure LC, which allows a good transmission of the bistable state (Lu et al. 2013).

Differential scanning calorimetry (DSC) is a valuable complementary method in PDLC characterization due to its accurate measurements, which are unaffected by the subjectivity of the human factor. The shape of the DSC curves and the variation of the transition temperatures are related to the dynamic of LC segregation, size distribution, and anchoring forces, and—most importantly—indicate the temperature range in which PDLCs can be used in practical applications. In addition, DSC establishes exactly the degree of miscibility of the two components (Graham et al. 1996). This is a distinctly important aspect that (i) affects the PDLC properties—dissolution of LC in the polymer matrix affects its transparency and decreases its glass/melting temperature, limiting the temperature range in which the PDLC can be used, and (ii) increases the PDLC manufacturing price because dissolution of the LC in the polymer matrix implies the use of larger amounts of LC, which usually is expensive.

Contact angle and surface free energy measurements offer the first insight into the potential use of the PDLC as biomaterial and also provide information about the physical interactions inside the material that determine the droplet anchoring. Usually, for a material to be considered as potentially biocompatible, the *water contact angle* value must be in the range of 60–90 degrees, which is considered within the domain of moderate wettability (Ikada 1994; Vasile and Pascu 2007). These values ensure an appropriate balance of hydrophobic/hydrophilic forces that will favor cellular adhesion and prevent the rejection of the implanted material.

Generally speaking, the *surface free energy* (SFE) reflects the chemical composition of the material and molecule orientation to its surface. In the case of PDLCs, orientation of the LC molecules inside the droplets and the interactions at the LC/polymer matrix interface determine the capability of the material to be electrically controlled and used in the design of biosensors as blood or sperm testers (Lin et al. 2012). Moreover, values of SFE higher than 22 mN/m proved to be beneficial in maintaining multicellular structure (Hallab et al. 2001).

Raman microspectroscopy provides a detailed view of the three-dimensional arrangement of the molecules inside the LC droplets. Because it proves the interaction of individual bond vibrations with laser light, the method allows the mapping of segregation as a function of chemical composition. The quantitative imaging of Raman intensity reveals information about the spatial organization of the molecules and their local environment. Thus, the method maps the director fields of the liquid crystal molecules in the horizontal plane and in the vertical direction (Blach et al. 2005).

7.4 POLYSULFONE AS POLYMER MATRIX FOR POLYMER DISPERSED LIQUID CRYSTALS ADDRESSED IN BIO-APPLICATIONS

Udel® polysulfone (PSU) is a high-performance polymer with outstanding mechanical properties and compressive strength; it can be processed from a melted state, is transparent, and is an insulator with a good resistance to moisture, chemical solvents, or aggressive chemical agents such as mineral acids, bases, electrolytes in the pH range 2–13, oxidant agents, surfactants, or aliphatic hydrocarbons (El-Hibri et al. 2015). It can be manufactured as films, membranes, or fibers that maintain their properties in a large range of temperatures (from −100°C to +150°C). Besides, PSU is a biologically friendly polymer, with good blood biocompatibility and a high drug clearance rate—better than the modified biopolymer cellulose, for example (Hoenich et al. 1995). Due to these properties, PSU is used in the automotive and electronic industry, in environmental protection, in the food industry, or in medical and healthcare domains; as an example, it is the most popular synthetic polymer used for hemodialyzers (Tsuruoka et al. 2000).

Polysulfone's capability to act as a matrix for PDLCs was tested for three distinct nematic liquid crystals, with the goal of proposing materials for application as smart packaging (Marin and Perju 2008, 2009, 2010; Perju et al. 2011, 2015). The polysulfone-based PDLCs were obtained by combining the SIPS and TIPS (thermal treating by heating/cooling between RT and T_g of the polysulfone) methods.

Using 4-cyano-4′-pentylbiphenyl (5CB) as nematogen and polysulfone (PSU) as a polymer matrix (Scheme 7.1) in different mass ratios, PDLC systems were obtained as free self-standing films.

POM confirmed the segregation of nematic droplets for the 5CB content higher than 30%, indicating that dissolution of the liquid crystal in the polysulfone matrix occurs under this value. The segregation of nematic droplets was observed by applying the TIPS treatment, although, in the case of the SIPS composites, a Schlieren nematic texture was observed, probably because of forming of an interpenetrating network with the polymer (Figure 7.1b and c). The droplets were uniformly distributed into the polymeric matrix and had a radial configuration with a hedgehog defect observed as a black point in the center, which is the signature of a homeotropic anchoring (Hamlington et al. 2007). This means that the long axis of the 5CB molecules lies perpendicular to the droplet walls. This type of alignment is different compared to the natural tendency of the 5CB to align planarly forming bipolar droplets (Figure 7.1a) and designates the constraining influence of the interphase forces at the droplet boundary. These forces were associated with an easier dipolar orientation of the polysulfone chains, gained at high temperature due the macromolecular mobility (Saxena et al. 2008), which further facilitates the interface interactions and, consequently, the LC anchoring. Microscopic observations revealed a close dependence of the droplet dimension related to the LC content and the ratio of interfacial tension/anchoring forces. The droplets were uniformly distributed into the matrix, almost equidistant from one each other.

The droplets have a spherical shape, as confirmed by SEM (Figure 7.1d), with submicrometric dimensions around 650 nm (in the case of the composites containing more than 30% LC). Interestingly enough, the PDLCs containing 30% LC, which did not show any droplets under polarized light, exhibited submicrometric droplets of around 250 nm in SEM, a size value that lies under the detection limit of the polarized light microscope. Thereby, the percentage limit at which the liquid crystal was dissolved in the polysulfone matrix was smaller than 30%.

The POM observations were confirmed by the DSC thermograms that revealed that the SIPS films, heated and cooled in the liquid crystal temperature range, show the specific transitions of the 5CB liquid crystal; the TIPS films exhibit a complex behavior reflecting the polymer matrix influence upon the droplet formation. Thus, multiple endotherms in the heating scan (TIPS treatment) were attributed to the melting of segregated liquid crystal in different amounts during the SIPS process, while the sole exotherm in the cooling scan corresponds to the occurrence of droplets of uniform size from the melted polysulfone remarkable is an endothermic shoulder registered on the DSC curve in the neighborhood of the glass transition, whose shape corresponds to a first order transition (a characteristic of crystalline melting).

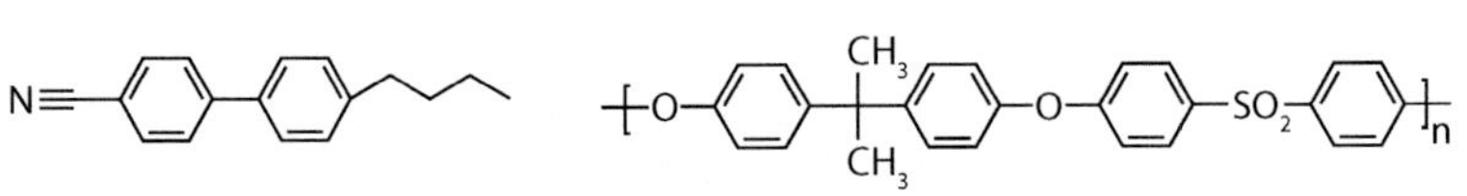

5CB PSU

SCHEME 7.1 Structure of 4-cyano-4′-pentylbiphenyl and polysulfone.

(a) (b) (c) (d)

FIGURE 7.1 Representative POM images of (a) 5CB, (b) SIPS-PDLC, (c) TIPS-PDLC, and SEM of (d) TIPS-PDLC.

This transition was attributed to the melting of the ordered polysulfone domains that occurred around the LC droplets. It was estimated that the forces at the interphase boundary guide the ordering of the liquid crystal and also that of the PSU polymer around it.

Surface analysis of the PDLC and pure polysulfone films showed that water contact angle values range in the domain of moderate wettability, which is specific for biomaterials (Ikada 1994). The SIPS films had better wettability compared to the pure polysulfone, which could be ascribed to the polysulfone oxygen atoms' preponderant exposure toward the film surface, making them capable of bonding with water via H-bonding. This was possible as a result of the segregation of methyl units of the polysulfone and alkyl groups of the liquid crystal. A slight increase in the hydrophobic feature was noted with thermal treatment (TIPS films), sustaining the hypothesis of the easier orientation of the polysulfone dipoles to the LC droplets, which results in a poorer exposure of the oxygen atoms to the surface and, consequently, a slight increase in the exposure of the methyl groups.

The evolution of the SFE for the SIPS and TIPS films was in agreement with the evolution of the contact angle. The SFE value of the pure polysulfone film (48.6 mN/m) results due to a major contribution from the London forces $\left(\gamma_S^{LW} = 46.2\right)$, while the contribution of the dipole–dipole and H-bonding forces is less significant ($\gamma_S^- = 1.45$ and $\gamma_S^+ = 0.95$, respectively). Microencapsulation of the liquid crystal by the SIPS method led to a slight increment in the SFE due to an increase in the basic contribution of the rich electron cyan groups (from 1.9 to 8.74, as a function of the liquid crystal content). The thermal treatment of the PDLC films is followed by a clear decrease in the polar contribution and a slight increase in the dispersive forces, in agreement with the anchoring phenomenon of nematic droplets via interphase interactions between polysulfone dipoles (gained at elevated temperature) and the electron deficient sites of the LC molecules, which are formed due to the presence of the electron withdrawing cyan groups.

The simultaneous analysis of the SIPS and TIPS systems allowed depiction of the droplet segregation mechanism into the polysulfone matrix.

In the first stage, the solution mixing of the 5CB liquid crystal with the polysulfone (SIPS method) led to a segregation of the two components by self-assembling their hydrophobic parts (aliphatic units) without an LC droplet forming. In the second

Az1

Az2

SCHEME 7.2 Structure of the cyano-containing azomethine liquid crystals.

stage, the thermal treatment up to the glass transition temperature (Tg) of the polysulfone (TIPS method) allowed the forming of mobile dipoles that gave the PSU matrix the ability to exert enough interfacial tension to constrain the liquid crystal to adopt the shape of round droplets, which were further stabilized by interfacial forces. The interphase interactions, combined with the tendency of the orientational alignment of the nematic LC molecules, force them to align with long axes perpendicular to the droplet wall, resulting in a homeotropic anchoring.

Photophysical investigation of the pure compounds and PDLC systems indicated a hypsochromic shift of the PDLC absorption once the LC amount in PDLC diminished. This is associated with the destruction of the dominating dimer structures in pure 5CB (Bezrodna et al. 2010) due to the interface forces that guide the 5CB molecules from the droplet boundary to its center. The emission spectra of the PDLCs registered the most intense UV luminescence for the composite with smaller droplets, attributed to a higher surface-to-volume ratio.

In a similar manner, the preparation and characterization of the 5CB/polysulfone composites were studied using another two PDLC systems based on a polysulfone matrix and two different azomethine liquid crystals (Scheme 7.2), containing a cyan and an aliphatic chain as end groups (Marin and Perju 2009; Perju et al. 2011). For these systems, a homeotropic anchoring was also evidenced, the most probably due to the similar structure of the liquid crystals.

These systems present the crystallization of the liquid crystalline droplets at a negative temperature, or close to 0°C, with the crystalline-nematic mesophase transition in a positive domain. This fact supports their use in the design of smart packaging for the food industry, when the change of the LC anisotropic properties as a function of temperature could indicate if the recommended storage temperature was exceeded and/or the alimentary products eventually were refrozen.

7.5 CHITOSAN—A BIOPOLYMERIC MATRIX FOR POLYMER DISPERSED LIQUID CRYSTALS

The application of PDLC systems for the next generation of biomaterials implies the development of composite films based on biocompatible polymers. In this context, chitosan is an excellent candidate. Extracted from the exoskeleton of crustaceans (such as crabs or shrimp), chitosan has plenty of good biological properties that seem to support its characterization as "too good to be true." Chitosan has been proven to be biocompatible, fungistatic, and anticarcinogenic, to be used as an immunoadjuvant, for wound healing and blood clotting, and as being capable of bonding the fatty acids to help in the fight against morbid obesity and related illnesses (Croisier and Jérôme 2013; Kumar et al. 2004). A survey of the chitosan physical properties revealed that this biopolymer meets the main features claimed by matrix polymers for obtaining PDLC: it is transparent, has the ability to form free-standing films, and is immiscible and inert with liquid crystals (Mucha 2003). Moreover, chitosan can be cross-linked with a large variety of agents, increasing its potential as a matrix for liquid crystal encapsulation.

PDLC systems having different 5CB contents (varying between 10 and 50%) were successfully prepared by using chitosan as a matrix (Scheme 7.3) and 4-cyano-4′-pentylbiphenyl (5CB) as a liquid crystal in different mass ratios (Marin et al. 2013). The segregation of the 5CB droplets was evidenced even for a low liquid crystal content (10%), proving the immiscibility of the two components (Figure 7.2a). The droplets were uniformly distributed into the matrix but with a large polydispersity of their diameter size. All the observed droplets were radial, as evidenced by polarized light microscopy (POM) (Figure 7.2a), indicating the ability of the chitosan matrix to homeotropically anchor the liquid crystal. Taking into consideration the polycationic nature of the chitosan, it can be surmised that strong interactions among the polycationic chitosan chains and electronegative C≡N groups of the liquid crystal are promoted.

The perpendicular arrangement of the long axes of the 5CB molecules on the droplet walls resulted in a homeotropic alignment, which was visualized under POM as radial droplets. In addition, when the PDLC samples were heated to the upper isotropization temperature of the liquid crystal, the isotropic droplets remained surrounded by a birefringent shadow—the signature of a polymer matrix ordering

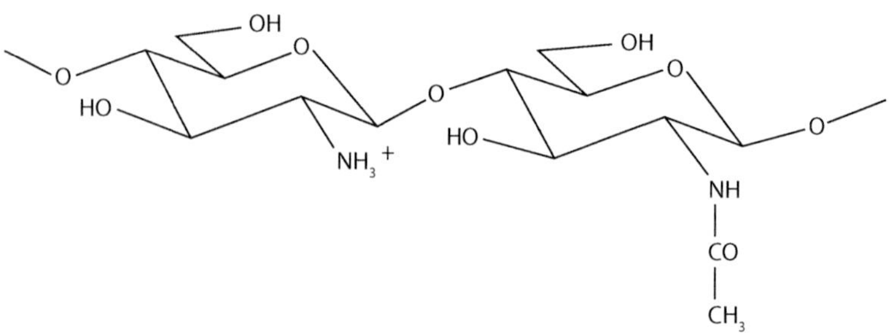

SCHEME 7.3 Chitosan in its polycationic state.

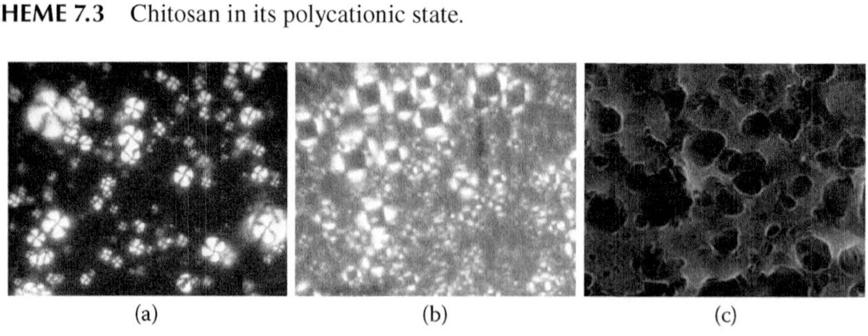

FIGURE 7.2 Representative images of chitosan-based PDLC: POM image of the PDLC containing 10% 5CB (a) at room temperature and (b) upper 5CB isotropization temperature; (c) SEM image of the PDLC containing 50% 5CB.

around the droplets (Figure 7.2b). This could be explained by an ordering coupling at interface, with the strong interface forces also generating the matrix alignment.

The droplets have a spherical geometry, and their integrity is kept up to 50% of liquid crystal (Figure 7.2c)—a value that constitutes the boundary for obtaining droplets—any greater, and the droplet density is too high and the interfacial tension not enough strong to further ensure the droplet integrity. The diameter size of the droplets was in the range of 300 nm–40 μm. No significant differences were observed in a comparison of the morphology of the PDLC systems containing different percentages of liquid crystal, indicating that the dynamic of the segregation was related to the conditions under which they were obtained and not to the intrinsic properties of the two components. By changing the preparation conditions (e.g., fast evaporation of the very thin films, under a vacuum at high temperature), a more uniform distribution of the droplet size was reached (Marin et al. 2013).

Mapping the stretching of the C≡N and C=C bonds of the liquid crystal when interacting with the polarized light by Raman microspectroscopy (Figure 7.3a) demonstrated that homeotropic anchoring of the liquid crystal has a three-dimensional character. The study relied on the theoretical premise that bond vibrations take place in the molecule plane and, consequently, have the same orientation as the director vector (**n**). It was observed that the vibration bands change their intensity when the electromagnetic vector of the polarized light (**P**) varies, indicating a preferential position of the director of the 5CB molecules (Figure 7.3b through d). Taking into consideration that the maximum intensity corresponds to the parallel arrangement of the two vectors ($\mathbf{n}||\mathbf{P}$) and the minimum intensity to a perpendicular position ($\mathbf{n} \perp \mathbf{P}$), it was concluded that the director field of the liquid crystal has a three-dimensional radial distribution inside the droplets.

The complete immiscibility of the two components was demonstrated by differential scanning calorimetry (DSC) where sharp phase transitions of the 5CB liquid crystal were registered (Marin et al. 2013). A distinct finding of the DSC measurements concerned the droplet polydispersity as evidenced by the shape of the transition bands. The narrow bands (isotropization endotherm and crystallization exotherm) in the case of the PDLCs containing lower percentage of liquid crystal correspond to the fast transitions of the preponderant small droplets while larger transitions in the case of the PDLCs containing higher percentage of liquid crystal correspond to a larger polydispersity with preponderant bigger droplets.

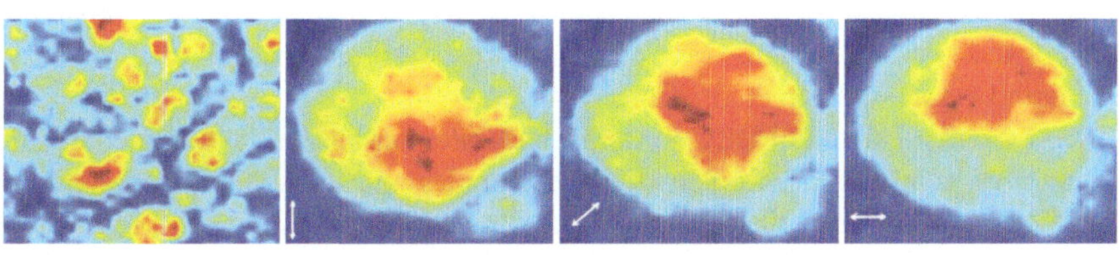

(a) (b) (c) (d)

FIGURE 7.3 Raman microspectroscopy of (a) the PDLC composite containing 30% of 5CB, and of a single droplet for 3 directions of the polarizer (b) **n** parallel to **P**, (c) 45° between **n** and **P** and (d) **n** perpendicular to **P** (Adapted from Marin Luminita et al. *Carbohyd. Polym.*, 95, no. 1, 16–24, 2013.)

The PDLCs emit UV light under irradiation, which intensity increased in the case of the composites containing smaller amount of liquid crystal distributed as smaller droplets—corresponding to the larger interphase surface.

7.6 POLYVINYL ALCOHOL BORIC ACID—A MATRIX WITH HIGH LIQUID CRYSTAL ANCHORING POWER

As stated earlier, applying PDLC composites in biomedical applications claims biocompatible polymers. The most appropriate polymer for such applications is polyvinyl alcohol (PVA)—one of the first polymers used as a carrying matrix when obtaining PDLCs (Mucha 2003). PVA is a nontoxic polymer that is water soluble, is completely biodegradable, has good chemical stability, and has a good film-forming ability—all good reasons to support its use in many biomedical applications (Baker et al. 2012). Obtaining PDLCs via encapsulation method resulted in polydisperse systems with planar anchoring (Dolgov et al. 2012). To decrease the polydispersity of LC droplets, a third component—a surfactant—was used, which also has the effect of changing the LC alignment into a homeotropic one (Teperek et al. 1995). The surfactant prompts strong interactions with the liquid crystal due to its ionic nature, determining a perpendicular alignment on the droplet walls. The strategy of surfactant use is efficient but has the drawback of difficult formulation and, consequently, higher costs. To overcome these disadvantages, the use of polyvinyl alcohol boric acid (PVAB) as carrying matrix was proposed. This polymer contains electron-deficient boron atoms capable of promoting attraction forces to electronegative sites of liquid crystal molecules (Marin et al. 2014). In this way, PVAB has the ability to overcome the difficulties of surfactant use. In addition, PVAB keeps the advantages of PVA and brings new advantages of the boric acid as an antiseptic activity; it promises to be a major player in biomedical applications.

Following this premise, PDLC systems were prepared using PVAB as a matrix and a nematic (4-cyano-4′-pentylbiphenyl: 5CB) or a smectic (butyl-*p*-[*p*′-*n*-octyloxy benzoyloxy]benzoate): Sm liquid crystal.

7.6.1 Polymer Dispersed Liquid Crystal Composites Based on Polyvinyl Alcohol Boric Acid Matrix and 5CB Nematogen

The 5CB nematogen was encapsulated into the PVAB matrix using various molar ratios of the two components. Freestanding films were obtained up to a content of 50% liquid crystal (Marin et al. 2014).

POM confirmed that PDLC systems had been successfully obtained by the presence of birefringent droplets for the samples with low liquid crystal content (Figure 7.4a), and a fine texture in the case of a higher content (Figure 7.4b), because of the droplets overlapping across the film thickness (Liu et al. 2011). The heating-cooling process of the PDLC films under POM revealed the radial nature of the droplets, indicating a homeotropic alignment of the liquid crystal molecules (Figure 7.4c). Multiple heating–cooling cycles did not affect thermotropic behavior, indicating thermal stability of these composite materials due to the immiscibility of the PVAB matrix with the nematic liquid crystal. The radial droplets present a central

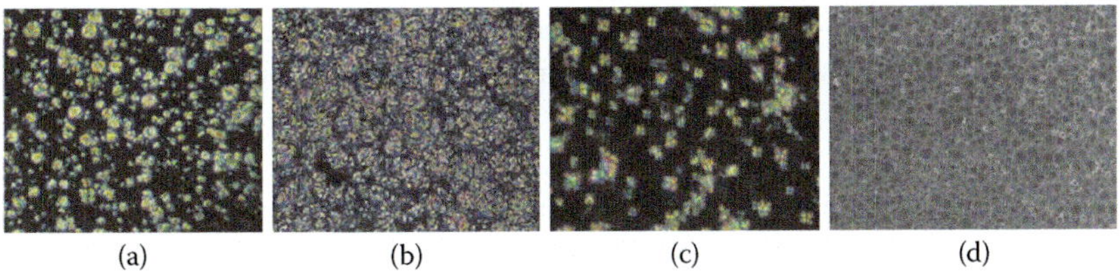

FIGURE 7.4 Representative POM (a,b,c) and SEM (d) PDLC images.

point defect (hedgehog) specific to a radial configuration of the director field of the liquid crystal (Hamlington et al. 2007), which is the signature of a homeotropic anchoring. This explanation was given after considering the strong attraction forces between the electron-rich cyan groups of liquid crystal and electron-deficient boron atoms of the PVAB polymer.

The droplets have spherical geometry, uniformly distributed in the polymer matrix, as demonstrated by SEM measurements (Figure 7.4d). The spherical shape of the droplets points to a proper balance of the superficial tension/interfacial forces, which are not able to transfer deforming stress from the matrix to the liquid crystal during the solvent evaporation. The droplet density increases along with the increase in LC content in the composite material, although their size polydispersity decreases (reaching monodispersity at 40% LC into the composite). Thus, although the droplet diameter ranges from 5–200 μm for the composite containing 10, 20, or 30% liquid crystal, a sharp dimension of 4–5 μm was registered for a content of 40%—ideal dimensions for good optical properties (Drzaic 1995). The pores of the droplets in the material are independent, showing the benefits brought by interface forces in keeping their integrity.

Calorimetric investigation of the phase separation process revealed the clear phase transitions of the liquid crystal encapsulated into the PVAB matrix—no matter how many heating/cooling cycles were applied, denoting a complete immiscibility of the two components and thus an excellent stability of the PDLC composites. Compared to the DSC curves of the pure components, those of the composites reflect their mutual influence. The symmetry and the narrow temperature range of the isotropization endotherm, in the case of the 40% content of liquid crystal, confirm the narrow polydispersity of the droplet size. On the other hand, the "left-tail" and "right-tail" and the large temperature range of the same transition for the other composites denoted the larger polydispersity and the evolution of the big/small ratio of the droplets along with the varying content of liquid crystal; the preponderant smaller droplets were obtained for samples with higher content of liquid crystal. This observation was confirmed by the evolution of the crystallization exotherm and crystalline–nematic endotherm, respectively. Although the composite containing 40% liquid crystal presents only a sharp peak corresponding to each transition, the others show many exotherms and endotherms, a signature of droplets of different size for which the transitions occur at different temperatures because of different intensities of the anchoring forces.

Encapsulation of micrometric droplets of hydrophobic liquid crystal into the hydrophilic PVAB matrix shifts the water contact angle into the moderate wettability domain, which is required by the biocompatible materials. Thus, although the water contact angle of the PVAB (58°) is less than the inferior boundary of the moderate wettability range (60–90°), that of the PVAB composites shifted within it, in the range 64–79°.

The value of the SFE of the PVAB film is given by a major contribution of the dispersive Lisfshitz–van der Waals forces $\left(\gamma_S^{LW} = 32{,}6\right)$ and a significant contribution of polar forces, electron donors $\left(\gamma_S^{-} = 12{,}6\right)$, and electron acceptors $\left(\gamma_S^{+} = 4{,}75\right)$, respectively, attributed to the electron-rich oxygen and electron-deficient boron, respectively. With the encapsulation of the polar liquid crystal containing electron-rich cyan groups, diminishing polar contribution (especially of the electron acceptor contribution) and increasing dispersive contribution take place, sustaining the premise of strong interface interactions among cyan groups and boron atoms. The SFE of the PDLC composites has values in the range 41–44 mN/m, a range that ensures good cell adhesion and proliferation (Hallab et al. 2001), once again pointing to the potential for bio-application of these composite materials.

The photoluminescence spectra of the PDLC indicated emission of UV light, the intensity of which increased when the LC content diminished and, consequently, the population of nanometric droplets increased. Interestingly enough, the emission spectra registered a new band compared to the emission profile of the two components, which is atributed to the interface interactions.

7.6.2 Polymer Dispersed Liquid Crystal Films Based on Polyvinyl Alcohol Boric Acid Carrying Matrix and a Smectic Liquid Crystal

Among the PDLC composites, those based on a smectic liquid crystal with a direct isotropic–smectic transition, and thus monotropic stability, are very important due to their capability of providing superior optic contrast and generating bistability—ideal properties for portable electronic devices. Moreover, the composites based on smectic liquid crystals demonstrated a reversible memory effect ascribed to the rigidity of the layered mesophase (Date et al. 1999; Lu et al. 2013; Soule et al. 2009; Yonezawa et al. 2004). Despite these benefits, only a few PDLC systems based on smectic liquid crystals have been reported due to the difficulty of obtaining smectic liquid crystals with direct isotropic–smectic transition and stable mesophase at room temperature and, in addition, the difficulty of finding a polymer matrix able to exert strong enough superficial tension to cause the smectic liquid crystal to grow as spherical droplets (Jeon et al. 2012; Kim et al. 2014; Lu et al. 2013; Tu et al. 2015).

In this context, the PVAB matrix that demonstrated an excellent ability to anchor nematic liquid crystals (Marin et al. 2014) also proved able to exert enough tension to constrain a smectic liquid crystal to adopt a spherical shape. In preparing PVAB–smectogen PDLCs, a smectic liquid crystal based on human friendly ester units, with a direct isotropic–smectic transition and mesophase stability range at human body temperature, has been used (Sm) (Scheme 7.4). The PDLC–smectogen composites were obtained by the encapsulation method using different mass ratios of the two components. Thin freestanding films with strong glass adhesiveness were obtained.

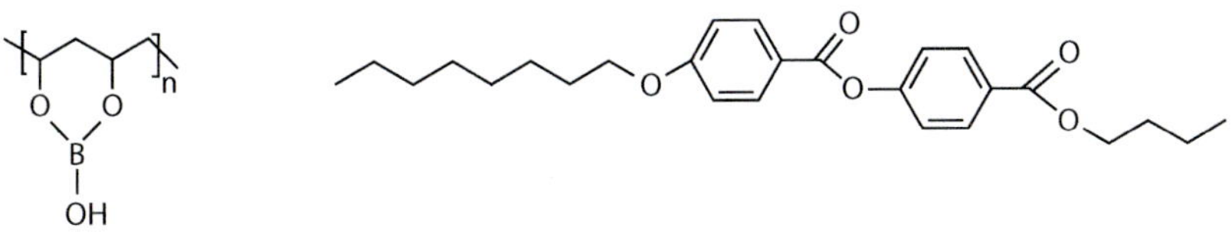

PVAB

Sm

SCHEME 7.4 Structure of the polyvinyl alcohol boric acid and benzoate smectogen.

Although pure liquid crystal presents a fan-shape texture characteristic of the smectic A mesophase, the PDLC films show round, birefringent droplets whose density grows with the content of Sm, resulting in a fine texture for the composites containing more than 30%. The droplet birefringence appeared as concentric circles of different colors (Figure 7.5a), similar to the polygonal textures of the smectogen compounds in a thick preparation (Demus and Richter 1978).

During heating, the crystalline–smectic transition could not be observed, and isotropization occurred slowly compared to that of pure liquid crystal, denoting the strong anchoring forces of the Sm smectogen in the PVAB matrix, which was consolidated by the interfacial forces. Moreover, when the PDLC samples were heated upper the isotropization temperature, the isotropic droplets remained surrounded by a milky birefringence characteristic of ordered polymers (Figure 7.5b). This was attributed to the extension of the ordering to the PVAB matrix via an ordering coupling at interface. The birefringent shadow intensified in time, suggesting that the ordering phenomenon evolved even in solid state. Considering the chemical structure of the two components (Sm and PVAB), it could be estimated that interfacial forces were based mainly on the physical attraction between the electron-deficient boron atoms and rich electron ester groups, which were able to develop a planar alignment of the director field of the smectogen consisting of a parallel disposition of the Sm molecules on the droplet walls.

Combining the POM observations and structural features resulted in the supposition that the droplet texture was polygonal and that the smectic layers were arranged as Duppin cyclides with the cone ellipses oriented on the plane of the droplet wall and the peak to its center.

This result was confirmed by Raman microspectroscopy. Following the vibrations characteristic of the smectogen (C=C at 1,600 cm^{-1} and COO at 1,720 cm^{-1})

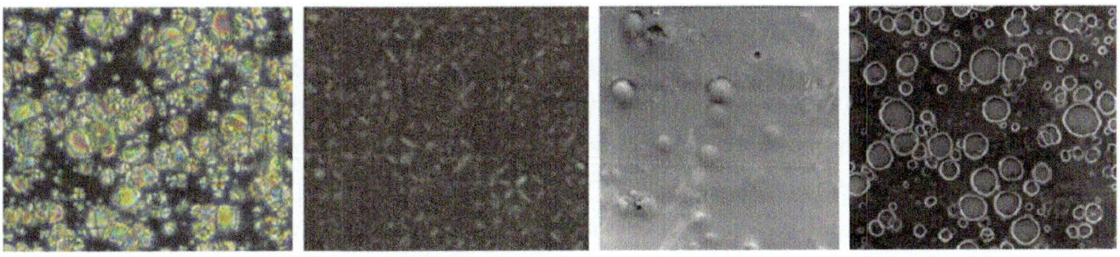

(a) (b) (c) (d)

FIGURE 7.5 Representative images of PDLCs (a,b) under POM and (c,d) under SEM.

and of the PVAB matrix (CH_2, CH_3 around 2,900 cm^{-1}), there was evidence of the droplet segregation on one hand and the preferential planar orientation of the smectogen molecules on the other.

It was established that the intensity of the Raman band characteristic of the liquid crystal, varies between a minimum and a maximum once the laser polarization of excitation was modified—a distinctive behavior of spherical droplets tilted in particular angles. Considering the theoretical premise that the maximum intensity is obtained for a parallel orientation of the director field on the electromagnetic vector of the incident plane of polarized light, while the minimum intensity is obtained for a perpendicular orientation, it was concluded that the director field of the liquid crystal has a radial direction inside the droplet, with the long axes parallel to the droplet wall. Thus, the Raman microspectroscopy indicated that the smectogen is tangentially anchored into the PVAB matrix, with the droplets consisting of focal cones in which the orientational director lies tangential at the interface.

Similarly with the encapsulation of the nematic liquid crystal into the PVAB matrix, the DSC data confirmed highly stable composites due to the component immiscibility, with a strong anchoring of the smectic droplets. A distinct observation of the smectic/PVAB system is the change in PVAB crystallinity. More specifically, the glass transition of the PVAB matrix shifted to higher temperature values compared to the pure PVAB polymer, which is in accordance with the increase of the PVAB crystallinity due to the ordering coupling at interface.

The spherical shape of the smectic droplets was revealed by the SEM measurements performed on the PDLC films (thermal, untreated, or treated). The untreated films showed the spherical shapes distributed on the film surface (Figure 7.5c), while the films heated to the upper isotropization temperature of the smectogen showed spherical pores (Figure 7.5d). Compared to other PDLC systems based on smectic liquid crystals, it is remarkable that PVABs allow the growth of smectic liquid crystal droplets instead of smectic batonnets, which are typical for the smectic mesophase. Considering the tendency of the Sm to grow as needles from solution, the forming of spherical droplets from the PVAB solution reflects the strong interfacial tension exerted by the matrix, which prevents the natural tendency of the smectic compound to grow as needles and determines its spherical encapsulation during the evaporation process.

The droplet diameter ranged from 4 to 11 μm, with a predominant population around 7 μm. As expected, the droplet density grows along with amount of liquid crystal, while the droplet dimension is not significantly affected.

In the case of the composite sample containing 40% of liquid crystal, the contact angle of the PVAB/Sm systems has values in the moderate wettability domain (65°). This value is close to that of the natural cornea (Durán-Sampedro et al. 2013), suggesting potential biomedical applications as contact lenses or other ocular devices, such as an artificial iris (Hsu et al. 2011; Lapointe et al. 2010).

The SFE has high values (36–38 mN/m) that facilitate maintaining the multicellular structure and also offer the possibility of electrical control in sensor applications. These values are close to those of a natural cornea, making these composites real candidates for ocular prostheses.

7.7 REMARKS

Well known for their applications in display manufacture, polymer dispersed liquid crystal (PDLC) composite materials return the researcher's attention to their possibility for a larger field of application, including biomedicine. This chapter highlights the newest contributions in the PDLC domain in terms of PDLC composite design, anchoring forces, properties, and the relationship between composite design and applications.

ACKNOWLEDGMENTS

The research leading to these results has received funding from the Romanian National Authority for Scientific Research, MEN–UEFISCDI, project number PN-II-PT-PCCA-2013-4-1861 (contract number 272/2014).

REFERENCES

Acton Ashton Q. *Acrylates—Advances in Research and Application.* Scholarly Editions, Atlanta, 2013.

Andrews Lester S, and Clary John J. Review of the toxicity of multifunctional acrylates. *J. Toxicol. Environ. Health.* 19 no. 2 (1986): 149–164.

Baker Maribel I., Walsh Steven P., Schwartz Zvi, et al. A review of polyvinyl alcohol and its uses in cartilage and orthopedic applications. *J. Biomed. Mater. Res. B. Appl. Biomater.* 100 no. 5 (2012): 1451–1457.

Bauer Michael, Hartmann Leo Afraneo, Kuschel Frank, et al. Display forming polymer/liquid crystal composite layers and their stability against external mechanical distortions. *J. Appl. Polym. Sci.* 117 no. 4 (2010): 1924–1933.

Bezrodna Tamara, Melnyk Victor, Vorobjev Vladimir, et al. Low-temperature photoluminescence of 5CB liquid crystal. *J. Lumin.* 130 no. 7 (2010): 1134–1141.

Blach Jean-Francoise, Daoudi Abdelaziz, Buisine Jean-Mark, et al. Raman mapping of polymer dispersed liquid crystal. *Vib. Spectrosc.* 39 no. 1 (2005): 31–36.

Bouteiller Laurent, and LeBarny Pierre. Polymer-dispersed liquid crystals: Preparation, operation and application. *Liq. Cryst.* 21 no. 2 (1996): 157–174.

Bronnikov Serghei, Kostromin Serghei, and Zuev Vjacheslav. Polymer-dispersed liquid crystals: Progress in preparation, investigation, and application. *J. Macromol. Sci. Phys.* 52 no. 12 (2013): 1718–1735.

Chen Lu Guang, and Shanks Robert. Thermoplastic polymer-dispersed liquid crystals prepared from solvent induced phase separation with predictions using solubility parameters. *Liq. Cryst.* 34 no. 12 (2007): 1349–1356.

Croisier Florence, and Jérôme Christine. Chitosan-based biomaterials for tissue engineering. *Eur. Polym. J.* 49 no. 4 (2013): 780–792.

Cupelli Daniela, Fiore Nicoletta P., De Filpo Giovanni, et al. Reverse mode operation polymer dispersed liquid crystal with a positive dielectric anisotropy liquid crystal. *J. Polym. Sci. B: Polym. Phys.* 49 no. 4 (2011): 257–262.

Date Motoo, Takeuchi Yasuhiro, and Kato Kiyoko. Droplet size effect on the memory-mode operating temperature of smectic-A holographic polymer dispersed liquid crystal. *J. Phys. D: Appl. Phys.* 32 no. 24 (1999): 3164–3168.

Demus Dietrich, and Richter Lothar. *Textures of Liquid Crystals.* Leipzig: VerlagChemie, Weinheim, 1978.

Dierking Ingo. Polymer network-stabilized liquid crystals. *Adv. Mater.* 12 no. 3 (2000): 167–181.

Dolgov Leonid, Yaroshchuk Oleg, Tomylko Sergiy, et al. Electro-optical memory of a nematic liquid crystal doped by multi-walled carbon nanotubes. *Condens. Matter. Phys.* 15 no. 3 (2012): 33401:1-8.

Drzaic Paul S. Putting liquid crystal droplets to work: A short history of polymer dispersed liquid crystals. *Liq. Cryst.* 33 no. 11–12 (2006): 1281–1296.

Drzaic Paul S. *Liquid Crystal Dispersion.* Singapore: World Scientific, 1995.

Durán-Sampedro Gonzalo, Agarrabeitia Antonia R., Cerdán Luis, et al. Carboxylates versus fluorines: Boosting the emission properties of commercial BODIPYs in liquid and solid media. *Adv. Funct. Mater.* 23 no. 34 (2013): 4195–4205.

El-Hibri Jamal M., and Axelrad Shari W. *Polyarylethersulfones.* In *Handbook of Thermoplastics*, Olabisi Olagoke, and Adewale Kolapo, eds., Boca Raton, FL: CRC Press, 2015.

Ellahi Mujtaba, Liu Fang, Song Ping, et al. Characterization and morphology of polymer-dispersed liquid crystal films. *Soft Mater.* 12 no. 3 (2014): 339–345.

Fergason James L. Polymer encapsulated liquid crystals for display and light control applications. *Tech. Digest. SID Int. Symp.* 16 (1985): 68–70.

Graham Russell M., Paterson Brian J.A., and Imrie Corrie T. Thermal characterization of polymer-dispersed liquid crystals by differential scanning calorimetry. *Chem. Mater.* 7 no. 11 (1996): 2185–2189.

Hallab Nadim J., Bundy Kirk J., O'Connor Kim, et al. Evaluation of metallic and polymeric biomaterial surface energy and surface roughness characteristics for directed cell adhesion. *Tissue. Eng.* 7 no. 1 (2001): 55–71.

Hamlington Benjamin D., Steinhaus Benjamin, Feng James J., et al. Liquid crystal droplet production in a microfluidic device. *Liq. Cryst.* 34 no. 7 (2007): 861–870.

Hoenich Nicholas A., Woffindin Celia, Mathews John N.S., et al. Biocompatibility of membranes used in the treatment of renal failure. *Biomaterials.* 16 no. 8 (1995): 587–592.

Hsu Tsung-Chun, Lu Ching-Heng, Huang Yu-Ting, et al. Concentric polymer-dispersed liquid crystal rings for light intensity modulation. *Sensor. Actuat. A-Phys.* 169 no. 2 (2011): 341–346.

Jeon Young Jae, Bingzhu Yin, Rhee June Tak, et al. Application and new developments in polymer-dispersed liquid crystal simulation studies. *Macromol. Theory. Simul.* 16 no. 7 (2007): 643–659.

Jeon Young Jae, Lee Gae Hwang, Jang Jae Eun, et al. Applications of multidirectional reflective light-control films on reflective polymer-dispersed liquid crystal displays for enhancement in image quality at lower viewing angles. *Liq. Cryst.* 39 no. 11 (2012): 1314–1319.

Kim Dae Soo, Yeo Dong Bin, Heo Kyong Chan, et al. Color polymer dispersed liquid crystal device with colored core-shell structure. *Mol. Cryst. Liq. Cryst.* 613 no. 1 (2015): 39–44.

Kim Jaeyong, and Han Jeong In. Effect of liquid crystal concentration on electro-optical properties of polymer dispersed liquid crystal lens for smart electronic glasses with auto-shading and auto-focusing function. *Electron. Mater. Lett.* 10 no. 3 (2014): 607–610.

Kitzerow Heinz S. Polymer-dispersed liquid crystals. From the nematic curvilinear aligned phase to ferroelectric films. *Liq. Cryst.* 16 no. 1 (1994): 1–31.

Kumar Ravi M.N.V., Muzzarelli Ricardo A.A., Muzzarelli Corrrado, et al. Chitosan chemistry and pharmaceutical perspectives. *Chem. Rev.* 104 no. 12 (2004): 6017–6084.

Lapointe Jerome, Durette Jean-Francoise, Harhira Aissa, et al. A 'living' prosthetic iris. *Eye.* 24 no. 11 (2010): 1716–1723.

Lin Yi-Hsin, Chu Ting-Yu, Tsou Yu-Shih, et al. An electrically switchable surface free energy on a liquid crystal and polymer composite film. *Appl. Phys. Lett.* 101 no. 23 (2012): 233502:1–4.

Liu Yan Jun, Ding Xiaoyun, Lin Sz-Chin Steven, et al. Surface acoustic wave driven light shutters using polymer-dispersed liquid crystals. *Adv. Mater.* 23 no. 14 (2011): 1656–1659.

Liu Yan Jun, and Sun Xiaowei W. Holographic polymer-dispersed liquid crystals: Materials, formation and applications. *Adv. Opto. Electron.* (2008): 684349.

Lu Yang, Wei Jie, Shi Yang, et al. Effects of fabrication condition on the network morphology and electro-optical characteristics of polymer-dispersed bistable smectic A liquid crystal device. *Liq. Cryst.* 40 no. 5 (2013): 581–588.

Marin Luminita, Ailincai Daniela, and Paslaru Elena. Monodisperse PDLC composites generated by use of polyvinyl alcohol boric acid as matrix. *RSC. Adv.* 4 no. 72 (2014): 38397–38404.

Marin Luminita and Perju Elena. New polymer-dispersed liquid crystals. Preparation and thermal characterization. *Metalurgia International, Special Issue: Exploring Romanian Resources in Materials Research* XIII, 2008.

Marin Luminita, and Perju Elena. Polysulfone as polymer matrix for new polymer dispersed liquid crystals. *Phase. Transit.* 82 no. 7 (2009): 507–518.

Marin Luminita, and Perju Elena. Optical response of cyanoazomethine liquid crystal droplets in PDLC films based on a polysulfone matrix. *J. Optoelectron. Adv. Mater.* 12 no. 6 (2010): 1378–1384.

Marin Luminita, Popescu Maria-Cristina, Zabulica Andrei, et al. Chitosan as matrix for biopolymer dispersed liquid crystal systems. *Carbohyd. Polym.* 95 no. 1 (2013): 16–24.

Mucha Maria. Polymer as important component of blends and composites with liquid crystals. *Progr. Polym. Sci.* 28 no. 5 (2003): 837–873.

Mucha Maria, Dunmur David A., Fukuda Atsuo, and Luckhurst Geoffrey R. *Physical Properties of Liquid Crystals: Nematics.* London: INSPEC, 2001.

Musevic Igor, and Humar Matjaz. Tunable liquid crystal optical microcavities. *Proc. SPIE.* 7955 (2011): 795509.

Perju Elena, Marin Luminita, Grigoras Vasile Cristian, et al. Thermotropic and optical behaviour of new PDLC systems based on a polysulfone matrix and a cyanoazomethine liquid crystal. *Liq. Cryst.* 38 no. 7 (2011): 893–905.

Perju Elena, Paslaru Elena, and Marin Luminita. Polymer-dispersed liquid crystal composites for bio-applications: Thermotropic, surface and optical properties. *Liq. Cryst.* 42 no. 3 (2015): 370–382.

Saxena Pooja, Gaur Mulayam Singh, Shukla Prashant, et al. Relaxation investigations in polysulfone: Thermally stimulated discharge current and dielectric spectroscopy. *J. Electrostat.* 66 no. 11–12 (2008): 584–588.

Sidiq Sumyra, Das Dibyendu, and Pal Santanu Kumar. A new pathway for the formation of radial nematic droplets within a lipid-laden aqueous-liquid crystal interface. *RSC Adv.* 4 no. 36 (2014): 18889–18893.

Smith Gregor W. Mixing and phase separation in liquid crystal/matrix systems. *Int. J. Mod. Phys. B.* 7 no. 25 (1993): 4187–4213.

Soule Ezequiel R., Abukhdeir Nasser M., and Rey Alejandro D. Thermodynamics, transition dynamics, and texturing in polymer-dispersed liquid crystals with mesogens exhibiting a direct isotropic/Smectic-A transition. *Macromolecules.* 42 no. 24 (2009): 9486–9497.

Srivastava Jagdeesh Kumar, Singh Rajendra Kumar, Dhar Ravindra, et al. Thermal and morphological studies of liquid crystalline materials dispersed in a polymer matrix. *Liq. Cryst.* 38 no. 7 (2011): 849–859.

Srivastava Jagdeesh Kumar, Singh Rajendra Kumar, Dhar Ravindra, et al. Phase diagrams and morphology of polymer-dispersed liquid crystals: An analysis. *Liq. Cryst.* 39 no. 11 (2012): 1402–1413.

Teperek Andrzej, Czajkowski Wojciech, and Fabianowski Wojciech. LC encapsulated in polymerized vesicles. *Proc. SPIE.* 2372 (1995): 408.

Tercjak Agnieszka, Serrano Elena, Larrañaga Maider, et al. Polymer dispersed liquid crystals based on poly(styrene-*b*-ethylene oxide), poly(bisphenol a carbonate) or poly(methylphenylsiloxane), and 4′-(hexyloxy)-4-biphenyl-carbonitrile: Analysis of phase diagrams and morphologies generated. *J. Appl. Polym. Sci.* 108 no. 2 (2008): 1116–1125.

Tsuruoka Shuichi, Sugimoto Kohichi, Hayasaka Tokie, et al. Ranitidine clearance during hemodialysis with high-flux membrane: Comparison of polysulfone and cellulose acetate hemodialyzers. *Eur. J. Clin. Pharmacol.* 56 no. 8 (2000): 581–583.

Tu Yingfeng, Gu Yan, Van Horn Ryan M., et al. A synthetic approach towards micron-sized smectic liquid crystal capsules *via* the diffusion controlled swelling method. *Polym. Chem.* 6 no. 13 (2015): 2551–2559.

Vasile Cornelia, and Pascu Mihaela. *Surfaces Properties of Polymers.* Trivandrum, India: Research Signpost, 2007.

White Timothy J., Natarajan Lalgudi V., Tondiglia Vincent P., et al. Polymerization kinetics and monomer functionality effects in thiol-ene polymer dispersed liquid crystals. *Macromolecules.* 40 no. 4 (2007): 1112–1120.

Woltman Scott J., Jay Gregory D., and Crawford Gregory P. Liquid-crystal materials find a new order in biomedical applications. *Nat. Mater.* 6 no. 12 (2007): 929–938.

Yonezawa Juniki, Martin Stephen M., Macosko Christopher W., et al. Rheology and morphology of smectic liquid crystal/polymer blends. *Macromolecules.* 37 no. 17 (2004): 6424–6432.

Yoshito Ikada. Surface modification of polymers for medical applications. *Biomaterials.* 15 no. 10 (1994): 726–736.

Zhang Tingting, Kashima Miki, Zhang Mingzhi, et al. Effects of the functionality of epoxy monomer on the electro-optical properties of thermally-cured polymer dispersed liquid crystal films. *RSC Adv.* 2 no. 5 (2012): 2144–2148.

8 From Macro- to Nanoscale Surface Morphological Features of Multiphase Polymer Films

Iuliana Stoica

CONTENTS

8.1 INTRODUCTION

The surface morphological features of multiphase polymer systems, including a wide range of materials such as composites, blends, and block copolymers, can be characterized at high magnification and resolution by atomic force microscopy (AFM), which is an excellent complimentary technique to the electron microscopy techniques, such as scanning electron microscopy (SEM) and transmission electron microscopy (TEM). Using the phase contrast mode, all these microscopy techniques can be used for composition imaging of the different phases in multiphase polymer films. Moreover, this chapter will present the manner in which AFM can also provide additional quantitative information regarding the differences in local mechanical properties by measuring the lateral forces (lateral force imaging), adhesion forces (adhesion force mapping), and Young's modulus (modulus mapping, indentation mapping).

8.2 MICROSCOPIC IDENTIFICATION METHODS OF THE SURFACE MORPHOLOGICAL FEATURES IN MULTIPHASE POLYMER FILMS

In order to develop new polymers with improved functionality that are suitable for a much wider range of applications, considerable importance has been given to the study of their structures and properties. Among many polymer characterization methods, electron microscopy techniques, divided into two different areas—scanning electron microscopy (SEM) and transmission electron microscopy (TEM), have the capability to provide direct pictures of the studied surfaces (Yao and Kung 2003). The images from a scanning electron beam instrument are built up by scanning a focused, highly convergent electron probe over an area of the sample and measuring a signal produced from the interaction of the electron beam with the specimen. Therefore, SEM images are referred to as virtual images because they are comprised of a signal different than that of the incident electron beam.

Image formation in SEM is dependent on the acquisition of signals produced from the electron beam and specimen interactions, which are divided into two major categories: elastic interactions and inelastic interactions. Elastic scattering results from the deflection of the incident electron by the specimen atomic nuclei or by outer shell electrons of similar energy. Incident electrons that are elastically scattered through an angle of more than 90° are called backscattered electrons (Figure 8.1). Inelastic scattering occurs through a variety of interactions among the incident electrons and the electrons and atoms of the sample, and it results in the primary beam electrons transferring substantial energy to that atom. As a result, the excitation of the specimen electrons during the ionization of specimen atoms leads to the generation of secondary electrons (Figure 8.1). When primary electrons interact with the sample, the primary electrons cause either secondary electrons or backscatter to be emitted from the sample, which are mainly used for image formation. A secondary electron image thus gives information about the topography of the sample.

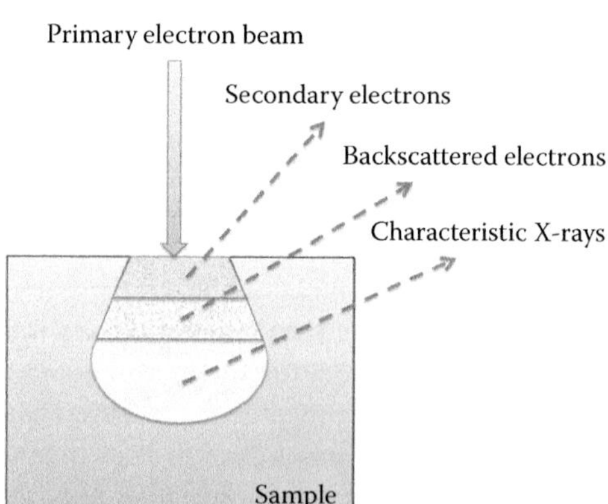

FIGURE 8.1 A schematic representation of processes taking place during electron–solid interaction.

Hence, a backscatter electron image gives information about the composition of the sample (Zhou et al. 2007).

Energy-dispersive X-ray spectroscopy (EDX) is an analytical technique used for the elemental analysis or chemical characterization of a sample. Its characterization capabilities are due in large part to the fundamental principle that each element has a unique atomic structure allowing a unique set of peaks on its X-ray emission spectrum (Goldstein 2003). To stimulate the emission of characteristic X-rays from a specimen, a high-energy beam of electrons is focused into the sample being studied. At rest, an atom within the sample contains ground state (or unexcited) electrons in discrete energy levels or electron shells bound to the nucleus. The incident beam may excite an electron in an inner shell, ejecting it from the shell while creating an electron hole where the electron was. An electron from an outer, higher-energy shell then fills the hole, and the difference in energy between the higher-energy shell and the lower energy shell may be released in the form of an X-ray (Figure 8.1). Because the energies of the X-rays are characteristic of the difference in energy between the two shells and of the atomic structure of the emitting element, EDX allows the elemental composition of the specimen to be measured (Goldstein 2003).

A transmission electron microscope utilizes instead a parallel stationary beam of electrons illuminating an area of the sample and forming an image using some portion of the electrons that pass through the specimen. In conclusion, TEM images can be classified as a real image. Depending on the mode used, TEM offers a variety of information. Bright field and dark field imaging are used to characterize defects and domain structures. Selected area diffraction combined with the tilting of the crystal in the microscope allows reconstruction of the reciprocal space and, in that way, obtains information about the crystal structure and identifies different phases. Finally, under certain conditions, high resolution lattice imaging allows direct visualization of the projected crystal's potential.

Atomic force microscopy (AFM) is another method widely used to investigate polymer surface topography on micrometric and nanometric scales to obtain roughness parameters or to map the qualitative differences of local surface properties such as chemical, adhesive, or elastic capabilities. This method is based on a principle that when a tip, integrated to the end of a spring cantilever, is brought within the interatomic separation between the tip and sample, interatomic potentials are developed between the atoms of the tip and the atoms of the surface. As the tip moves across the surface, the interatomic potentials force the cantilever to bounce up and down with the changes in surface contours. Therefore, by measuring the deflection of the cantilever, the topographic features of the surface can be mapped out. Among all the AFM modes (contact mode, tapping mode, and non contact mode), the tapping mode is suitable for the imaging of soft samples (Van den Berg et al. 1994; Tsukruk and Singamaneni 2012) and can be effectively applied in the field of polymers to study surface characteristics. Although the topographic image shows some corresponding features, surface roughness hinders the identification of domains in multiphase polymer films. Thus, it is necessary to apply phase imaging, which allows precise resolution of the different material phases. In this mode, the probe oscillating vertically near its mechanical resonance frequency is brought into proximity with the surface. As the probe lightly taps the surface, the amplitude of oscillation is reduced and the AFM uses this change in amplitude to track the surface topography. In addition to its amplitude, the probe motion can be characterized by its phase relative to a driving oscillator. The phase signal changes when the probe encounters regions of different composition (Figure 8.2). Phase shifts are registered as bright and dark regions in phase images, comparable to the way height changes are indicated in height images.

Although AFM contact mode is not as commonly used in polymer investigations, there still are some contact techniques very useful in distinguishing regions with different local mechanical properties on the same sample surface. One of them is lateral force microscopy (LFM), in which the lateral deflection of the cantilever is a result of the force applied to the cantilever when it moves horizontally across the sample surface, and the magnitude of this deflection is determined by the frictional coefficient, the topography of the sample surface, the direction of the cantilever movement, and the cantilever's lateral spring constant. As seen in Figure 8.3, when scanning in the contact mode perpendicular to the longitudinal axis of the cantilever, in addition to the cantilever's deflection in the normal direction, an added torsional bending of the cantilever occurs. This is caused by the movement of forces acting on the tip. With minor deflections, the angle of torsion is proportional to the side (lateral) force. When moving over a flat surface with zones of different friction factors, the angle of torsion will change in every new zone. This allows measuring of the local friction force. If the surface is not absolutely flat, such an interpretation is complicated. To distinguish zones of different friction and relief influence, a second pass on the same line in opposite direction can be utilized (Magonov and Whangbo 1996). In this way, LFM is very useful for studying a sample whose surface consists of areas with different friction and also for obtaining edge-enhanced images of any surface.

Another contact technique that can be used to analyze multiphase polymer films (i.e., local variations in the elastic properties of the surface) is force–distance spectroscopy. In this mode, the force–distance curves are plots of distance dependent on the forces that act on the tip in the vicinity of the surface. They are registered when the tip approaches the surface or is retracted from it (Maver et al. 2013). Typical approach/retract force–distance curves and their stages are presented in Figure 8.4: (1) the cantilever starts to approach the surface; (2) the tip approaches the surface;

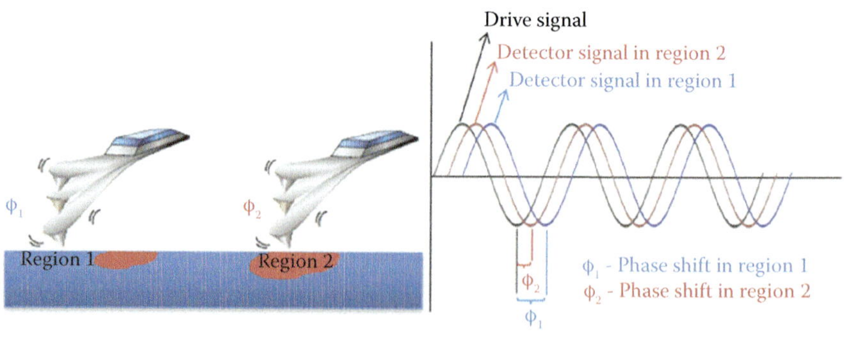

FIGURE 8.2 A schematic representation of AFM phase imaging.

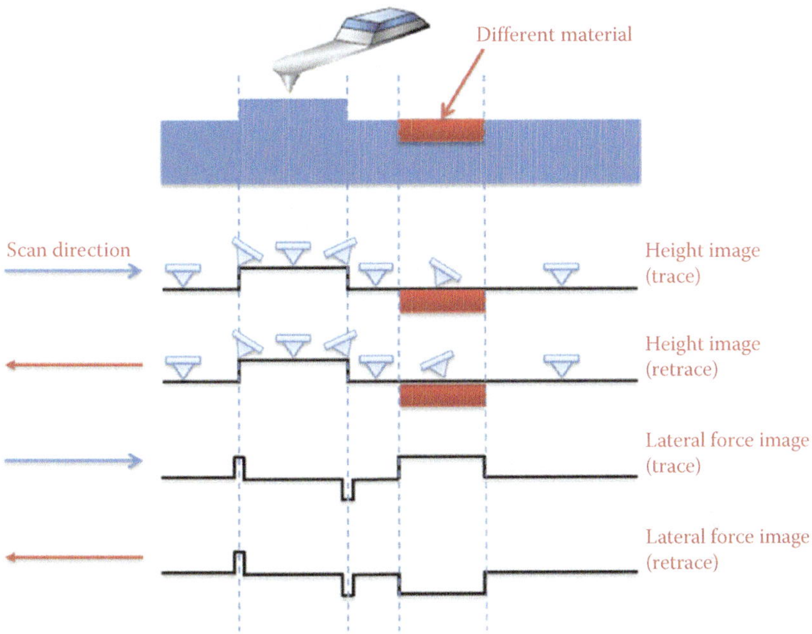

FIGURE 8.3 A schematic representation of AFM and LFM imaging.

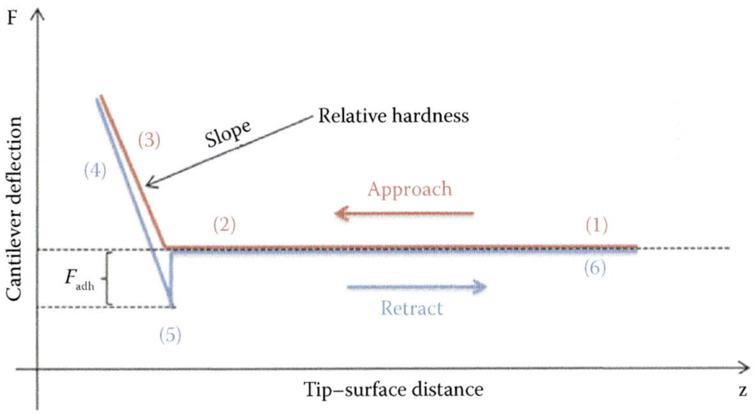

FIGURE 8.4 Stages of force–distance curves.

(3) the cantilever is deflected upward, and the measured force sharply increases; (4) the tip starts to retract, and the deflection starts to decrease; (5) the tip snaps in, and the cantilever is deflected downwards; and (6) the tip snaps out and returns to its equilibrium position.

In the linear region of the force–distance curves, the slope is related to the elastic modulus of the system. As shown in Figure 8.4, when the cantilever is much softer

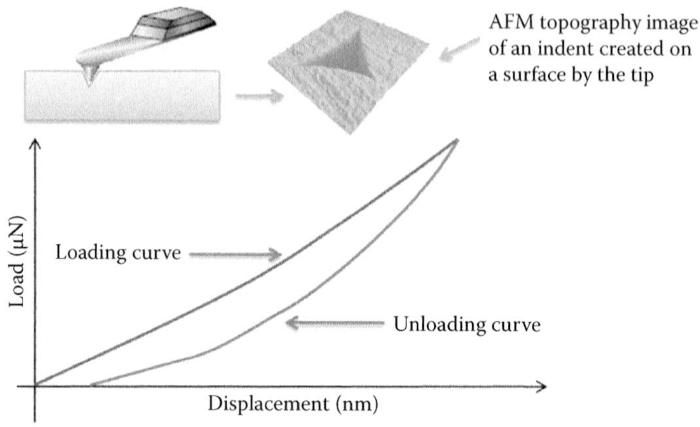

FIGURE 8.5 Loading and unloading displacement curves during nanoindentation.

than the sample surface, the slope of the curve mostly reflects the spring constant of the cantilever. Instead, when the cantilever is much stiffer than the sample surface, the slope of the force–distance curve allows investigation of the elastic properties of the sample.

Elastic modulus and hardness can be also evaluated by analysis of the force–displacement curve, using AFM nanoindentation. The important factors in this technique are the force calibration and the depth (or so-called displacement). The force constant of a cantilever, the shape of the tip, and other mechanical properties such as tip glue determine the total force exerted onto a sample.

The nanoindentation pit, created by the indenter stressing the surface of a sample at a specified force that is sufficiently high enough to deform the sample for about 1–2 s, has a distinct geometry that can be characterized using, for example, tapping mode AFM, as shown in Figure 8.5. Typical loading and unloading displacement curves during AFM nanoindentation in cycle are also depicted in Figure 8.5. In this case, the hysteresis indicates that the deformation is not fully elastic and is partially inelastic. However, for some materials, if the deformation mainly is elastic, the loading and unloading curves would be closely identical. Hardness can be calculated by dividing the loading force by the projected residual area of the indentation. Besides hardness, Young's modulus of elasticity can be obtained from the slope of the unloading curve. That being the case, a comparison of atomic force microscopy and scanning and transmission electron microscopy is presented in Table 8.1.

8.3 MORPHOLOGY AND MECHANICAL CHARACTERISTICS IN MICRO- AND NANOSTRUCTURED MULTIPHASE POLYMER BLENDS

Polymer blends mainly are created using the blending process of two or more different polymers. The preparing of polymer blends is necessary to reduce costs by combining high-quality polymers with cheaper materials and to create a polymer

TABLE 8.1
Comparison of AFM with Other Microscopy Methods

Characterization Technique		
Scanning Electron Microscopy	**Transmission Electron Microscopy**	**Atomic Force Microscopy**
• Conducted in a vacuum or ambient environment	• Conducted in a vacuum environment	• Conducted in a vacuum, ambient, or fluid environment
• Based on scattered electrons	• Based on transmitted electrons	• Based on probing the surface features with a sharp tip by raster scanning
• Lateral resolution: 1–5 nm	• Lateral resolution: 0.1 nm	• Lateral resolution: 2–10 nm
• Vertical resolution: N/A	• Vertical resolution: N/A	• Vertical resolution: 0.1 nm
• Magnification: 20 X–65 × 10^4 X	• Magnification: 60 X–15 × 10^6 X	• Magnification: 5 ×10^2 X–10^8 X
• Provides a two-dimensional image	• Provides a two-dimensional image	• Provides a two- and three-dimensional image
• Provides information about topography, composition imaging, elemental composition	• Provides real image of the sample, domain structures, identity of different phases	• Provides information about topography, composition imaging, differences in local mechanical properties

that has a desired combination of the different properties of its components. When producing blends, the main issue is to enhance the compatibility and connection among the polymeric phases in order to obtain an optimized morphology and, thus, favorable properties for the intended purpose.

According to Michler, the morphology of polymer blends is determined by the incompatibility of the polymer components (which cause phase separation) and the processing (intensity of blending, shear forces during mixing, viscosities, temperature). Incompatible polymers—such as polystyrene with polypropylene, polyethylene, or poly(methyl methacrylate)—usually show strong phase separation with the formation of large particles. In this context, the contact among polymeric components can be improved by introducing compatibilizers (Michler 2008)—for example, a block copolymer premade or generated in situ by reaction that preferentially locates at the polymer–polymer interface (Baker et al. 2001). The addition of such compatibilizers can lead to more stable and finer scale morphologies (Harrats et al. 2003).

In general, morphologically speaking, in pure polypropylene/polystyrene (PP/PS) blends, large PS particles are visible in the PP matrix. The spherical shapes of the PS particles and debonding from the matrix are good indicators of low compatibility (Michler 2008). You and Jia investigated the effects of a styrene–ethylene/propylene (SEP) diblock copolymer on the morphology of 80/20, 60/40, 40/60, and 20/80 PP/PS blends. The scanning electron microscopy results showed that a SEP diblock copolymer, acting as a compatibilizer in PP/PS immiscible blends, can diminish the coalescence of the dispersed particles, reduce their average particle size, and change their phase morphologies significantly. It was found that SEP has better compatibilization effects on the PP/PS (20/80) blends. In this case, at 5 wt% and 10 wt% of SEP, the particles of the dispersed phase were distributed evenly in the matrix, and their sizes were reduced significantly, while the interphases between PP and PS were blurred (You and Jia 2003).

In order to reduce the interfacial tension, to stabilize the morphology, and to increment the interfacial adhesion, Diaz et al. tried to compatibilize PP (80 wt%)/PS (20 wt%) blends by adding a tailor-made copolymer (TCM), which was prepared from the pure homopolymers via Friedel–Crafts alkylation reaction, in different concentrations, varying the mixing time from 24 to 36 min. The improvements in blend morphological aspects after compatibilization were studied by SEM. For blends mixed for 24 min, a gradual decrease of particle size with the TMC addition was noted. Still, the interfacial adhesion did not show any improvement until 10 wt% TMC was added. For the 36-min mixed blends, both particle size reduction and interfacial adhesion improvement were clearly seen, starting from 0.3 wt% up to 0.7–1.0 wt% of TMC added (Diaz et al. 2006).

Brostow et al. prepared various concentrations of PP/PS blends with styrene -ethylene/butylene–styrene block copolymer (SEBS) added as a compatibilizer between the two main polymers. Initially, as a result of high interfacial tension between the components, two-phase morphology with PS particles dispersed in the PP matrix was observed. By adding the compatibilizing agent, the blend morphology was improved, and smaller particles of the dispersed PS phase in the PP matrix blends were obtained. By adding more than 5% SEBS, no further effect on the blends was highlighted (Brostow et al. 2008).

In recent years, an original concept of compatibilization by using rigid nanoparticles, such as silica nanoparticles, has been proposed (Elias et al. 2007). The morphologies of polymer blends generated during processing are usually unstable, and morphology coarsening often occurs in the melt state; therefore, suppressing the morphology coarsening is crucial for obtaining polymer blends with tailored and stable structure and properties. Recently, Liu et al. reported the morphology coarsening behavior of co-continuous PP/PS (50/50) blends, pure and filled with 1.01, 3.09, 4.17 wt% nanosilica particles (SiO_2 np), subjected to quiescent annealing in the melt state for various times (5, 30, and 120 min) at 190°C. For a pure blend, with increased annealing time, the phase size increases significantly. Compared with a pure blend, for 1.01 wt% and 3.09 wt% nanosilica particle filled PP/PS (50/50) blends, the phase size increase slows down. When the nanosilica concentration is as high as 4.17 wt%, the coarsening process is greatly suppressed, and the morphology stability is greatly improved (Liu et al. 2014).

In addition to the morphology evolution data obtained for pure 50/50 PP/PS blend and 1 wt% SiO_2 np filled 50/50 PP/PS blend during annealing for various periods of time at 190°C, Feng et al. compared those obtained for pure 70/30 PP/PS blend and 1 wt% SiO_2 np filled 70/30 PP/PS blend during annealing for various periods of time at 210°C. First, it was observed that the SiO_2 particles prefer to locate at the interface in any blend composition. In addition, in the beginning of the annealing process, the PS phase looks like highly elongated fibers in the case of a pure 70/30 PP/PS blend. With an increase in annealing time, these fibers break up, retract, become spherical droplets, and finally merge. For a 1 wt% SiO_2 np filled 70/30 PP/PS blend, most

PS fibers can retract in 30 min but, after that, the retracted spherical PS droplets do not collide, immerge, and form bigger spherical PS droplets as does that in a pure 70/30 blend; the shape and the size of the droplets are fixed, indicating that a further coalescence process is highly suppressed (Feng et al. 2015).

Using a new AFM tool to rapidly map the viscoelastic properties from the frequency and quality factor of the contact resonance peak, Killgore et al. investigated a blend containing PS domains in a PP matrix. The obtained maps indicate that there is relatively little difference in the stiffness (storage modulus) of the two polymers, but the damping (loss modulus) shows greater contrast (Killgore et al. 2011).

Polystyrene/polyethylene (PS/PE) blends consist of incompatible polymers that show strong phase separation. Using SEM investigations, Lyu et al. have shown that the addition of block copolymer to PS/PE blends reduces coalescence and stabilizes droplet morphologies (Lyu et al. 2002). In addition, blending two immiscible polymers of near 50/50 composition leads to co-continuous morphology. Galloway et al. studied the effect of PS–PE block copolymers, with molecular weights varying from 6 to 200 kg/mol, on the co-continuous morphology of 50/50 (w/w) polystyrene/high-density polyethylene (PS/HDPE) blends. Using an intermediate molecular weight PS–PE (40 kg/mol), SEM investigations showed remarkable results in reducing the phase size and stabilizing the blend morphology during annealing. The existence of an optimal molecular weight block copolymer is due to a balance between the ability of the block copolymer to reach the interface and its relative stabilization effect at the interface (Galloway et al. 2005).

Polystyrene (PS) and poly(methyl methacrylate) (PMMA) are frequently used as binary model systems to study phase separation dynamics in polymer thin films and in polymer–polymer and polymer–substrate interactions (Zhu et al. 1999; Winesett et al. 2000). Using scanning force microscopy (SFM), Zhu et al. showed that, after the PS/PMMA bilayer samples were annealed for seven days at 180°C and washed in cyclohexane to selectively dissolve the PS homopolymer, the phase of PS was differently organized on the PMMA layer surface due to the content of a 30% dPS-PMMA diblock copolymer, namely in undissolved micelles (for a thickness of 100 nm for PS layer), coexisting micelles and microemulsion structures (for 80 nm), and in microemulsion (for 50 nm).

The location of the diblock copolymer at the interface between the PS/PMMA domains was confirmed using the lateral force mode of the SFM. The regions with higher friction contrast due to the PS block of the diblock copolymer (not readily dissolved in cyclohexane) are located at the edges of the PMMA domains (Zhu et al. 1999).

Walheim et al. reported PS droplets after annealing a 50/50 w/w PS/PMMA sample for 12 h at 190°C (Walheim et al. 1997), and Ade et al. have also found using scanning force microscopy surface phase-separated PS and PMMA domains with PS droplets after annealing at 180°C (Ade et al. 1999). Based on AFM microscopy investigations, Harris et al. showed that PS and PMMA are phase-separated on the surface; only in PS-*b*-PMMA diblock copolymer films is the PS block always at the surface and the PMMA block never at the surface (Harris et al. 2003).

Pohjakallio et al. brought new insights for the preparation of bicomponent Langmuir–Schaefer (LS) films, especially those containing blends of immiscible polymers such as PS and PMMA; 10:1 and 5:1 (mass ratio) PMMA/PS blend films deposited on silica substrates have been investigated by AFM (topography and phase imaging), and a complete phase separation of PMMA and PS at the air/water interface (a PMMA monolayer with PS islands) has been observed. The LS dipping leads to morphology when the continuous PMMA layer is located at the air-exposed surface (Pohjakallio et al. 2011). Recently, the effect of carbon black (CB) concentration and mixing protocol on morphology development of PS/PMMA co-continuous blends during quiescent annealing was investigated using TEM (Scherzer et al. 2015). CB was preferentially localized in the dark domains of the PS phase in the studied blends. Also, the morphology of the blends is strongly influenced by the mixing protocol.

As discussed, PS and PMMA form highly incompatible blends, but increasing the length of the ester alkyl chain gives more compatibility (Affrossman et al. 2000). Therefore, Cappella measured the topography, dissipated energy, adhesion, stiffness, and elastic modulus of a 50:50 blend of polystyrene (PS) and poly(*n*-butyl methacrylate) (PnBMA) with an atomic force microscope (AFM) in tapping mode and with force–distance curves. As seen in Figure 8.6, PS and PnBMA components can be distinguished in all images.

From the histograms of the adhesion, stiffness, and logarithm of the Young's modulus, the mean adhesion force, stiffness, and Young's modulus for PnBMA and PS regions were calculated. It was concluded that, in the investigated 50/50 PnBMA/PS blends, PnBMA has a higher adhesion and a smaller stiffness than PS and, hence, a higher dissipated energy (Cappella 2011).

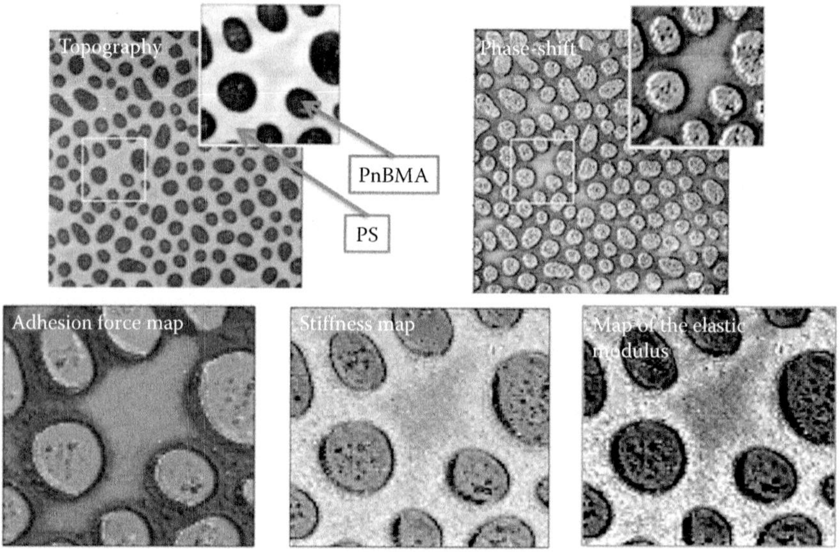

F_{adh}(PnBMA) = 0.71 ± 0.04 μN S (PnBMA) = 0.81 ± 0.04 E (PnBMA) = 3.2 GPa

F_{adh}(PS) = 0.42 ± 0.08 μN S (PS) = 0.89 ± 0.03 E (PS) = 10.7 GPa

FIGURE 8.6 Topography, phase-shift, adhesion force map, stiffness map, and map of the elastic modulus of the PnBMA/PS blend and calculated values for mean adhesion force, stiffness, and Young's modulus for PnBMA and PS regions. (Adapted from Cappella Brunero. *Polymers*., 3, no. 3, 1091–1106, 2011.)

8.4 PHASE MORPHOLOGY AND LOCAL MECHANICAL PROPERTIES IN BLOCK COPOLYMER SYSTEMS

Over the years, electron microscopies and scanning or atomic force microscopy were found to be the most powerful analytical tools in block copolymer research because they are able to highlight the type, size, and arrangement of the different blocks induced by microphase separation (Hasegawa and Hashimoto 1996).

Electron microscopy (especially TEM) can be used in the characterization of block copolymer nanostructures because of its ability to image the highly local structure of a bulk sample, enabling a straightforward elucidation of the specimen morphology. In this way, it was possible to collect TEM images of a highly asymmetric styrene/butadiene block (SB) copolymer (Adhikari and Michler 2004). An SB block copolymer has an essentially lamellae-like morphology, with each polybutadiene (PB) lamella embedding a thin, discontinued polystyrene (PS) layer.

Mixing the SB block copolymer with low molecular weight PS, with an average molecular weight of about 15,000 g/mol (PS015), leads to partial thickening of PS lamellae as the PS015 chains are accommodated by the PS block chains of SB (at 20 wt% of PS015), coexisting cylindrical and lamellar morphologies (at 40 wt% of PS015), and co-continuous morphology (at 60 wt% of PS015) (Michler 2008). The contrast during the TEM investigation of the polystyrene-*block*-polybutadiene-*block*-polystyrene (SBS) triblock copolymers depends on the nature of the monomers present. By changing the polystyrene/polybutadiene (PS/PB) volume ratio, a wide variety of morphologies can be produced, namely PS cylinders (for 28% PS), lamellae (for 45% PS), and PB cylinders (for 74% PS) (Michler 2008). Adhikari et al. studied morphology formation and deformation behavior of asymmetric styrene/butadiene triblock copolymers (total polystyrene (PS) content ~70%) consisting of PS outer blocks, each held apart by a styrene-*co*-butadiene random copolymer block (PS-*co*-PB). They observed a significant shift in the phase behavior compared to pure symmetric triblock copolymers, which can be attributed to the asymmetric architecture and the presence of PS-*co*-PB as a soft block. Whatever the morphology type, when compared with the hard phase, the soft phase was found to deform to a significantly higher degree of orientation (Adhikari et al. 2004). Recently, Roache et al. developed a simple technique to allow for the lift-off and subsequent transfer of poly(styrene-*block*-ethyleneglycol) films to TEM grids. This method offers better control over film processing than dip coating the TEM grid (Roache et al. 2015).

Together with TEM, SEM investigations have been useful in the description of diblock copolymer polystyrene-*b*-poly(acrylic acid) (PS-*b*-PAA) particle formation (Figure 8.7), namely from vesicles and lamellae to unclosed and closed porous particles and on until final porous particles. Thus, Yu et al. have highlighted that these self-assembled microspheres have monodispersed nanopores that contribute to their high sorption capacity and sustained release behavior (Yu et al. 2014).

SEM also has been used, for example, to establish the orientation of lamellar formations for poly(styrene-*block*-4-trimethylsilylstyrene) (PS-*b*-PTMSS) and

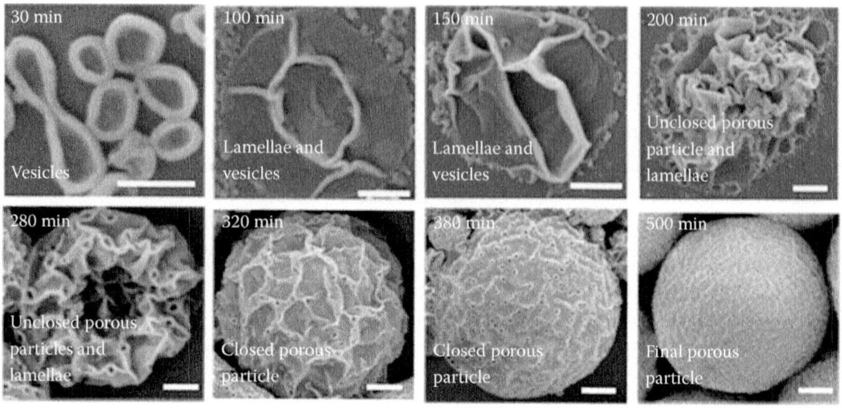

FIGURE 8.7 SEM images of a single PS-*b*-PAA nanoporous particle formation. (Adapted from Yu Haizhou et al. *Nature Communications*. 5: 4110(1–10), 2014.)

poly(styrene-*block*-methyltrimethylsilylmethacrylate) (PS-*b*-PTMSM) (Maher et al. 2014) or to study the effect of polymer size on the structures of some membranes fabricated from four poly(isoprene-*b*-styrene-*b*-4-vinylpyridine) triblock terpolymers with similar block volume fractions but different molar mass (Dorin et al. 2014).

In addition, atomic force microscopy can be successfully used to map the structure and properties of block copolymers by evaluating the sample topography or differences in the viscoelastic properties. This technique is also suited for the study of thin films deposited onto flat surfaces. In general, the AFM contrast among the components of the block copolymer arises from differences in the local nanomechanical properties. Using AFM tapping-mode phase imaging, systems were successfully imaged where one of the components is much stiffer than the other at room temperature, such as in styrene/butadiene, methyl alkyl acrylate/methacrylate, and styrene/isoprene systems. Thus, using AFM, Silverstein et al. described the post-polymerization modifications of polystyrene-*block*-poly(1,2-butadiene) (PS-*b*-PB) diblock and PS-*b*-PB-*b*-PS triblock copolymers. Disordered cylindrical nanostructures were observed for the parent PS-*b*-PB copolymer (film cast from CH_2Cl_2). Similar cylindrical patterned nanostructures also were highlighted for BOC-cysteamine modified PS-*b*-PB block copolymers (film cast from DMF) or for thioglycolic acid modified PS-*b*-PB block copolymers (film cast from DMF) (Silverstein et al. 2012).

With atomic force microscopy, Tong et al. compared the morphology of some poly(methylmethacrylate)-*b*-poly(alkylacrylate)-*b*-poly(methylmethacrylate) triblock copolymers, namely poly(methylmethacrylate)-*b*-poly(isooctyl acrylate)-*b*-poly(methylmethacrylate) (MIM), poly(methylmethacrylate)-*b*-poly(*n*-butylacrylate)-*b*-poly(methylmethacrylate) (MnBM), poly(methylmethacrylate)-*b*-poly(*n*-propylacrylate)-*b*-poly(methyl methacrylate) (MnPM), and poly(methylmethacrylate)-*b*- and poly(ethylacrylate)-*b*-poly(methylmethacrylate) (MEM), using a procedure designed to image microdomains of different mechanical properties. The amplitude and phase of the oscillating tip were recorded simultaneously at each point of the surface, with a lateral resolution of a few nanometers, which provided a topographic image and cartography of the mechanical properties of the sample surface. This capability relies on the difference in modulus of the thermoplastic microdomains of PMMA and the rubbery phases of poly(alkyl acrylate). The phase separation was observed for all the investigated triblock copolymers, except for the MEM and the MnPM triblocks of 10,000–50,000–10,000 MW (Tong et al. 2001).

Schricker et al. characterized the surface morphology of poly(methyl methacrylate)–poly(acrylic acid) (PMMA-*b*-PAA) and poly(methyl methacrylate)–poly(2-hydroxyethyl methacrylate) (PMMA-*b*-PHEMA) block copolymers using AFM. It was observed that block copolymers have a distinct morphology that is absent in homopolymer films, the images collected using both a phase and height mode demonstrating that the structure is due to the phase separation of the block copolymer. Also, comparing the results obtained for a PMMA-*b*-PAA diblock copolymer with those for a PAA-*b*-PMMA-*b*-PAA triblock copolymer such that for both the MMA to AA ratio is 1:1, distinct morphologies can be distinguished, with the triblock displaying a ring structure. In order to find the influence of block copolymer morphology on protein adhesion, the AFM tips were functionalized with proteins (fibronectin, bovine serum albumin, and collagen), and the adhesion of that particular protein to polymer surfaces was measured in phosphate buffer solution (PBS). According to height and corresponding adhesive force maps, there was a large increase in the average adhesive force of the fibronectin and bovine serum albumin tips from the random copolymer to the diblock and triblock. Also they use antibody-functionalized atomic force microscopy tips to determine how the surfaces will affect the conformation of fibronectin, an important parameter in evaluating surface biocompatibility (Schricker et al. 2012).

AFM can even be employed for block copolymers with components that have very similar properties, such as polystyrene and poly(methyl methacrylate). Both tapping-mode AFM height image and corresponding phase image of a compositionally asymmetric PS/PMMA diblock copolymer (70/30) showed a clearly visible hexagonal arrangement of the PMMA cylinders in the PS matrix (Michler 2008).

Film thicknesses influence the arrangement of PMMA and PS components in PS-*b*-PMMA copolymers. Ham et al. (2008) observed a loss of orientation or order in thicker films of PS-*b*-PMMA deposed on substrates modified with random copolymers (PS-r-PMMA). For values of the block copolymer thickness close to polymer domain spacing (35.4 nm), distinct PMMA cylinders in the PS matrix can be observed (Albert and Epps 2010).

8.5 CONCLUSIONS

Multiphase polymer systems—including a wide range of materials such as composites, block copolymers, blends, alloys, and gels—have gained much interest due to their special properties. In this context, the characterization of the morphology, composition of the different phases, and local mechanical properties (friction force, adhesion force, stiffness, Young's modulus) in multiphase polymer films is crucial for understanding the behavior of these materials under different conditions. The most effective methods that can be used for these purposes are electron microscopy techniques (such as scanning electron microscopy and transmission electron microscopy) and atomic forces microscopy (height imaging, phase imaging, lateral force imaging, adhesion force mapping, modulus mapping, and indentation mapping).

ACKNOWLEDGMENTS

This work was supported by a grant from the Romanian National Authority for Scientific Research, CNCS-UEFISCDI, project no PN-II-ID-PCE-2011-3-0937, 2016 stage.

REFERENCES

Ade Harald Werner, Winesett Donald Andrew, Smith A.P., Qu S., Ge S., Sokolov Jonathan, and Rafailovich Miriam. Phase segregation in polymer thin films: Elucidations by X-ray and Scanning Force Microscopy. *Europhys. Lett.* 45 no. 4 (1999): 526–532.

Adhikari Rameshwar, Huy Trinh An, Buschnakowski Matthias, Michler Goerg H., and Knoll Konrad. Asymmetric PS-block-(PS-co-PB)-block-PS block copolymers: Morphology formation and deformation behavior. *New J. Phys.* 6 no. 1 (2004): 28(1–20).

Adhikari Rameshwar, and Michler Goerg. Influence of molecular architecture on morphology and micromechanical behavior of styrene/butadiene block copolymer systems. *Prog. Polym. Sci.* 29 no. 9 (2004): 949–986.

Affrossman Stanley, Jerome Robert, O'Neill Scott, Schmitt T., and Stamm Manfred. Surface structure of thin film blends of polystyrene and poly(n-butyl methacrylate). *Colloid Polym. Sci.* 278 no. 10 (2000): 993–999.

Albert Julie, and Epps Thomas. Self-assembly of block copolymer thin films. *Mater. Today.* 13 no. 6 (2010): 24–33.

Baker W., Scott C.E., and Hu G.H. *Reactive polymer blending.* Cincinnati: Hanser, 2001.

Brostow Witold, Holjevac Grguric Tamara, Olea-Mejia Oscar, Rek Vesna, and Unn Jaykumar. Polypropylene + polystyrene blends with a compatibilizer. Part I. Morphology and thermophysical properties. *e-Polymers.* no. 033 (2008): 1–9.

Cappella Brunero. Mechanical properties and adhesion of a micro structured polymer blend. *Polymers.* 3 no. 3 (2011): 1091–1106.

Diaz Monica, Barbosa Silvia, and Capiati Numa. Addition compatibilization of PP/PS blends by tailor-made copolymers. *Polym. Eng. Sci.* 46 no. 3 (2006): 329–336.

Dorin Rachel Mika, Phillip William, Sai Hiroaki, Werner Jörg, Elimelech Menachem, and Wiesner Ulrich. Designing block copolymer architectures for targeted membrane performance. *Polymer.* 55 no. 1 (2014): 347–353.

Elias L., Fenouillot Françoise, Majesté Jean-Charles, and Cassagnau Philippe. Morphology and rheology of immiscible polymer blends filled with silica nanoparticles. *Polymer.* 48 no. 20 (2007): 6029–6040.

Feng Jian-Ming, Liu Xi-Qiang, Bao Rui-Ying, Yang Wei, Xie Bang-Hu, and Yang Ming-Bo. Suppressing phase coarsening in immiscible polymer blends using nano-silica particles located at the interface. *RSC Adv.* 5 no. 91 (2015): 74295–74303.

Galloway Jeffrey, Jeon Hyun, Bell Joel, and Macosko Christopher. Block copolymer compatibilization of cocontinuous polymer blends. *Polymer.* 46 no. 1 (2005): 183–191.

Goldstein Joseph. *Scanning Electron Microscopy and X-ray microanalysis: Third Edition.* New York: Springer, 2003.

Ham Sujin, Shin Changhak, Kim Eunhye, Ryu Du Yeol, Jeong Unyong, Russell Thomas, and Hawker Craig. Microdomain orientation of PS-b-PMMA by controlled interfacial interactions. *Macromolecules.* 41 no. 17 (2008): 6431–6437.

Harrats Charef, Fayt Roger, Jérôme Robert, and Blacher Silvia. Stabilization of a cocontinuous phase morphology by a tapered diblock or triblock copolymer in polystyrene-rich low-density polyethylene/polystyrene blends. *J. Polym. Sci. Polym. Phys.* 41 (2003): 202–216.

Harris Mark, Appel Guenter, and Ade Harald. Surface morphology of annealed polystyrene and poly(methyl methacrylate) thin film blends and bilayers. *Macromolecules.* 36 no. 9 (2003): 3307–3314.

Hasegawa Hirokazu, and Hashimoto Takeji. Chapter14. Self-assembly and morphology of block copolymer systems. In *Comprehensive polymer science: Suppl. 2*, Sundar L. Aggarwal and Saverio Russo (eds.), pp. 497–539. London: Elsevier, 1996.

Killgore Jason, Hurley Donna, Yablon Dalia, Turner Joseph, Yuya Philip, Proksch Roger, Gannepalli Anil, and Tsou Andy. Viscoelastic property mapping with contact resonance force microscopy. *Langmuir.* 27 no. 23 (2011): 13983–13987.

Liu Xi-Qiang, Wang Qian-Yu, Bao Rui-Ying, Yang Wei, Xie Bang-Hu, and Yang Ming-Bo. Suppressing phase retraction and coalescence of co-continuous polymer blends: Effect of nanoparticles and particle network. *RSC Adv.* 4 no. 90 (2014): 49429–49441.

Lyu SuPing, Jones Todd, Bates Frank, and Macosko Christopher. Role of block copolymers on suppression of droplet coalescence. *Macromolecules.* 35 no. 20 (2002): 7845–7855.

Magonov Sergei and Whangbo Myung-Hwan. *Surface analysis with STM and AFM. Experimental and theotetical aspects of image analysis.* Weinheim: Wiley-VCH, 1996.

Maher Michael, Bates Christopher, Blachut Gregory, Sirard Stephen, Self Jeffrey, Carlson Matthew, Dean Leon, Cushen Julia, Durand William, Hayes Colin, Ellison Christopher, and Willson Grant. Interfacial design for block copolymer thin films. *Chem. Mater.* 26 no. 3 (2014): 1471–1479.

Maver Uroš, Maver Tina, Peršin Zdenka, Mozetič Miran, Vesel Alenka, Gaberščček Miran, and Stana-Kleinschek Karin. Polymer characterization with the atomic force microscope. In *Polymer science*, Faris Yılmaz (ed.), pp. 113–132. Rijeka: InTech, 2013.

Michler H. Goerg. *Electron microscopy of polymers.* Berlin: Springer-Verlag, 2008.

Pohjakallio Maija, Aho Timo, Kontturi Kyösti, and Kontturi Eero. Morphology of poly(methyl methacrylate) and polystyrene blends upon Langmuir–Schaefer deposition. *Soft Matter.* 7 no. 2 (2011): 743–748.

Roache Fergus, Radjaini Mazdak, Williams David, Gerrard Juliet, Travas-Sejdic Jadranka, and Malmstrom Jenny. Novel lift-off technique for transmission electron microscopy imaging of block copolymer films. *Ultramicroscopy.* 150 (2015): 71–73.

Scherzer Selina, Pavlova Ewa, Esper Julian, and Starý Zdeněk. Phase structure, rheology and electrical conductivity of co-continuous polystyrene/polymethylmethacrylate blends filled with carbon black. *Compos. Sci. Technol.* 119 no. 23 (2015): 138–147.

Schricker Scott, Palacio Manuel, and Bhushan Bharat. Designing nanostructured block copolymer surfaces to control protein adhesion. *Philos. Trans. Roy. Soc. A.* 370 no. 1967 (2012): 2348–2380.

Silverstein Joshua, Casey Brendan, Natoli Mary, Dair Benita, and Kofinas Peter. Rapid modular synthesis and processing of thiol–ene functionalized styrene–butadiene block copolymers. *Macromolecules.* 45 no. 7 (2012): 3161–3167.

Tong J.D., Leclère Philippe, Doneux C., Brédas Jean-Luc, Lazzaroni Roberto, and Jérôme Robert. Morphology and mechanical properties of poly(methylmethacrylate)-b-poly(alkylacrylate)-b-poly(methylmethacrylate). *Polymer.* 42 no. 8 (2001): 3503–3514.

Tsukruk Vladimir, and Singamaneni Srikanth. *Scanning probe microscopy of soft matter: Fundamentals and practices.* Weinheim: Wiley-VCH, 2012.

Van den Berg R., De Groot H., Van Dijk M.A., and Denley D.R. Atomic force microscopy of thin triblock copolymer films. *Polymer.* 35 no. 26 (1994): 5778–5781.

Walheim Stefan, Boltau Martin, Mlynek Jürgen, Krausch Georg, and Steiner Ullrich. Structure formation via polymer demixing in spin-cast films. *Macromolecules*. 30 no. 17 (1997): 4995–5003.

Winesett Donald Andrew, Ade Harald Werner, Sokolov Jonathan, Rafailovich Miriam, and Zhu Shenghe. Substrate dependence of morphology in thin film polymer blends of polystyrene and poly(methyl methacrylate). *Polym. Int.* 49 no. 5 (2000): 458–462.

Yao Nan and Kung Eugene. Polymer characterization using electron microscopes. In *Industrial applications of electron microscopy*, Zhigang R Li (ed.), Chapter 13, pp. 357–386. New York: Marcel Dekker, 2003.

You Chang-jiang and Jia De-min. Effects of styrene-ethylene/propylene diblock copolymer (SEP) on the compatibilization of PP/PS blends. *Chinese J. Polym. Sci.* 21 no. 4 (2003): 443–446.

Yu Haizhou, Qiu Xiaoyan, Nunes Suzana, and Peinemann Klaus-Viktor. Biomimetic block copolymer particles with gated nanopores and ultrahigh protein sorption capacity. *Nat. Commun.* 5 (2014): 4110(1–10).

Zhou Weilie, Apkarian Robert, Wang Zhong Lin, and Joy David. Fundamentals of scanning electron microscopy (SEM). In *Scanning microscopy for nanotechnology. Techniques and applications*, Weilie Zhou and Zhong Lin Wang (eds.), pp. 1–40. New York: Springer, 2007.

Zhu Shenghe, Liu Ying, Rafailovich Miriam, Sokolov Jonathan, Gersappe Dilip, Winesett Donald Andrew, and Ade Harald Werner. Confinement-induced miscibility in polymer blends. *Nature*. 400 no. 6739 (1999): 49–51.

9 Insights into Depth of Structural Organization Using X-ray Methods

Magdalena Aflori

CONTENTS

9.1 SHORT HISTORY

Small-angle X-ray scattering (SAXS) and X-ray diffraction (XRD) are important techniques for resolving the structure of matter down to atomic resolution. Many papers describing X-ray methods are valuable for pure science or practical applications, and their results are used in various branches of science and technology. Phase and microphase separation induced by different factors (crosslinking, hydration, stretching, and temperature) are studied by small-angle X-ray scattering, while X-ray powder diffraction is a powerful method used in the identification of multiphase systems.

Following the work of Wilhelm Conrad Röntgen, who received the first Nobel Prize in Physics (1901) for discovering X-rays, three major branches of science have developed from the use of this radiation method. The relationship between the relative absorption of the X-ray by matter and the average atomic number and density of the matter formed the basis of *X-ray radiography*, from which various diagnostic methods for medical or industrial use were developed. The second branch, *X-ray crystallography*, is concerned with structure and was developed following the classic work of Max Von Laue (1912) in which diffraction experiments from a single crystal confirmed the wave character of X-rays. The third branch, *X-ray spectrometry*, has been developed over the past 30 years as a routine application for elemental analysis.

The first SAXS observations on amorphous materials, including colloidal solutions and liquid mixtures, were reported in late 1920 (Krishnamurti 1929). Over the years, there have been several X-ray collimation system designs: the four pinhole (or slit) systems, the Kratky collimation system (Kratky 1954), and the use of Bonse-Hart (Bonse and Hart 1965) channel-cut crystals for very small scattering angles.

In 1935, Le Galley (1935) developed the first X-ray powder diffractometer, followed by Parrish and Gordon (1945) and Friedman (1945), who developed Geiger counter spectrometers. In 1947 in North America, Philips introduced the first commercial equipment based on the Geiger counter spectrometers (Jenkins and Snyder 1996).

Every crystalline material gives a unique XRD pattern, a "fingerprint" that enables qualitative analyses to be performed in a multiphase system. Those patterns are listed in a database used for single-phase recognition on a search-match method basis. Early work on quantitative phase analysis by X-ray powder diffraction was reported by Navias (1925) and by Clark and Reynolds (1936). The first attempts to list standard diffraction patterns as a qualitative analysis tool were published in 1936 (Hanawalt and Rinn 1936). Soon afterward (Hanawalt et al. 1938), a file of approximately 1,000 diffraction patterns was published that contained an indexing system based on the use of three-dimensional spacings having the strongest intensities (Jenkins and Snyder 1996). The most common methods based on those works are the *absorption–diffraction method* (Alexander and Klug 1948; Lennox 1957) and the *internal standard method* (Klug and Alexander 1974).

Today, the digital diffraction patterns acquired with well-developed instruments and computer software offer the possibility of performing quantitative phase analysis using all data rather than considering only a few reflections of the strongest lines. The *full-pattern method* fits the entire experimental diffraction pattern with a calculated pattern obtained from a database as a combination of standard diffraction patterns (Smith et al. 1987). The best example of using the full-pattern method in quantitative measurements is the Rietveld method (Rietveld 1969), which minimizes the sum of the weighted squared differences between the observed and calculated intensities at every step in a digital powder pattern. The crystal structure of all phases must be present in the database in order to use this method (Hill and Howard 1987; Bish and Howard 1988).

A review of current X-ray methods applied in multiphase polymer systems is presented in this chapter.

9.2 MULTIPHASE SYSTEMS STUDIED BY X-RAY METHODS

Wide- and small-angle scattering are fully developed X-ray techniques for studying structural features. The theory is complete, the experimental devices are well developed, laboratory systems are commercially available, and dedicated X-ray facilities for special applications exist in several international research centers. Data analysis techniques have improved substantially in the last two decades through computers' use of numerical methods. By nature, SAXS is used in the study of relatively large-scale structures, while wide-angle X-ray diffraction (WAXD) deals mainly

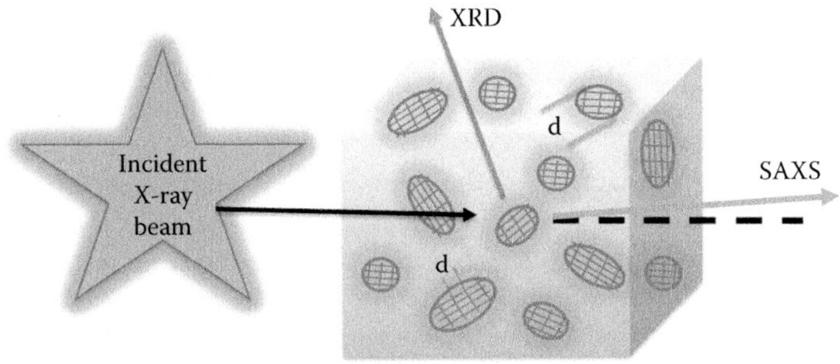

FIGURE 9.1 SAXS and XRD are complementary analysis methods that help in better understanding samples or processes.

with the atomic structure of crystals. SAXS includes not only the diffraction of large lattice spacing—on the order of tens, hundreds, or even thousands of interatomic distances—but also the scattering by perturbed or nonperiodic structures of amorphous and mesormorphic materials. Although the Bragg equation remains the fundamental means of describing WAXD, the scattering (or diffraction) of semicrystalline or amorphous materials is often described in terms of electron density at point x, in reciprocal or Fourier space. A proper structure and morphology analysis of many materials requires information using a combination of SAXS and WAXD (Figure 9.1), even though the WAXD measurements can appear in the normal SAXS q (the magnitude of the scattering wave vector) range because of larger lattice spacing. Usually, in WAXD where crystal lattice planes' diffraction angles are in the range of 3°–170°, Bragg's law is used to derive information concerning molecular structure and atomic arrangement. In this case, the distance between crystal lattice planes is determined using the Ångstrom (Å) scale. SAXS (with angles in the range 0°–3°) is measured in transmission scattering at particles or by electron density changes (Chu and Hsiao 2001).

SAXS is now a well-established standard technique for the investigation of nonperiodic structures with colloidal dimensions from about 10 Å up to several thousand Å, which are enormously large compared to the X-ray wavelength (e.g., the most frequently used CuKα line of 1.54 Å), the angular range of observable scattering being correspondingly small. Any scattering process is characterized by a reciprocity law, which gives an inverse relationship between particle size and scattering angle. Because X-rays are primarily scattered by electrons, we can summarize the situation as follows: SAXS is always observed, and only observed, when electron density inhomogeneities of colloidal size exist in the sample. Here, we will confine our considerations to coherent scattering because incoherent scattering is negligibly weak at very small angles (Chu and Hsiao 2001).

9.2.1 Multiphase Systems—Current Knowledge

In the study of cultural heritage, XRD at small and wide angles is an ideal tool for the characterization of phases present in artifacts from all over the world

(Simionescu et al. 2009). Usually the analysis is carried out using XRD (Bragg–Brentano geometry), which requires artwork sampling (Mahnke et al. 2009; Mannes et al. 2015). However, this destructive technique is not suitable for certain cultural heritage objects. Some scientists (Duran et al. 2008) have used XRD configuration with Göbel mirrors to study rough artifacts without sampling. For bent samples or for obtaining information about the depth profile of some samples, grazing incidence attachments can be used. A simple and inexpensive procedure for attaching self-cleaning and antimicrobial properties is the restoration of stone built heritage items using metallic particle-like nanosized silver or titanium dioxide, due to their photocatalytic features (Simionescu et al. 2010; Bordianu et al. 2015). Investigations of those multiphase systems using XRD resulted in useful techniques such as a method for binding metallic nanoparticles to stone surfaces (Simionescu et al. 2012; La Russa et al. 2016).

Powder XRD and SAXS are considerably advantageous in pigment microanalysis of clay structures in the color layers of paintings and provide reliable quantitative analyses of their relative contents. A large amount of work (Bandyopadhyay and Sinha Ray 2010; Bartczak et al. 2014; Lanson et al. 2015; Hradil et al. 2016) has been carried out that differentiates pure smectites, interstratified illite/smectite structures, and chlorites in paint and preparatory layers of actual paintings. (The results were used to differentiate among earth pigments of various provenances for Baroque paintings and for Early Renaissance preparation layers for gilding.) These studies have also been used to establish general rules for proper discrimination of clay structures (on the base of various characteristic shifts of basal diffraction lines that depend on the type of organic binder and on the relative content of expandable layers in the clay structure).

The use of X-ray techniques in pharmaceutical science has gained more interest through increased collaborations with the materials science community and through increased availability of instrumentation, particularly synchrotron sources. In modeling pharmaceutical systems, X-ray powder diffractometry is used for quantitative characterization, while structural insights into amorphous and nanocrystalline materials have been enabled by coupling X-ray scattering with the pair distribution function. Some authors (Laity et al. 2015) studied a multiphase system containing a commercial granulated clay (as a potential compaction excipient for pharmaceutical tableting in contrast with polymeric pharmaceutical excipients such as spheronized microcrystalline cellulose) and they proposed a model system demonstrating the capabilities of SAXS for investigating compaction in other situations. Using X-ray methods, the variations in compaction around the debussed notch (with a small region of high density below and low density along the flanks) enabled the authors to identify extensive cracking as the cause of many problems in commercial tableting.

The progress of pharmaceutical science also relies on the ability to understand the structural aspects of drugs and drug delivery systems. The preparation method, in addition to sample history, can dictate the nature and the stability of the amorphous phase. A consequence of the physical instability of amorphous phases is their crystallization propensity, and recent X-ray instrumental advances have substantially enhanced our ability to detect and quantify crystallization in a variety of complex matrices (Dong and Boyd 2011).

Although novel antimicrobial drugs, such as antimicrobial peptides, are urgently needed due to the existence of multidrug resistant bacteria, the identification of new modes of action is expensive and time consuming. In addition, current approaches are not conveniently available, so the exploration of how the SAXS method can contribute to classifying the mode of action for novel antimicrobials supports rapid decision making in drug development. Some authors (von Gundlach et al. 2015) used SAXS as a structure sensitive tool that provides rapid feedback on drug-induced ultrastructural alterations in whole *E. coli* cells. By using several well-characterized antimicrobial compounds and the analysis of resulting SAXS curves by principal component analysis, they demonstrated that ultrastructural changes depend on the antibiotics used and on their specific mode of action. The SAXS data were correlated with transmission electron microscope (TEM) images. In contrast to real-space imaging techniques, SAXS allows one to obtain nanoscale information averaged over approximately one million cells. The antimicrobial peptide showed a different mode of action than all tested antibiotics and, therefore, is a good candidate for further drug development.

X-ray crystallography has become increasingly precise and today can provide specific volumes and their differences as a function of temperature, pressure, humidity, and time. The phase transition mechanism obtained by X-ray crystallography assigned heat-exchange data to the transition of one form into another (Céolin and Rietveld 2016). X-ray wide- and small-angle scattering are commonly used for the study of another multiphase system that frequently occurs in polymers: lamellar superstructures. Lamellar to nonlamellar membrane transitions play an important role not only in cell life but also in food processing and in the design of novel drug nanocarriers. Using time-resolved synchrotron X-ray scattering techniques, the formation pathways of these nanostructures were elucidated in great detail, and three different lamellar to nonlamellar phase transitions were detailed (Rappolt 2013).

9.2.2 Theoretical Background

If we consider a two-phase system consisting of two different substances, a and b, respectively, each substance has a constant electron density and is filling the volume, V_a and V_b, respectively, from the total volume V. If the phases are homogenous and have a distinct, sharp interface (small particles embedded in a polymer matrix), the SAXS data analysis consists of resolving the equation:

$$I(\vec{q}) = K(\rho_a - \rho_b)^2 \frac{1}{V} \left| \int_V \sigma(\vec{r}) e^{i\vec{q}\bullet\vec{r}} d^3r \right|^2 \tag{9.1}$$

where K is a constant determined from absolute intensity calibration, $(\Delta\rho)^2 = (\rho_a - \rho_b)^2$ is the scattering contrast that can be calculated if the phases are known, and $\frac{1}{V} \left| \int_V \sigma(\vec{r}) e^{i\vec{q}\bullet\vec{r}} d^3r \right|^2$ is the structure function that contains complete information about the structure of the system (size, shape, and arrangement of the two phases); the value of $\sigma(\vec{r})$ is 1 for $\vec{r}$ in phase a, and its value is 0 for $\vec{r}$ in phase b. The differential cross section per unit volume $\frac{d\Sigma}{d\Omega}\left(cm^{-1}\right)$ is given by the quantity $(\Delta\rho)^2 \frac{1}{V}\left|\int_V \sigma(\vec{r}) e^{i\vec{q}\bullet\vec{r}} d^3 r\right|^2$.

A single particle scattering involves interference from the waves emanating from two points in that particle, separated by a distance r = 1/q. For an isotropic system, one must first consider averaging any starting point in a particle and then averaging any direction for the vector r. One can obtain a double summation that is identical to the determination of the moment of inertia for a particle. When the electron density is used as the weighting rather than the mass density, this moment of inertia is called the *radius of gyration* (R_g) of the particle. The radius of gyration reflects a second moment in the distribution of the shape and size of a particle (domain) about the mean (Glatter and Kratky 1982).

The parameter R_g is obtained in two steps—the first is averaging all possible positions in the particle from which a vector "r" can start and be within the particle, and the second step is determining the probability that a randomly directed vector "r" from an arbitrary starting point in the particle will fall in the particle (Glatter and Kratky 1982). The probability p(r) in the vicinity of r (the particle size) can be graphically represented by a Gaussian probability cloud created by the summation of all possible positions of the particle where the center of the probability cloud is in the particle phase.

At very low q, this corresponds to the volume fraction particles squared. At sizes r = 1/q close to the average particle size or radius or gyration, this probability is reflected by a decaying exponential function. The decaying exponential function can be written in terms of r or in terms of q. (The Fourier transform of a Gaussian distribution is a Gaussian distribution.) Such an analysis leads to Guinier's law (Guinier and Fournet 1955), where the average size is reflected in the radius of gyration, Rg. Rg is the moment of inertia for a particle using the electron density rather than the mass as a weighting factor.

$$I(q) = N_p n_e^2 \exp\left(\frac{-q^2 R_g^2}{3}\right) \tag{9.2}$$

9.2.3 Simultaneous SAXS/WAXD

Nowadays, one of the most frequently used methods for real-time study of the structure and morphology changes during phase transformation is the simultaneous SAXS and WAXD technique. The principle of this combined method is quite simple: two position-sensitive detectors are placed in different locations, covering a wide angular range, such as 100 μrad < θ < 0.5 rad, with a maximum four orders of magnitude in q being obtained (Bark et al. 1992; Mathot et al. 1996). The structure, morphology, and orientation of the studied samples determine the positioning of the detectors. Usually, when a minimal blank space between the SAXS and WAXD angular ranges exists, the two detectors are operating synchronously. Many multiphase system problems can be resolved by the use of simultaneous SAXS/WAXD techniques (Martorana et al. 1999). These problems could include crystallization and melting, phase transformation of polymers (Bark and Zachmann 1993; Krueger and Zachmann 1993; Wang et al. 1998), polymerization (Herzberg et al. 1999), and the formation of colloids and gels (Dokter et al. 1994), just to name a few.

Two geometrical configurations can be achieved in simultaneous SAXS/WAXD techniques: one for isotropic systems (using two one-dimensional position-sensitive detectors) and the other one for anisotropic systems (using two area detectors). In a complete review, Bras and Ryan (1998) stated the importance of these techniques for understanding fundamental aspects of phase transformations in the fields of polymer processing. Amenitsch and coworkers reported the design of a wiggler beamline for SAXS/WAXD measurements (Amenitsch et al. 1998) developed specifically for time-resolved structure studies on gels, liquid crystals, (bio)polymers, amorphous materials, muscles, and proteins in solutions. For anisotropic systems, simultaneous two-dimensional SAXS/WAXD measurements have been demonstrated using two area detectors (Bras et al. 1995), but this arrangement is not widely used because the WAXD images collected this way often have limited values for quantitative data analysis.

9.3 CASE STUDY: SILVER-EMBEDDED POLYLACTIC ACID

Some authors have seen improved specific properties of polylactic acid (PLA) films (Rhim 2013) through thin silver (Ag) coating. Meanwhile by changing coating thickness or the amount of coating, some specific properties of the polymers were controlled to make them suitable for a specific application such as packaging for snack foods. The deposited Ag nanoparticles have an efficient reinforcing effect on the thermal, mechanical, and water vapor barrier properties, as well as the antibacterial activity of PLA nanocomposite films for fresh food packaging applications (Yu et al. 2015). However, further research on large-scale production of the film is needed for commercial applications (mainly in the food industry). As an example of a multiphase system studied by X-ray methods, a two-step treatment of polylactic acid for incorporating silver at the polymer surface was carried out in this chapter.

9.3.1 Experimental Procedure

In order to create new functionalities that facilitate the silver-based solution binding on the polymer surface, the film surface was exposed to a plasma treatment. The experiments were performed on an EMITECH RF plasma device using helium and oxygen as background gas at a pressure, $p = 10^{-5}$ mbar. The power was fixed at 50 W, and the film was immersed for 5 min and 10 min, respectively, into the plasma that fills the gas vessel by diffusion. The sample was immersed for 14 days at room temperature in a sodium hydroxide/silver nitrate ($NaOH/AgNO_3$) solution, protected from light. After two weeks, the sample was rinsed with deionized water and subsequently investigated by different X-ray characterization techniques. The mechanisms involved in this kind of treatment have been described elsewhere (Aflori 2014;

Miron et al. 2014). These X-ray methods are useful tools in characterizing the treated polymer because the antibacterial activity of silver nanoparticles depends not only on their quantity and size (Carlson et al. 2008) but also on their shape (Pal et al. 2007) and level of aggregation (Lok et al. 2007) because the generation of large, spacious aggregates leads to a loss in antibacterial activity.

9.3.2 XRD Measurements

Diffraction data were obtained on a thick layer of powder samples using a Bruker D8 Advance Bruker AXS instrument (Bragg-Brentano geometry, diffractometer circle radius 250 mm) with CuKα (1.5406 Å) radiation and a scintillation detector at room temperature. Scans were collected under the following conditions: divergence slit 0.6 mm fixed, antiscatter slit 0.6 mm fixed, receiving slit 0.2 mm fixed, step width 0.01°, counting time per step 1 s, and instrument parameters 36 KV, 40 mA. The beam was collimated, compressed, and frequency filtered by a Göbel mirror. Phase identification was performed using DiffracPlus EVA and TOPAS (Coelho 2005) software from the Bruker DiffracEvaluation Package (Bruker AXS 2005) and the International Centre for Diffraction Data (ICDD) PDF2-2005 database (ICDD 2005).

The results of the XRD measurement for the PLA and the silver-treated PLA films are shown in Figure 9.2. For comparison, the pattern of XRD peaks of the $AgNO_3$ powder was shown as a reference in the same figure. The first high-intensity broad peak was observed at 2 q of 16.56° in the pristine polymer and, in silver-treated films, is attributed to the amorphous structure of PLA base film (Shameli 2010).

At the bottom of Figure 9.2, the reference peaks from the ICDD database for $AgNO_3$ are marked with red and, for Ag, with blue (ICDD PDF-2005 Database: Files 00-043-0649 for $AgNO_3$ and 00-001-1167 for Ag). Based on the quantitative

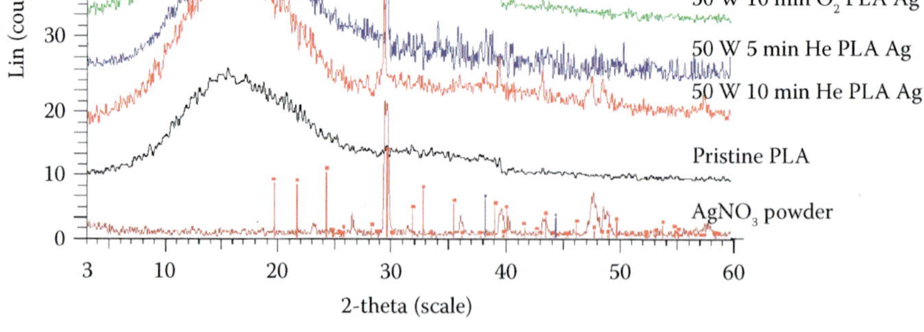

FIGURE 9.2 Phase identification from XRD patterns of all PLA samples.

results of the diffraction peaks, it can be concluded that a higher amount of silver was present in the polymer-treated samples for 50 W 10 min He plasma PLA Ag. In the patterns of the treated samples, two crystalline peaks at 38.17° and 44.37° are present; these are attributed to (111) and (200) crystallographic planes of face-centered cubic silver crystals, respectively (Martínez-Castañón et al. 2008; Shameli 2010).

9.3.3 SEM and EDAX Measurements

SEM investigations of all samples were performed with a QUANTA 200 instrument, the specimens being fixed using adhesive paste on aluminum (Al) conducting supports of cylindrical shapes. Prior to SEM analysis, untreated and modified PLA substrates were sputter-coated with an ultra-thin layer of gold (JEOL-JFC) for 30 s at 20 mA. The SEM system was equipped with a dispersive X-ray spectroscopy (EDAX) module in order to obtain qualitative elemental information.

From SEM pictures (Figure 9.3a), the pristine polymer shows homogenous, flat, uniform surfaces without any important imperfections. Following the two-step treatments, significant changes in the surface topography induced by the silver embedding were evidenced. The two kinds of plasmas (in helium and in oxygen) produced different morphologies of the polymer surfaces. Helium–plasma treatments followed by $AgNO_3$ treatments (Figure 9.3b) showed a relatively homogenous distribution of the silver nanoparticles on granulated surfaces. For the oxygen–plasma (Figure 9.3c) treatments, the surfaces were homogenous and the silver nanoparticles embedded on the polymer matrix.

From EDAX measurements, the highest amount of silver (Figure 9.3 and Table 9.1) was obtained from a helium–plasma treatment for 50 W 10 min and the lowest one from an oxygen–plasma treatment for 50 W 5 min.

9.3.4 SAXS Measurements

The silver-embedded PLA films were studied by SAXS measurements using a Bruker Nanostar instrument having a microfocus X-ray source with copper (Cu) radiation and operation power 30 W and using a VÅNTEC-2000 two-dimensional X-ray detector with a 140 mm × 140 mm detector window. Precise sample positioning in the X-ray beam is achieved by determining the sample transmission as a function of position (nanography). The sample is scanned perpendicular to the beam, and the transmission is determined at each x,y position and displayed as a color-coded map; the user can then select a proper position(s) for the data collection. The system also contains spot focus optics for sealed tube, three-pinhole collimation for the best signal-to-noise ratio. The sample-to-detector distance is 1,070 mm.

The two-dimensional images acquired for all studied samples and processed in Bruker Leptos software are present in Figure 9.4.

The integrated two-dimensional data from Figure 9.4 transformed into a one-dimensional pattern in a logarithmic scale are presented in Figure 9.5. In this figure, well-defined peaks were observed for the treated samples. The peaks can be either the effect of well-dispersed particles in size (form factor) or in a good arrangement (structure factor).

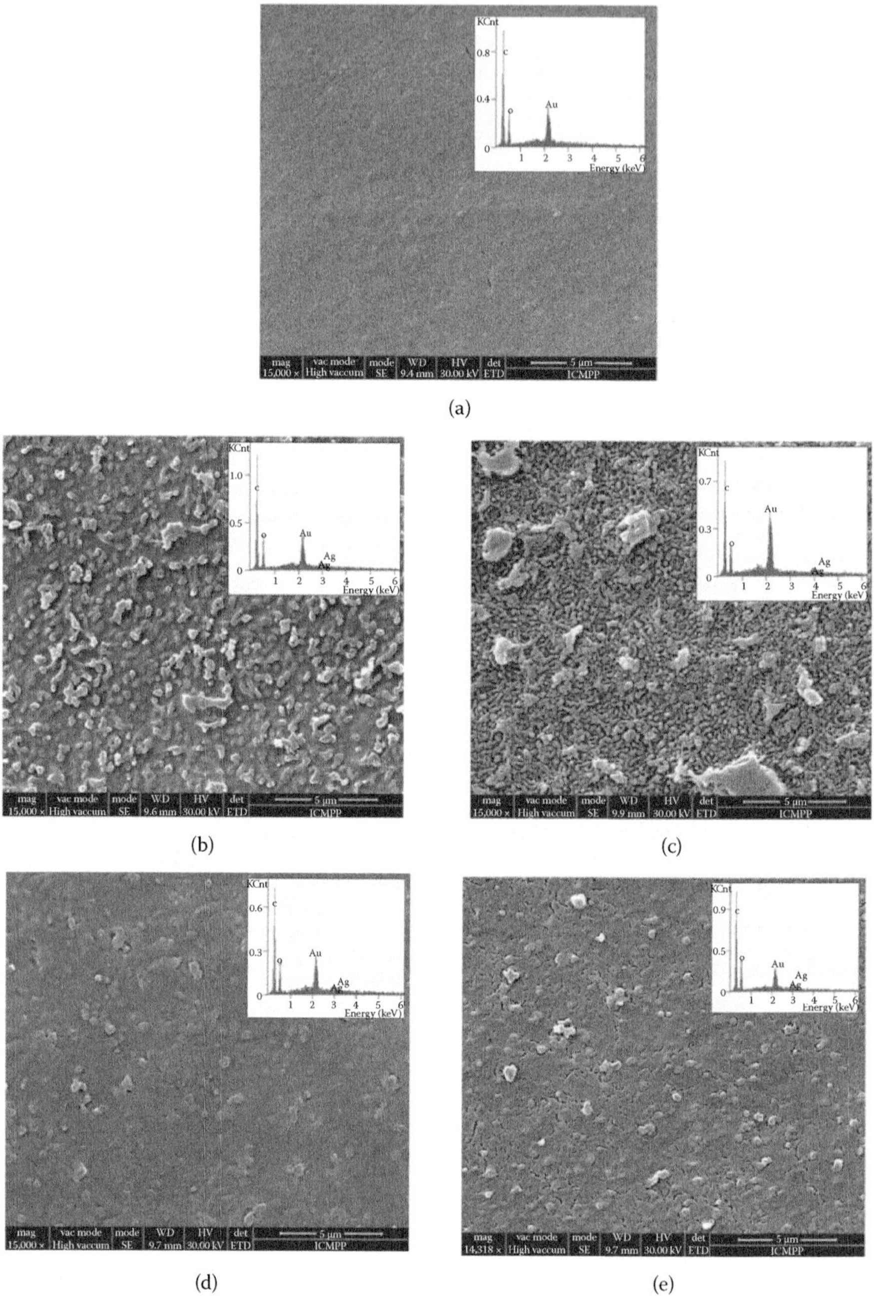

FIGURE 9.3 SEM–EDAX measurements for (a) pristine PLA; (b) 50 W 5 min He plasma PLA Ag; (c) 50 W 10 min He plasma PLA Ag; (d) 50 W 5 min O_2 plasma PLA Ag; and (e) 50 W 5 min O_2 plasma PLA Ag.

TABLE 9.1
Results of Elemental Analysis from EDAX Measurements for All PLA Samples (Original Data)

Sample	C (%)	O (%)	Ag (%)	Au (%)
Pristine PLA	68.37	21.08	–	10.55
50 W 5 min He plasma PLA Ag	69.74	21.62	00.19	8.45
50 W 10 min He plasma PLA Ag	64.40	19.21	00.32	16.07
50 W 5 min O_2 plasma PLA Ag	68.01	20.23	00.23	11.53
50 W 10 min O_2 plasma PLA Ag	67.86	23.29	00.16	8.69

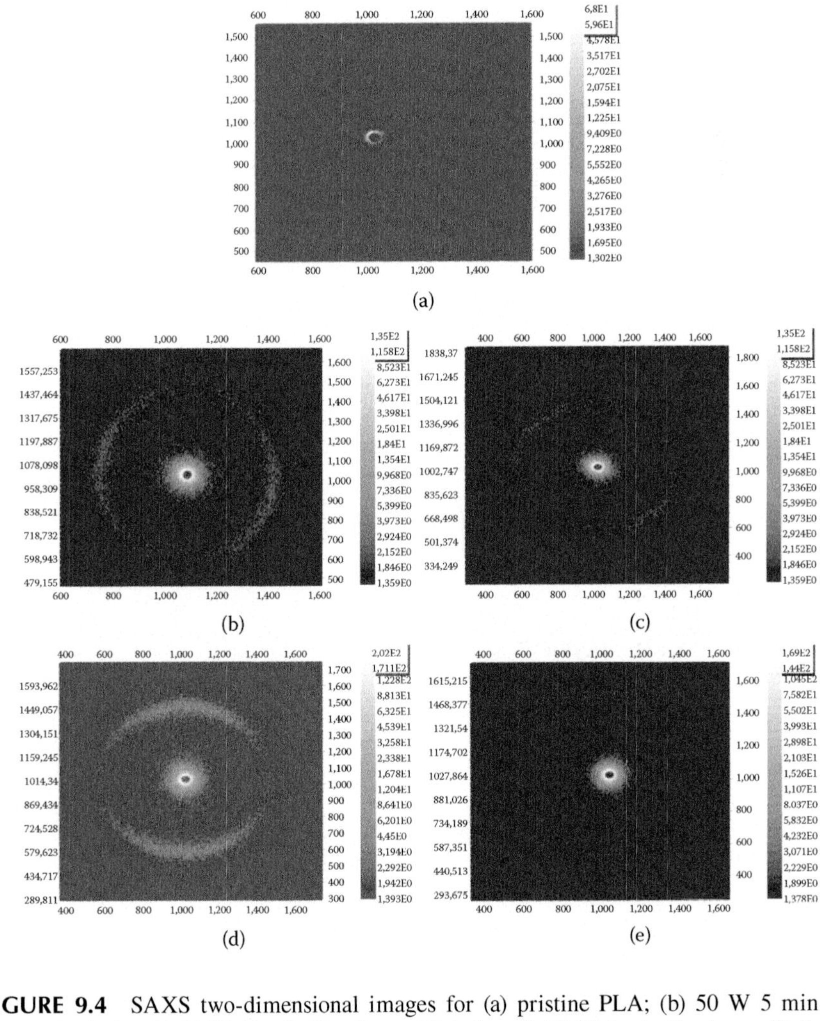

FIGURE 9.4 SAXS two-dimensional images for (a) pristine PLA; (b) 50 W 5 min He plasma PLA Ag; (c) 50 W 10 min He plasma PLA Ag; (d) 50 W 5 min O_2 plasma PLA Ag; and (e) 50 W 5 min O_2 plasma PLA Ag.

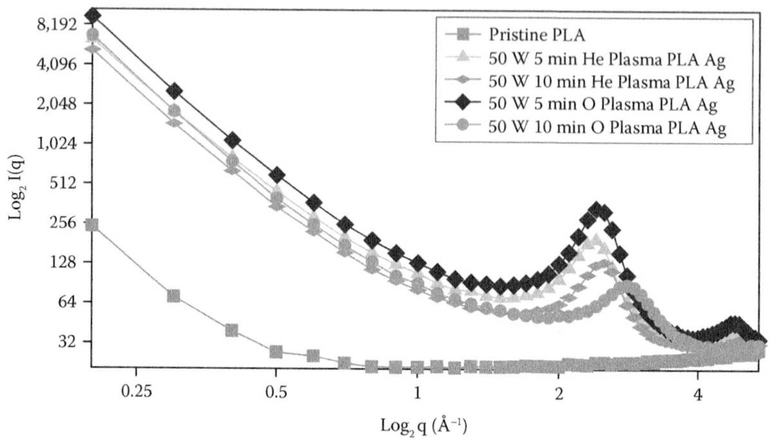

FIGURE 9.5 The X-ray intensity profile for all PLA samples from SAXS-integrated two-dimensional data.

A Guinier–Porod empirical model can be used to fit shapes such as spheres—in our case, silver nanoparticles. Figure 9.6 and Table 9.2 illustrate the results of the empirical model application used to calculate size and dimensionality of scattering objects for a generalized Guinier/power law object. The Guinier–Porod fit was obtained from Igor Pro software (Igor Pro).

Using the Igor Pro program, the returned value of this empirical model is P(Q) as written in Equations 9.3 and 9.4, plus the incoherent background term. The result is in the units of [$cm^{-1}sr^{-}$] absolute scale. The following functional form is used:

$$I(q) = \frac{G}{q'} \exp\left(\frac{-q^2 R_g^2}{3-s}\right) \quad \text{for } q \leq q_1 \tag{9.3}$$

$$I(q) = \frac{D}{q^m} \exp\left(\frac{-q^2 R_g^2}{3-s}\right) \quad \text{for } q \geq q_1 \tag{9.4}$$

where m is the Porod exponent. This is based on the generalized Guinier law for such elongated objects (Glatter and Kratky 1982). For three-dimensional globular objects (such as spheres), s = 0 and one recovers the Guinier formula (Equation 9.2). For two-dimensional symmetry (such as for rods), s = 1, and for one-dimensional symmetry (such as for lamellae or platelets), s = 2. A dimensionality parameter 3-s is defined, and it is 3 for spherical objects, 2 for rods, and 1 for plates.

Enforcing the continuity of the Guinier and Porod functions and their derivatives yields

$$q_1 = \frac{1}{R_g}\sqrt{\frac{(m-s)(3-s)}{2}} \tag{9.5}$$

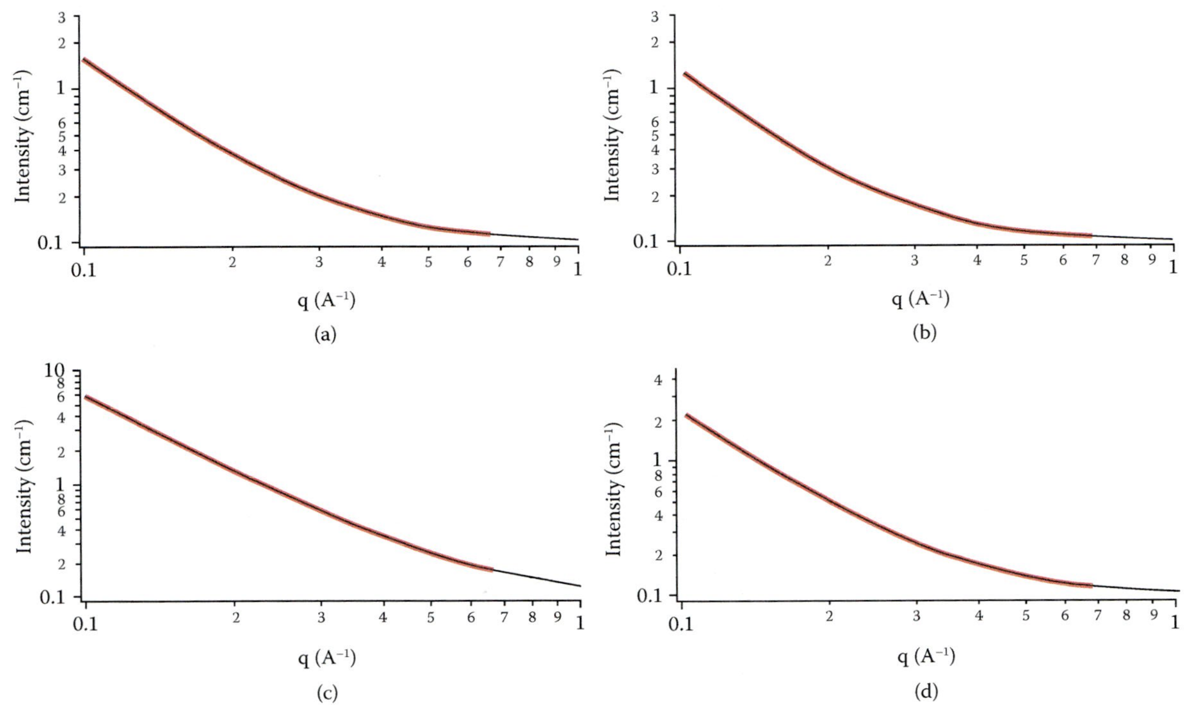

FIGURE 9.6 The Guinier–Porod model for (a) 50 W 5 min He plasma PLA Ag; (b) 50 W 10 min He plasma PLA Ag; (c) 50 W 5 min O_2 plasma PLA Ag; and (d) 50 W 5 min O_2 plasma PLA Ag. The experimental data is in red, and the fitting model is in black.

TABLE 9.2
SAXS Parameters Obtained by Fitting Experimental Data with Guinier–Porod Model (Original Data)

Sample	R_g (Å)	Porod Exponent
50 W 5 min He plasma PLA Ag	59.39	2.50
50 W 10 min He plasma PLA Ag	64.27	2.56
50 W 5 min O_2 plasma PLA Ag	41.70	2.28
50 W 5 min O_2 plasma PLA Ag	56.43	2.47

and

$$D = G\exp\left(\frac{-q_1^2 R_g^2}{3-s}\right) q_1^{(m-s)} = \frac{G}{R_g^{m-s}} \exp\left[-\frac{(m-s)}{2}\right]\left(-\frac{(m-s)(3-s)}{2}\right)^{\frac{(m-s)}{2}} \quad (9.6)$$

Table 9.2 presents the parameters obtained by using the described Guinier–Porod model. Using Equations 9.2, 9.3, and 9.4, the radius for spheres (Baur and Gerold 1964) can be obtained by using the formula $R = \sqrt{\frac{5}{3}} R_g$.

9.4 REMARKS AND FUTURE DIRECTIONS

The tunability of X-ray wavelength over a range near the K or L edge of an element has made SAXS/WAXD measurements a feasible approach for structure and morphology investigations of a specific element in the presence of other elements. The data collected by scattering and diffraction methods must be analyzed with varying mathematical operations to construct the information from reciprocal to real space, and very often the solution of this transformation is not unique. Therefore, a modeling approach has to be developed to extract the structural information. Today, the advantage of performing synchrotron X-ray experiments has advanced research opportunities in the fields of materials science, chemistry, physics, biology, and engineering. The field of nanoscience and nanotechnology (nanoparticles, nanocomposites, and biological systems) has become a major research thrust in all of the aforementioned domains whereby X-ray techniques are proven to be powerful tools for characterizing the structure, morphology, and dynamics of these materials. In time-resolved scattering experiments, the data collection rate can be overwhelming due to the presence of more intense sources (synchrotron X-rays) and detectors with a better resolution ((large charge-coupled device [CCD] area detectors). Most scattering software programs are lacking in transportability or generality, and software development for scattering analysis has fallen behind the experimental capability. The processing of very large scattering data sets needs to be a continuous effort, from an interdisciplinary standpoint (computer science, physicists, chemists, biologists, and so on), in order to tackle the software needs. Many dynamic experiments on transient properties that were not possible before now can be carried out in real time (with resolution in seconds to milliseconds) and/or in situ. The interpretation of SAXS/WAXD could be improved by the coupling of additional variables such as mechanical or thermodynamic parameters. For resolving complex problems, it is desirable to combine SAXS/WAXD with other experimental techniques, such as microscopy, Fourier transform infrared spectroscopy, laser light scattering, Raman spectroscopy, thermal analysis, and different forms of rheological techniques (Chu and Hsiao 2001).

REFERENCES

Aflori, M. Surface characterization of peritoneal dialysis catheter containing silver nanoparticles. *Rev. Roum. Chim.* 59 no. 6–7 (2014): 523–526.

Alexander, L.E., and Klug, H.P. X-ray diffraction analysis of crystalline dusts. *Anal. Chem.* 20 (1948): 886–894.

Amenitsch, H., Bernstorff, S., Kriechbaum, M., et al. The small-angle X-ray scattering beamline ateElettra: A new powerful station for fast structural investigations on complex fluids with synchrotron radiation. *Nuovo Cimento Soc. Ital. Fis. D.* 20 no. 12 (1998): 2181–2190.

Bandyopadhyay, J., and Sinha Ray, S. The quantitative analysis of nano-clay dispersion in polymer nanocomposites by small angle X-ray scattering combined with electron microscopy. *Polymer.* 51 no. 6 (2010): 1437–1449.

Bark, M., and Zachmann, H.G. Simultaneous measurements of small-angle X-ray-scattering, wide-angle X-ray-scattering and heat-exchange during crystallization and melting of polymers. *Acta Polym.* 44 no. 6 (1993): 259–265.

Bark, M., Zachmann, H.G., Alamo, R., and Mandelkern, L. Investigations of the crystallization of polyethylene by means of simultaneous small-angle and wide-angle X-ray-scattering. *Makromol. Chem.* 193 no. 9 (1992): 2363–2377.

Bartczak, Z., Rozanski, A., and Richert, J. Characterization of clay platelet orientation in polylactide–montmorillonite nanocomposite films by X-ray pole figures. *Eur. Polym. J.* 61 (2014): 274–284.

Baur, R., and Gerold, V. Comparative X-ray and electron-microscopic measurement of the size of Guinier-Preston zones in an aluminum-silver alloy. *Acta Metall.* 12 no. 12 (1964): 1449–1453.

Bish, D.L., and Howard, S.A. Quantitative phase analysis using the Rietveld method. *J. Appl. Crystallogr.* 21 no. 2 (1988): 86–91.

Bonse, U., and Hart, M. Principles and design of Laue-case X-ray interferometers. *Z. Phys.* 188 no. 2 (1965): 154–164.

Bordianu, I.E., David, G., Simionescu, B., et al. Functional silsesquioxane-based hierarchical assemblies for antibacterial/antifungal coatings. *J. Mater. Chem. B.* 3 (2015): 723–727.

Bras, W., Mant, G.R., Derbyshire, G.E., et al. Real-time simultaneous wide-angle and small-angle fiber diffraction. *J. Synchrotron Radiat.* 2 (1995): 87–92.

Bras, W., and Ryan, A.J. Sample environments and techniques combined with small angle X-ray scattering. *Adv. Colloid Interface Sci.* 75 (1998): 1–43.

Carlson, C., Hussain, S.M., Schrand, A.M., et al. Unique cellular interaction of silver nanoparticles: Size-dependent gencration of reactive oxygen species. *J. Phys. Chem. B.* 112 no. 43 (2008): 13608–13619.

Céolin, R., and Rietveld, I.B. X-ray crystallography, an essential tool for the determination of thermodynamic relationships between crystalline polymorphs. *Ann. Pharm. Fr.* 74 no. 1 (2016): 12–20.

Chu, B., and Hsiao, B.S. Small-angle X-ray scattering of polymers. *Chem. Rev.* 101 no. 6 (2001): 1727–1761.

Clark, G.L., and Reynolds, D.H. Quantitative analysis of mine dusts-an X-ray diffraction method. *Ind. Eng. Chem. Anal. Ed.* 8 no. 1 (1936): 36–40.

Dokter, W.H., Beelen, T.P.M., van Garderen, H.F., et al. Simultaneous monitoring of amorphous and crystalline phases in silicalite precursor gels. An *in situ* hydrothermal and time-resolved small- and wide-angle X-ray scattering study. *J. Appl. Crystallogr.* 27 no. 6 (1994): 901–906.

Dong, Y.-D., and Boyd, B.J. Applications of X-ray scattering in pharmaceutical science. *Int. J. Pharmaceutics.* 417 (2011): 101–111.

Duran, A., Herrera, L.K., Jimenez de Haro, M.C., Justo, Á., and Perez-Rodriguez, J.L. Non-destructive analysis of cultural heritage artefacts from Andalusia, Spain, by X-ray diffraction with Göbel mirrors. *Talanta.* 76 no. 1 (2008): 183–188.

Friedman, H.A. Geiger counter spectrometer for industrial research. *Electronics.* 18 (1945): 132–137.

Glatter, O., and Kratky, O. *Small angle X-ray scattering.* London: Academic Press, 1982.

Guinier, A., and Fournet, G. *Small angle scattering of X-rays.* New York: Wiley, 1955.

Hanawalt, J.D., and Rinn, H.W. Identification of crystalline materials. *Ind. Eng. Chem. Anal. Ed.* 8 no. 4 (1936): 244–247.

Hanawalt, J.D., Rinn, H.W., and Frevel, L.K. Chemical analysis by X-ray diffraction—Classification and use of X-ray diffraction patterns. *Ind. Eng. Chem. Anal. Ed.* 10 no. 9 (1938): 457–512.

Herzberg, O., Gehrke, R., and Epple, M. Combined in-situ small and wide-angle synchrotron X-ray scattering (SAXS-WAXS) applied to a solid-state polymerization reaction. *Polymer.* 40 no. 2 (1999): 507–511.

Hill, R.J., and Howard, C.J. Quantitative phase analysis from neutron powder diffraction data using the Rietveld method. *J. Appl. Crystallogr.* 20 no. 6 (1987): 467–474.

Hradil, D., Bezdička, P., Hradilová, J., and Vašutová, V. Microanalysis of clay-based pigments in paintings by XRD techniques. *Microchem. J.* 125 (2016): 10–20.

ICDD PDF-2 Database. Kabekkodu Soorya, International Centre for Diffraction Data, Newton Square, PA, 2005.

Igor Pro Software (2013). https://www.wavemetrics.com/products/igorpro/manual.htm

Jenkins, R., and Snyder, R.L. *Introduction to X-ray powder diffractometry.* New York: Wiley, 1996.

Klug, H.P., and Alexander, L.E. *Xray diffraction procedures.* New York: Wiley, 1974.

Kratky, O. Neues verfahren zur herstellung von blendenstreuungsfreien Röntgen-kleinwinkelaufnahmen. *Z. Elektrochem.* 58 (1954): 49–53.

Krishnamurti, P. X-ray diffraction in liquid mixtures. *Proc. Indian Assoc. Cultiv. Sci.* 12 (1929): 331–355.

Krueger, K.N., and Zachmann, H.G. Investigation of the melting behavior of poly(aryl ether ketones) by simultaneous measurements of SAXS and WAXS employing synchrotron-radiation. *Macromolecules.* 26 no. 19 (1993): 5202–5208.

Laity, P.R., Asare-Addo, K., Sweeney, F., Šupuk, E., and Conway, B.R. Using small-angle X-ray scattering to investigate the compaction behaviour of a granulated clay. *Appl. Clay Sci.* 108 (2015): 149–164.

Lanson, B., Ferrage, E., Hubert, F., et al. Experimental aluminization of vermiculite interlayers: An X-ray diffraction perspective on crystal chemistry and structural mechanisms. *Geoderma.* 249–250 (2015): 28–39.

La Russa, M.F., Rovella, N., de Buergo, M.A., et al. Nano-TiO_2 coatings for cultural heritage protection: The role of the binder on hydrophobic and self-cleaning efficacy. *Progr. Org. Coating.* 91 (2016): 1–8.

Le Galley, D.P. A type of Geiger-Müller counter suitable for the measurement of diffracted Mo K X-rays. *Rev. Sci. Instrum.* 6 (1935): 279–283

Lennox, D.H. Monochromatic diffraction absorption technique for direct quantitative X ray analysis. *Anal. Chem.* 29 no. 5 (1957): 766–772.

Lok, C.N., Ho, C.M., Chen, R., et al. Silver nanoparticles: Partial oxidation and antibacterial activities. *J. Biol. Inorg. Chem.* 12 no. 4 (2007) 527–534.

Mahnke, H.E., Denker, A., and Salomon, J. Accelerators and X-rays in cultural heritage investigations. *Comptes Rendus Phys.* 10 no. 7 (2009): 660–675.

Mannes, D., Schmid, F., Frey, J., Schmidt-Ott, K., and Lehmann, E. Combined neutron and X-ray imaging for non-invasive investigations of cultural heritage objects. *Phys. Proc.* 69 (2015): 653–660.

Martínez-Castañón, G.A., Niño-Martinez, N., Martínez-Gutierrez, F., et al. Synthesis and antibacterial activity of silver nanoparticles with different sizes. *J. Nanopart. Res.* 10 no. 8 (2008): 1343–1348.

Martorana, A., Piccarolo, S., and Sapoundjieva, D. SAXS/WAXS study of the annealing process in quenched samples of isotactic poly (propylene). *Macromol. Chem. Phys.* 200 no. 3 (1999): 531–540.

Mathot, V.B.F., Scherrenberg, R.L., Pijpers, M.F.J., and Bras, W. Dynamic DSC, SAXS and WAXS on homogeneous ethylene-propylene and ethylene-octene copolymers with high comonomer contents. *J. Therm. Anal.* 46 no. 3–4 (1996): 681–718.

Miron, C., Sava, I., Aflori, M., et al. Surface modifications of aromatic polyimides by RF-oxygen plasma and pulsed electrical discharges in water. *J. Optoelectron. Adv. Mater.* 16 no. 9–10 (2014): 1206–1211.

Navias, A.L. Quantitative determination of the development of in fired clays by X-ray method. *J. Am. Ceram. Soc.* 8 (1925): 296–302.

Pal, S., Tak, Y.K., and Song, J.M. Does the antibacterial activity of silver nanoparticles depend on the shape of the nanoparticle? A study of the gram-negative bacterium *Escherichia coli. Appl. Environ. Microbiol.* 73 no. 6 (2007): 1712–1720.

Parrish, W., and Gordon, S.G. Precise angular control of quartz-cutting by X-rays. *Am. Mineral.* 30 (1945): 326–346.

Rappolt, M. Chapter two—Formation of curved membranes and membrane fusion processes studied by synchrotron X-ray-scattering techniques. *Adv. Planar Lipid Bilayers Liposomes.* 17 (2013): 29–54.

Rhim, J.-W. Preparation and characterization of vacuum sputter silver coated PLA film. *LWT—Food Sci. Technol.* 54 (2013): 477–484.

Rietveld, H.M. A profile refinement method for nuclear and magnetic structures. *J. Appl. Crystallogr.* 2 no. 2 (1969): 65–71.

Shameli, K., Ahmad, M.B., Yunus, W.M.Z.W., et al. Silver/poly(lactic acid) nanocomposites: Preparation, characterization, and antibacterial activity. *Int. J. Nanomed.* 5 (2010): 573–579.

Simionescu, B., Aflori, M., and Olaru, M. Protective coatings based on silsesquioxane nanocomposite films for building limestones. *Constr. Build. Mater.* 23 no. 11 (2009): 3426–3430.

Simionescu, B., Bordianu, I.E., Aflori, M., et al. Hierarchically structured polymer blends based on silsesquioxane hybrid nanocomposites with quaternary ammonium units for antimicrobial coatings. *Mater. Chem. Phys.* 134 no. 1 (2012): 190–199.

Simionescu, B., Olaru, M., Aflori, M., and Cotofana, C. Silsesquioxane-based hybrid nanocomposite with self-assembling properties for porous limestones conservation. *High Perform. Polym.* 22 no. 1 (2010): 42–55.

Smith, D.K., Johnson, G.G., Scheible, A., Wims, A.N., Johnson, J.L., and Ulman, G. Quantitative X-ray powder diffraction using the full diffraction pattern. *Powder Diffr.* 2 (1987): 73–77.

von Gundlach, A.R., Garamus, V.M., Gorniak, T., et al. Small angle X-ray scattering as a high-throughput method to classify antimicrobial modes of action. *Biochim. Biophys. Acta (BBA) – Biomembranes* 1858 no. 5 (2015): 918–925.

Wang, W., Schultz, J.M., and Hsiao, B.S. Time-resolved simultaneous SAXS/WAXS studies of peek during isothermal crystallization, melting, and subsequent cooling. *J. Macromol. Sci. Phys.* B37 no. 5 (1998): 667–682.

Yu, H.-Y., Yang, X.-Y., Lu, F.-F., Chen, G.-Y., and Yao, J.-M. Fabrication of multifunctional cellulose nanocrystals/poly(lactic acid) nanocomposites with silver nanoparticles by spraying method. *Carbohydr. Polymer.* 140 (2015): 209–219.

10 Dynamic Mechanical Analysis in Polymeric Multiphase Systems

Mariana Cristea

CONTENTS

10.1 INTRODUCTION

Polymer chemists that work in basic or applied research have knowledge and skills relating to the creation of polymeric structures with controlled characteristics or to the development of end products based on polymers. In the world of polymer chemistry, when dealing with dynamic mechanical analysis (DMA) in polymer investigation, there often are initial misunderstandings. On one side, the method is only thought of as an excellent technique for determining the glass transition temperature of polymers, which is obvious if some requirements related to the appearance of the sample are fulfilled. This is partially true because DMA is a method to determine the relaxation in polymers and, usually, α-relaxation is associated with glass transition. However, DMA offers much more information than just a temperature value. On the other side, the stress–strain curve is sometimes erroneously considered to be the method's expected result. The stress–strain curve is the result of a simple tensile test, with the stress that is applied to the sample increasing linearly in time.

There is an accessible DMA reference for chemists and material scientists that does offer easy familiarization with the method (Menard 1999). However, it is important to stress that DMA belongs to a group of methods that take advantage of the time-dependent properties of polymers. This means that, with an externally input action, the polymer structure undergoes changes, and it takes time to return to equilibrium. When the input is oscillatory, the output is also oscillatory-shaped. Because of the time-dependent properties of polymers, there always is a lag between the input and the output signal. The output signal is analyzed in terms of shift and amplitude. Besides DMA, this group of methods includes dielectric spectroscopy (Runt and Fitzgerald 1997), frequency domain fluorimetry (Lakowicz 2006), and frequency domain magnetic investigations (Fannin 1998).

As with many other techniques, DMA instruments have been optimized and updated to the Digital Age, making them approachable after some basic training. However, a comprehensive survey simply named "dynamic mechanical analysis" would be largely confined to the theoretical aspects of DMA, with few tangible examples. Therefore, most overviews of DMA focus on a particular area: glasses (Liu et al. 1978; Roling 2001; Yoon and Eisenberg 1978), polymers (Gradin et al. 1989; Menard and Menard 2015; Sepe 1998), pharmaceuticals (Jones et al. 2012), gels (Warren et al. 2012; Xie et al. 2008), coatings (Schlesing et al. 2004), food (Jeckle and Becker 2015; Vélez-Ruiz et al. 1997), or adhesives (Caussé et al. 2013).

Polymeric multiphase systems, as their name implies, are composed of several phases (at least two) (Sperling 2006). Also, at least one of the components of the system is a polymer. It is important to mention that even if in the scientific literature the terms multiphase and multicomponent can be found in the same context, they do not have exactly the same meaning (Thomas et al. 2011). A multicomponent system includes more than one component. A multiphase system supposes that phase separation occurs as a result of incompatibility among the components or among different parts of the component structure, for a one-component system. DMA is an important method for studying multiphase polymer systems of different types, such as polymer blends (Badia et al. 2015), interpenetrating polymer networks (Sperling and Hu 2014), (nano)composites (Robertson 2014; Spitalsky et al. 2010), or gels (Choi et al. 2015; Nicotera et al. 2007).

This chapter first will present the principles of dynamic mechanical analysis and some of its facets in polymer investigations and then give an overview of trends in viscoelastic parameters of multiphase polymeric systems in the glassy region, glass transition region, and the rubbery plateau.

10.2 DYNAMIC MECHANICAL ANALYSIS

10.2.1 Theoretical Background

Polymers have viscoelastic properties; they contain both elastic and viscous components. When the solicitation submitted to polymeric materials is mechanical, their time-dependent properties can be examined by creep, stress relaxation, and DMA. In DMA, a sample is subjected to a sinusoidal load (stress or strain), with the frequency ν [Hz], and the response (strain or stress) is recorded. The oscillating load has an amplitude as small as possible to keep the behavior in the linear viscoelastic range. Therefore, the response signal will be also sinusoidal. However, because of the viscoelastic character of the sample, always there is a phase shift between the input load and the response (Figure 10.1).

When the sinusoidal load is the stress, it can be expressed as

$$\sigma = \sigma_0 \sin\omega t \tag{10.1}$$

where σ is the stress at the time t, σ_0 is the stress amplitude, and $\omega = 2\pi\nu$ is the angular frequency.

The strain response signal is

$$\gamma = \gamma_0 \sin(\omega t + \delta) \tag{10.2}$$

where γ is the strain amplitude at the time t and δ is the phase difference between the stress signal and the strain signal.

For a perfect elastic solid, $\delta = 0$, but for a perfect viscous fluid, $\delta = 90^{\circ}$. In the case of a viscoelastic material, $0^{\circ} < \delta < 90^{\circ}$.

The resulting strain signal is analyzed in terms of its amplitude (γ) and shift δ. Due to linear viscoelasticity, the stress amplitude is proportional to the strain amplitude, and a complex modulus is defined as the ratio of the sinusoidal stress to the strain. The complex modulus E* comprises the contribution of both the elastic component and the viscous component:

$$E^* = E' + iE'' \tag{10.3}$$

$$E' = E^*\cos\delta \text{ and } E'' = E^*\sin\delta$$

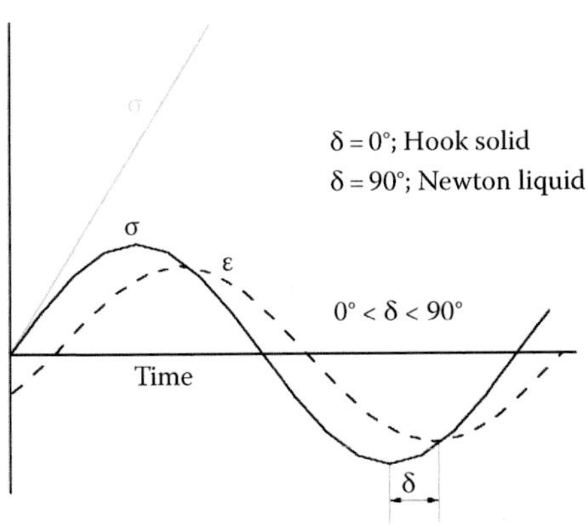

FIGURE 10.1 Schematical representation of the sinusoidal stress load subjected to a polymer sample and of the corresponding sinusoidal strain response. The linear gray line represents the variation of the load in the case of a quasistatic mechanical experiment.

where E′ is the storage (elastic) modulus that is related to the rigidity of the sample, and E″ is the loss (viscous) modulus, correlated with the flowing properties. The ratio between the two moduli represents the loss factor (damping):

$$\tan\delta = \frac{E''}{E'} \tag{10.4}$$

The loss factor tan δ expresses which component is dominant—the elastic or the viscous one. It defines the ratio of the energy dissipated to the energy stored elastically. Some refer to this as internal friction (Blanter et al. 2007). Details on the mathematical data treatment performed to deduce E′, E″, and tanδ can be found in certain references (Chartoff et al. 2009; Mano et al. 2002).

The references related to DMA that have been mentioned so far have reviewed many facets of the procedure. However, the key point when dealing with DMA is to understand it as being part of rheology, defined briefly as the study of the relation between the stress applied on materials and the resulting deformation (De Kee and Wissbrun 1998). Samples are investigated in rheology by using rotational tests and oscillatory tests. Oscillatory tests are also referred to as dynamic mechanical analysis (Mezger 2006). Other names also are used to refer to DMA, such as dynamic mechanical thermal analysis (Badia 2015; Jones 2012) or mechanical spectroscopy (Blanter et al. 2007; Razzaq et al. 2007). Table 10.1 includes the basic characteristics of rheometers and DMA instruments, presented comparatively.

Rheometers are used in a much larger frequency range than DMA, for a large variety of materials, from liquids to soft solids. Their configurations are plate–plate and cone–plate, whereas torsion and tension attachments are optional. Dynamic mechanical analyzers perform only oscillatory experiments on load-bearing samples (i.e., those that have a shape with well-defined dimensions, such as films, fibers, or bars). The samples can be fixed in specific attachments, such as tension, bending, shear, or compression. A new attachment has been developed for DMA that is intended for non–self-supporting samples—a material pocket (Pinheiro and Mano 2009). The material pocket behaves elastically over the studied temperature

TABLE 10.1
DMA Instruments and Rheometers, Similarities and Differences

	Rheometers	DMA Instruments
Measurements	Rotational and oscillatory	Oscillatory
Frequency range	10^{-6} Hz to 1000 Hz	10^{-2} Hz to 100 Hz
Samples	Liquids, soft samples	Load-bearing samples
Fixtures	Plate–plate, cone–plate	Tension, bending, compression, shear
Options	Accessories for tension, torsion	Accessories for liquids, pastes

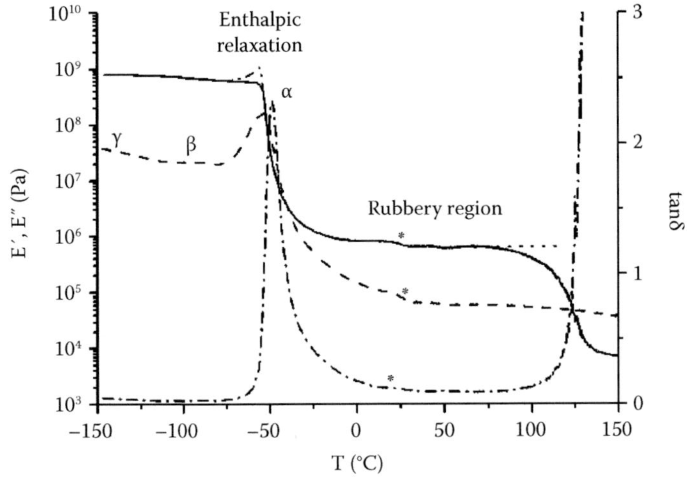

FIGURE 10.2 Schematical representation of the viscoelastic behavior of an elastomer in a single frequency DMA experiment performed by temperature sweep. The star marks a melting peak, and the black dotted line indicates the viscoelastic behavior for a more constrained polymer.

and frequency range, and its modulus is much higher than the sample (by several orders of magnitude). The abbreviations E′ and E″ are used if the sample is subjected to tension, bending, or compression load. When the attachment of choice is shear, the abbreviations for storage modulus and loss modulus are G′ and G″. DMA is sometimes called the rheology of the solid. For a larger perspective, readers are directed to a history of the method and its operating principles (Menard 1999).

Despite the fact that a comprehensive viscoelastic characterization of a polymer means the study of molecular motions over a large range of temperatures and frequencies, the most utilized experiment is an isochronal temperature sweep. Figure 10.2 schematically depicts the result of such an experiment performed on a thermoplastic elastomer.

In the decreasing order of temperatures, the relaxations are denoted α, β, γ, and so on. They are assessed from the drop of E′ modulus or peaks of E″ modulus and loss factor tan δ.

10.2.2 Sub-Glass Transition Region

At low temperature, in the glassy region, the chains are immobilized into a rigid structure due to a very small free volume. Relaxation is the time required to respond to a change in temperature or pressure (Sperling 2006). A relaxation process is detected when its characteristic times are close to the time scale of the experiment. In the glassy region, the polymer has difficulties in responding to the rapid change of the sinusoidal deformation ($\omega \cdot \tau >> 1$, τ: relaxation time). They are very rigid ($E' > 10^9$ Pa) and only local mobility is displayed, represented by the motions of side groups or by backbone movements of bending or stretching (β and γ secondary relaxations). An important observation is that DMA is peculiarly sensitive to secondary relaxations, as opposed to differential scanning calorimetry (DSC). The caloric capacity c_p associated with these relaxations is too small to be sensed by the calorimeters in use. Due to the cooperativity of secondary relaxations, their activation energy (E_a, J/mol) can be determined using the Arrehenius equation, from the slope of the semilogarithmic plot ln ν versus the reciprocal peak temperature 1/T.

$$\nu = A\exp\left(-\frac{E_a}{RT}\right) \tag{10.5}$$

where ν is the frequency (Hz), A is the pre-exponential factor, R is the gas constant ($8.314\ J \cdot mol^{-1} \cdot K^{-1}$), and T is the relaxation absolute temperature (K).

Apart from the secondary relaxations that result from the structural particularities of the investigated polymers, there are some relaxations with unambiguous origins. A faint γ-relaxation close to −150°C is associated with aliphatic methylene segments (Willbourn 1958). Its presence was reported in polyurethanes and polyimides (Cristea et al. 2009, 2011). The presence of traces of water in the polymer is ascertained by a specific γ_w relaxation around −100°C. There is evidences that the position of this relaxation is not connected with the content of water but with microstructural features such as chain packing (Bas et al. 2003). The activation energy of these relaxations is under 50 kJ/mol. However, as the structures become more constrained, due to the chain packing or because the chain is part of a network, $E_a(\gamma_w, \gamma)$ can raise above 100 kJ/mol (Cristea et al. 2009, 2011).

10.2.3 Glass Transition Region

As the temperature increases, long-range coordinated molecular motions are activated, triggering the α-relaxation. The relaxation time of the chains closely follows the rate of oscillatory deformation ($\omega \cdot \tau \cong 1$), and both the moduli and the loss factor greatly depend on the temperature. A considerable drop in E′ modulus takes place and tan δ peak registers a high value (Figure 10.2). Usually, α-relaxation is associated with glass transition.

A few aspects should be pointed out regarding the nomenclature. A transition represents a change of state induced by changing the temperature or pressure (Sperling 2006). When working with DMA, it is more appropriate to use the term "relaxation" instead of "transition." Nevertheless, both are used in routine work, without paying attention to particularities. In addition, the term "dispersion" (in the sense of emission and absorption of energy) is found frequently in the literature (Da Silva et al. 2000). Three approaches are possible for the estimation of the glass transition temperature: the onset temperature (the temperature where E′ starts to decrease as a result of the increased mobility) and two peak temperatures, E″ and tan δ. Despite the fact that the three values are not equivalent, all of them are correct because a polymer does not relax suddenly at a fixed temperature but in a range of temperatures. However, a temperature value also should include an explanation relative to the method of determination and the experimental conditions. Chartoff et al. advocated that E″ peak temperature is the best choice because it is closer to the initial drop of E′ than tan δ peak, where the polymer did not completely lose its structure (Chartoff et al. 2009).

In practice, the temperature associated with the region of glass transition is avoided because of numerous fast transformations occurring there. On the contrary, from a scientific point of view, the glass transition region may be regarded as a fingerprint of the polymer. The qualitative and quantitative trends of the E′, E″, and tan δ curves are well suited for discovering important clues about the structure–property relation.

During glass transition, a drop of E′ modulus of three orders of magnitude is an attribute of a linear amorphous polymer. A smaller drop suggests that there may be some constraints in the structure of the polymer, such as crystalline domains or crosslinks. The broadness of the tan δ peak indicates the structural homogeneity of the polymer. When the tan δ peak is broader (and the E′ drop is less abrupt), there is little structural homogeneity in the polymer. This occurs because the presence in the same macromolecular chain of segments with various mobilities (relaxation times) implies consecutive relaxations at successive temperatures. The temperature range associated with the transition spreads out.

Caution should be exercised when comparing the glass transition temperatures (T_g) obtained by DSC and DMA for at least three reasons. First, the two methods work according to different operating principles. Second, the heating rate used in DSC is much higher than that of DMA. Third, it is compulsory that the sample has a shape in the DMA experiments. The polymers analyzed by DSC mostly are powders such as they resulted from synthesis. All the procedures needed to shape them (film, bar) involve operations (dissolution, casting, high temperature heating, melting, extrusion, etc.) that confer a thermal history to the polymer. From a practical standpoint, it is the same basic compound but not the same material. The hump that is noticed frequently on an E′ curve, just before the onset point that marks the beginning of the glass transition (the dotted gray line in Figure 10.2), is associated with the rearrangement occurring in the polymer chains as soon as they sense some mobility. The stress accumulated during sample processing is relived in this region (Menard 1999; Sikorska et al. 2014). It corresponds to the relaxation enthalpy noticed in DSC.

10.2.4 Experimental Conditions

As stated earlier, the behavior during glass transition is very sensitive to experimental conditions: heating rate, frequency, and the nature of the attachment. DMA is a time-consuming method. In order to avoid the heterogeneity of the sample temperature in bulk and a too high thermal lag between the sample temperature and the thermocouple temperature, a small heating rate is recommended (Lacík et al. 2000). Usually, a value of 2°C/min is a good compromise for fulfilling both requirements; however, a correct value should be chosen according to the given circumstances. Figure 10.3 displays the result of a DMA experiment on a poly(amic acid) performed at 1 Hz and at three heating rates: 0.5°C/min, 2°C/min, and 5°C/min.

The shoulder corresponds to the glass transition of the amic acid chain segments that is shifted to a higher temperature at the highest heating rate (5°C/min), probably because of the thermal lag. The second peak is the relaxation associated with the glass transition of the imide formed by thermal imidization in the DMA oven (Cristea et al. 2008). The height of tan δ peak is higher at 0.5°C/min than at 2°C/min, suggesting less-constrained molecular movements in the first case.

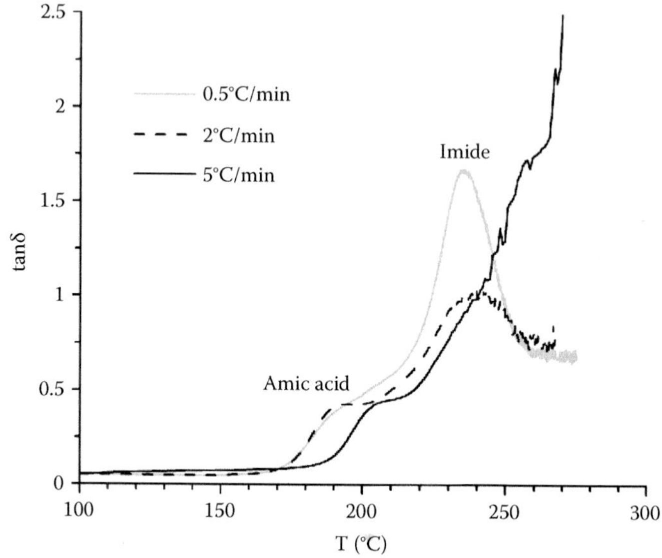

FIGURE 10.3 Variation of tan δ with temperature for a poly(amic acid) at 1 Hz and different heating rates.

At a lower heating rate, intramolecular imidization prevails over intermolecular imidization, which generates crosslinks and therefore less mobility. At a very high heating rate (5°C/min), the imidization reaction does not have enough time to proceed. The coordinated movements of chain segments characteristic of the glass transition become so intense that the polymer chains start to slip past each other and flow. The tan δ increases continuously until the film breaks.

In a DMA experiment, there is also a strong dependence between the heating rate and the frequency. Being a time-dependent experiment, the heating rate should be correlated with the frequency of the experiment. This requirement is expressed by the relation (Jäckle 1986):

$$\omega >> \frac{\frac{dT}{dt}}{\Delta t} \tag{10.6}$$

where ω is the angular frequency, rad · s^{-1}; Δt is the time range wherein the variation of temperature is ΔT.

10.2.5 Structure–Property Relation During the Glass Transition

Because tan δ incorporates the combination of storage and loss moduli variation with T and ν, its variation can be useful in a broader context:

- To assess that tan δ peak can be associated with a true relaxation
- To appreciate whether other phenomena occur during α-relaxation
- To determine the activation energy of α-relaxation, under some conditions

When evaluating the results of a DMA experiment and trying to understand it, the tendency is to associate any tan δ peak with a relaxation. However, not every tan δ peak or shoulder corresponds to a relaxation. In any relaxation region, tan δ peak shifts to a higher temperature as the frequency is increased because, as the rate of deformation is increased, the macromolecular chains have less time to respond to the applied stress. Therefore, a higher energy (temperature) is needed to initiate the molecular movements associated with the relaxation. By performing an experiment at multiple frequencies and carefully choosing the experimental conditions, the evolution of tan δ peak assesses the occurrence of relaxation (as was noted earlier in this chapter). Practically, the definition of the relaxation process provides the answer—the process of going from a state where the experimental frequency is too fast for a certain mode of molecular motion to a state where it is too slow (Lee and Goldfarb 1981). A schematical representation of this criterion is represented in Figure 10.4, and concrete examples representing the difference between α-relaxation and crystallization or between α-relaxation and curing can be found in some of the references (Cristea et al. 2010; Stark 2013).

The highest value of tan δ is attained when the frequency of molecular relaxation harmonizes with the experimental frequency. As the frequency is progressively raised, it becomes difficult for the polymer to respond to oscillating load, and it behaves more elastically than it really is. This fact is evidenced by the decrease of tan δ peak height with increasing frequency because the increase of E′ modulus takes place at the expense of the E″ modulus (Figure 10.4). When other phenomena that confer additional mobility to polymer chains overlap α-relaxation, an increase of tan δ peak with increasing frequency will take place (Cristea et al. 2009).

Due to the cooperative character of molecular movements associated with this relaxation, the linearity of the semilog plot ν versus 1/T is not fulfilled (Fried 2007). The activation energy of α-relaxation (glass transition) can be determined by using the equation (10.5) only when it is utilized in a limited frequency range.

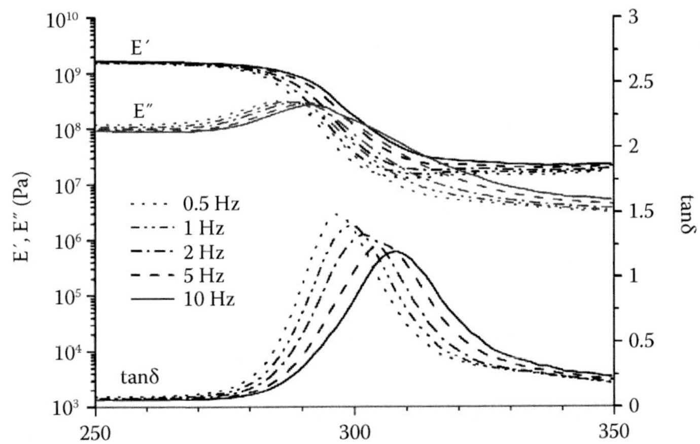

FIGURE 10.4 Variation of E′, E″, and tan δ for a polyimide at 2°C/min and five frequencies.

In this case, an apparent activation energy is determined, and it has higher values than secondary relaxations: 100–300 kJ/mol for urethane–soybean oil networks (Cristea et al. 2013), 200–300 kJ/mol for polyurethane–epoxy interpenetrating polymer networks (Cristea et al. 2009), more than 400 kJ/mol for semicrystalline poly(ethylene terephtalate) (Cristea et al. 2010), and more than 600 kJ/mol for polyimides (Cristea et al. 2008, 2011). The glass transition temperature is the most appropriate reference temperature when applying the time–temperature correspondence in a multifrequency experiment. The procedure allows estimation of the viscoelastic behavior of a polymer in time, in certain conditions, and is based on the fact that the viscoelastic properties at a certain temperature can be shifted along the frequency scale to obtain the variation on an extended time scale (Brostow 2007; Williams et al. 1955). The shift factor is described by the Williams–Landell–Ferry (WLF) equation:

$$\log a_T = -\frac{C_1(T - T_r)}{C_2 + (T - T_r)} \tag{10.7}$$

where a_T is the shift factor, C_1 and C_2 are the values of the viscoelastic coefficients, and T_r is the reference temperature (the glass transition temperature).

The value E_a at the reference temperature is given by the relation:

$$E_a = -\frac{2.303 R C_1 T_r^2}{C_2} \tag{10.8}$$

10.2.6 Rubbery and Terminal Region

In this region, the rate of the oscillatory load is too high for the segmental motion ($\omega\tau$<<1); both E′ and E″ stay in a plateau region. Situations can be encountered where there is no rubbery plateau because the structure of the polymer induces a very high mobility gained during α-relaxation. The increase of E″ at the expense of E′ during the glass transition is very high; flowing process overlaps α-relaxation and tan δ does not drop back to pretransition values (Damaceanu et al. 2014; Sava et al. 2015). Also, when the structure of the polymer is too constrained, flowing does not take place (Damaceanu et al. 2010). The linearity of the rubbery plateau may be interrupted by other phenomena: melting, cold crystallization, or orientation (Cristea et al. 2010, 2011; Ionita et al. 2015; Smola et al. 2014). Melting will be marked by a small decrease of E′ and a small tan δ peak that does not depend on frequency (Figure 10.2). A DSC experiment exhibits a melting peak in this range of temperature. The increase of E′ modulus is apt to occur for the last two phenomena. Also, when the structure belongs to a flexible elastomer, the flowing region is followed by another plateau due to the permanent crosslinks between the chains (Figure 10.2).

10.3 POLYMERIC MULTIPHASE SYSTEMS

The focus of intense fundamental and applied research in polymer science is on the continuous development of new materials oriented to practical need. Very often the required properties are not obtained by designing new polymer structures but by mixing polymers with other polymers (polymer blends) or with various reinforcements (composites). The interfacial properties determine the characteristics of the new system (i.e., the proportion and the manner of connection among various phases). Because of high interfacial tension and low interfacial adhesion, polymers are (partially) immiscible, and interpenetration at their surface is very low. In the case of the composites, the polymer can only bind or adhere to the nonpolymer surface (Sperling 2006). The interphase region may be strengthened by formation of bonds among polymers such as in copolymers or by using compatibilizers. In interpenetrating polymer networks (IPN), the polymers are combined in a network form with at least one of them being crosslinked, and no interaction occurs among the polymers (Sperling and Hu 2014). The simplest multiphase system is that of a semicrystalline polymer (Boyd 1985; Struik 1987).

DMA is a glass transition temperature-based analysis for detecting phase separation (Xavier 2014). In multiphase systems, each phase has its own glass transition temperature (T_g). Accordingly, the proof for miscibility is the existence of a single T_g, and the degree of phase separation is evaluated as the shift of T_gs relative to the T_gs of the component polymers or of the homopolymers. DMA is considered to be an indirect method for the characterization of multiphase systems because it allows only macroscopic physical characterization. Therefore, phase-separated systems present quite complex morphologies, whether the system is single- or multiphase should be confirmed by morphological (SEM, TEM, AFM) or scattering (SAXS, neutron) techniques (Thomas et al. 2011).

10.3.1 Behavior of Polymeric Multiphase Systems in the Glassy Region

The issues that can be addressed in the glassy region are the effects of multiphase formation on the values of the storage modulus and on the fate of secondary relaxations. Theoretically, it is expected that the moduli of the multicomponent polymeric materials are situated among the moduli of the involved components, but the phase morphology and the continuity in the system have their own roles in this behavior. When polypropylene (PP), a semicrystalline polymer, is put together with polystyrene (PS), they form an immiscible mixture ($E'_{PS} \approx 2E'_{PP}$) (Paramesvaranpillai et al. 2015). The microscopy indicates that for low PS content, small domains of PS are dispersed in the PP matrix. However, when PS is the major component, big islands of PS are dispersed in the PP matrix. For situations when the PS content in the mixture is lower than 40% and PP is the continuous phase, there is no increase of the E′ modulus as the PS proportion is raised. When PS is in combination with polyamide 12 (PA12) or with the copolymer ethylene/vinyl acetate (EVA), the blends with less stable morphology (co-continuous structure) have unfavorable interfacial interactions (Aghjeh et al. 2015; Jose et al. 2006). Therefore, a monotonous variation of the E′ modulus with an increase in the more rigid compound content does not take place. The compatibility among some polymers can be substantially enhanced when their structures are complementary (i.e., they can interact). Polyamide 66 (PA66) is compatible with poly(hydroxyl ether of bisphenol A) (PHE) because of the formation of hydrogen bonding between hydroxyl groups of PHE and the amide groups of PA66 (An et al. 2014). In the glassy region, the E′ moduli of the mixtures with different compositions PA66/PHE are very close to each other, although there is a significant difference between these values and the E′ of PA66. This compatibility is also reflected in the secondary relaxations. The β-relaxation of PA 66, associated with the segmental motions of amide groups, vanishes as PHE is included in the mixture and γ-relaxation shifts to a higher temperature. Similarly, in ethylene-norbornene copolymers, the copolymer with higher norbornene content has a less visible γ-relaxation, and the β-relaxation, attributed to the norbornene, is more distinct (Makrocka-Rydzyk et al. 2010).

The rigidity of the random copolymer butylacrylate/acrylonitrile (BA/AN) increases in the whole range of the investigated temperature as the content of the hard component (AN) is increased (Suresh et al. 2003). With regard to the origin of secondary relaxations in semicrystalline polymers, reported conclusions do not match. However, most of them concluded that the crystallinity has no effect on β-relaxation because they have localized character (Boyd 1985; Boyd and Boyd 2001).

The presence of reinforcement in polymer (nano)composites (Thomas et al. 2011; Wen 2007) generally increases the value of E′ modulus due to restrictions of vibrations and short-range rotational motions (Bindu and Thomas 2013; Dufresne 2000; Liu et al. 2005; Musto 2006). Exceptions to this behavior exist, and one of them was reported on the FeS_2/polyimide composite (Sava 2009). Incorporation of low quantities of pyrite microparticles restrains the possible physical interactions among polyimide chains, decreasing the rigidity of the composite relative to the pristine polyimide. Another situation was mentioned for an all-polymer composite obtained from a pair of incompatible polymers: polysulfone (PSF) (the matrix polymer) and cross linked polydimethylsiloxane (PDMS) submicron particles (the disperse phase) (Racles et al. 2013). All-polymer composites contain polymer nanoparticles as the disperse phase in a polymer matrix and, from the point of view of properties, these are situated between polymer blends and polymer composites. The PDMS particles act like a plasticizer for the PSF matrix (E'_{PSF} = 2 GPa, $E'_{PSF/PDMS}$ = 0.7 GPa).

10.3.2 Behavior of the Polymeric Multiphase Systems in the Glass Transition and Rubbery Regions

In semicrystalline polymers, the stiffening crystalline phase limits the mobility of the amorphous phase and the anticipated behavior would be an increase in the glass transition temperature with the increase of crystallinity, simultaneous with the reduction of the height of the tan δ peak. However, this statement is a very simplified, even incomplete, description of semicrystalline polymers. Although not generally accepted, a convention was established for their relaxations: α-relaxation is associated with the slippage of the crystallites past each other, while the movement of coordinated segments in amorphous phase is responsible for β-relaxation (Boyd 1985; Menard 1999). Moreover, an increase of T_g with the degree of crystallinity was reported for poly(ethylene terephthalate) when the degree of crystallinity was less than 30%. Thereafter, T_g diminished with the degree of crystallinity (Boyd 1985). This phenomenon was explained later in terms of dual amorphous phases (Wunderlich 2005). As long as the coupling between the polymer molecules that belong to the amorphous phase and the crystalline phase is weak, the result will be a broadening of the temperature to higher values. When the coupling is strong, a new rigid amorphous fraction is formed in addition to the mobile amorphous phase. In this sense, DMA was used to investigate the influence of the rigid amorphous phase in the segmental mobility of the mobile amorphous regions for semi-crystalline poly(L-lactic acid) (Picciochi 2007).

The inward shift of the glass transition temperatures for each phase denotes there is some degree of miscibility between two phases (Hara and Sauer 1998; Wetton and Corish 1989). Usually, the DSC technique is the first choice for the determination of the glass transition temperature. The various possibilities of connections that may happen in a polymer blend (interfacial adhesion, molecular interpenetration, chain packing, ion-dipole interactions) broaden the glass transition region and make the change of the specific heat capacity (Δc_p) too small to be detected. Instead, DMA turned out to be sensitive enough. For example, in mixtures of ionic poly(*p*-phenylene terephthalamide) and poly(4-vinylpyridine), DMA displayed transition in the temperature range where DSC did not show any information (Chen et al. 2003). If each phase maintains its own T_g, the two polymers are incompatible. However, an experienced eye may notice details in the evolution of the tan δ peak amplitude with the composition of the polymer mixture that can reflect the occurrence of interfacial adhesion between phases (Jose et al. 2015). A random copolymer should display a single tan δ peak, associated with the glass transition temperature, that is located between the T_gs of the homopolymers. A narrow tan δ peak indicates that the copolymer is homogeneous; all the chains have nearly the same composition. A wide tan δ peak means that the copolymer is heterogeneous. In the BN/AN random copolymer, mentioned in the previous paragraph, the glass transition temperature is displayed between the T_g of poly(butyl acrylate) (–55°C) and the T_g of poly(acrylonitrile) (105°C) (Suresh et al. 2003). Also, there is only a small shift of tan δ peak toward higher temperature with increasing AN content, but the amplitude of this peak has a smaller value as the AN content is higher (i.e., lower mobility). Moreover, the sample with the highest AN content presents a bimodal shape because of microscopical heterogeneity. A homogeneous random PVC–PBA copolymer was synthesized by single-electron transfer degenerative transfer living radical polymerization (SET-DTLRP) (Calafel et al. 2015). Although the triblock copolymer 30PVC–40PBA 30PVC has two tan δ peaks, approaching the T_gs of the homopolymers, a single T_g is noticed in the random copolymer 60PVC–40PBA. The tan δ peaks in the triblock copolymer are broad and their amplitude is smaller than the correspondent homopolymer. The single, sharp tan δ peak of the homopolymer that did not reduce its amplitude confirms its homogeneity.

The rubbery plateau also represents an indicator of miscibility (Badia et al. 2015). Linear triblock copolymers composed of poly(*n*-butyl acrylate) and polystyrene (PS–PBA–PS) have phase-separated morphologies exhibiting two rubbery plateaux (Pakula et al. 2011). The first plateau lies between the T_g of the PBA blocks and the T_g of the PS blocks; in this region, the glassy PS microdomains act like physical crosslinks for the flexible PBA blocks. The second rubbery plateau starts after the T_g of PS and continues until the order–disorder transition, where the microdomain structure transforms into a disordered homogeneous phase (Han et al. 1989). The length and the storage modulus of the rubbery plateau depend on the molecular architecture and on the composition of the copolymers. For the same composition, the three-arm star PBA–PS copolymers exhibit superior properties compared to linear triblock PS–PBA–PS copolymers in terms of phase separation and the storage modulus value of the first rubbery plateau (three times higher). When the three-arm star copolymers incorporate beside PBA, poly(methyl methacrylate) (PMMA), or polyacrylonitrile (PAN), the second rubbery plateau is extended up to very high ordered–disordered temperatures (i.e., the range of microdomain morphology is extended) (Dufor et al. 2008a, 2008b). The SAXS does not show evidence of the bottle brush PBA–PS copolymers' significant phase separation, and DMA confirms this by showing a single glass transition temperature (Pakula et al. 2011).

The IPN systems display dual-phase continuity. This means that, in the situation of a bicomponent system, it will be characterized by phase separation, but each phase presents phase continuity throughout the material. Each phase has its own glass transition region, with the particularity that the transitions are very broad and have moved toward each other (Lipatov and Alekseeva 2007; Sperling and Hu 2014). Efforts are focused on increasing the damping properties: raising of the tan δ peak that is simultaneous with the broadening of the transition temperature range. In an attempt to obtain organic glasses by using the favorable characteristics of poly(methyl metacrylate) (PMMA) (refractive index close to that of glass, good light transmission, etc.), it was combined with a polycarbonate (PCR39®) in the form of an IPN in order to minimize its sensitivity to abrasion (Berrebi et al. 2015). The resulted IPN takes advantages from the features of both components: only one α-relaxation is revealed, the intensity of which increases with the increase of PMMA proportion, and the storage modulus of the rubbery plateau rises with the proportion of PCR39® as a result of co-continuous morphology. Further progress in developing a tougher IPN has advanced by adding molecular units that allow the formation of supramolecular associations (Monemian and Korley 2015). DMA combined with TEM investigations demonstrated that only the strong interfacial coupling between the matrix and the supramolecular phase confer toughness enhancement.

It is difficult to outline rigorous rules on how reinforcement influences the viscoelastic behavior in the glass transition region. Huang et al. reviewed the effect of filler particles on the dynamics of elastomers (Huang et al. 2015). Due to the reduced chain mobility of the polymer upon addition of reinforcement, the height of tan δ usually decreases. This effect was noticed for natural rubber–ZnO nanocomposites (Bindu and Thomas 2013), glass fiber-reinforced poly(trimethylene tereftalate) (Liu et al. 2005), bacterial polyester/cellulose whisker natural composites (Dufresne 2000), and in flax-reinforced ethylene-propylene-diene (EPDM) composites (Cristea et al. 2015). Also, the value of the glass transition temperature does not modify very much. However, a pronounced strain–amplitude dependency can be noticed in an investigation of viscoelastic behavior versus strain (Payne effect). This is the result of the debonding of the polymeric chains from the filler surface (Gradin et al. 1989; Gauthier et al. 2004).

10.4 CONCLUDING REMARKS

Dynamic mechanical analysis belongs to the macroscopic physical characterization methods of polymeric multiphase systems. Because polymers are very sensitive to temperature in the region of the glass transition, any change in the glass transition temperature of the phases gives information about the miscibility of the polymeric systems. The advantages of DMA come from two directions. On one side, when considering DMA as a thermal analysis method, it provides excellent results where other thermal methods fail. It has been proven to be much more sensitive than DSC in revealing the transitions. On the other side, DMA is a rheological method, and it offers information about the interconnection among phases when the variations with temperature and frequency of the viscoelastic parameters E′, E″, and tan δ are examined in detail. Nevertheless, for an accurate description of a multiphase system, the DMA results should be supported by morphological investigations (SEM, TEM, AFM), interface/interphase investigations (SAXS, SANS), or by spectroscopical research (UV, FTIR, NMR, XPS).

REFERENCES

Aghjeh R. Masoud, Khonakdar A. Hossein, and Seyed H. Jafari. Application of mean-field theory in PP/EVA blends by focusing on dynamic mechanical properties in correlation with miscibility analysis. *Compos. B. Eng.* 79 (2015): 74–82.

An Jingjing, Cao Xinyu, Ma Yongmei, et al. Preparation and characterization of polyamide 66/poly(hydroxyl ether of bisphenol A) blends without compatibilizer. *J. Appl. Polym. Sci.* 131 no. 13 (2014): 40437.

Badia José-David, Santonja Blasco Laura, Martínez-Felipe Alfonso, and Ribes-Greus Amparo. Dynamic mechanical thermal analysis of polymer blends. In *Characterization of Polymer Blends*, Sabu Thomas, Yves Grohens, Parameswaranpillai Jyotishkumar (eds.), 365–392. Weinheim, Germany: Wiley-VCH, 2015.

Bas Corine, Tamagna Cécile, Pascal Thierry, and Alberola N. Dominique. On the dynamic mechanical behavior of polyimides based on aromatic and alicyclic dianhydrides. *Polym. Eng. Sci.* 43 no. 2 (2003): 344–355.

Berrebi Mickaël, Fabre-Francke Isabelle, Lavédrine Bertrand, and Fichet Odile. Development of organic glass using interpenetrating polymer networks with enhanced resistance towards scratches and solvents. *Eur. Polym. J.* 63 (2015): 132–140.

Bindu Patanair, and Thomas Sabu. Viscoelastic behavior and reinforcement mechanism in rubber nanocomposites in the vicinity of spherical nanoparticles. *J. Phys. Chem. B.* 117 no. 41 (2013): 12632–12648.

Blanter S. Mikhail, Golovin S. Igor, Neuhäuser Hartmut, and Sinning Hans-Rainer. In *Internal friction in metallic materials. A handbook.* Berlin: Springer Verlag, 2007.

Boyd H. Richard. Relaxation processes in crystalline polymers: Experimental behavior—A review. *Polymer.* 26 no. 3 (1985): 323–347.

Boyd U. Sylke, and Boyd H. Richard. Chain dynamics and relaxation in amorphous poly(ethylene terephtalate): A molecular dynamics simulation study. *Macromolecules.* 34 no. 20 (2001): 7219–7229.

Brostow Witold. Mechanical properties. In *Physical properties of polymers handbook*, James E. Mark (ed.), 424–445, New York: Springer, 2007.

Calafel Itxaso, Muñoz María Eugenia, Santamaría Anton, Boix Miquel, Conde Jose Ignacio, and Pasqual Belén. PVC/PBA random copolymers prepared by living radical polymerization (SET-DTLRP): Entanglements and chain dimensions. *Eur. Polym. J.* 73 (2015): 202–211.

Caussé Nicolas, Quiroga Cortes Luis, Dantras Éric, et al. New bonded assembly configuration for dynamic mechanical analysis of adhesives. *Int. J. Adhes. Adhes.* 46 (2013): 1–6.

Chartoff P. Richard, Menczel D. Joseph, and Dillman H. Steven. Dynamic mechanical analysis. In *Thermal analysis of polymers. fundamentals and applications,* Joseph D. Menczel, R. Bruce Prime (eds.), 387–495. Hoboken, NJ: Wiley, 2009.

Chen C. Wen, Sauer A. John, and Hara Masanori. Melt processable blends of ionic poly(p-phenylene terephthalate) and poly(4-vinylpyridine): Relaxation behavior. *J. Polym. Sci. Pol. Phys.* 41 no. 13 (2003): 1468–1475.

Choi Bogyu, Loh Xian Jun, Tan Aloysius, et al. Introduction to in situ forming hydrogels for biomedical applications. In *In-situ gelling polymers for biomedical applications*, Xian Jun Loh (ed.), 5–35. New York: Springer, 2015.

Cristea Mariana, Airinei Anton, Ionita Daniela, and Stelescu Maria Daniela. Relaxation behavior of flax-reinforced ethylene-propylene-diene rubber. *High Perform. Polym.* 27 no. 5 (2015): 676–682.

Cristea Mariana, Ibanescu Sorin, Cascaval N. Constantin, and Rosu Dan. Dynamic mechanical analysis of polyurethane-epoxy interpenetrating polymer networks. *High Perform. Polym.* 21 no. 5 (2009): 608–621.

Cristea Mariana, Ionita Daniela, Doroftei Florica, and Simionescu C. Bogdan. Effect of long-term and short-term dynamic mechanical evaluation of networks based on urethane and soybean oil. *J. Mech. Behav. Biomed. Mater.* 17 (2013): 317–326.

Cristea Mariana, Ionita Daniela, Hulubei Camelia, Timpu Daniel, Popovici Dumitru, and Simionescu C. Bogdan. Chain packing versus chain mobility in semialiphatic BTDA-based polyimides. *Polymer.* 52 no. 8 (2011): 1820–1828.

Cristea Mariana, Ionita Daniela, and Simionescu C. Bogdan. A new insight in the dynamo-mechanical behavior of poly(ethylene tereftalate). *Eur. Polym. J.* 46 no. 10 (2010): 2005–2012.

Cristea Mariana, Ionita Gheorghiu Daniela, Bruma Maria, and Simionescu C. Bogdan. Thermal behavior of aromatic polyamic acids and polyimides containing oxadiazole rings. *J. Therm. Anal. Calorim.* 93 no. 1 (2008): 63–68.

Damaceanu Mariana-Dana, Musteata Valentina-Elena, Cristea Mariana, and Bruma Maria. Viscoelastic and dielectric behaviour of thin films made from siloxane-containing poly(oxadiazole-imide)s. *Eur. Polym. J.* 46 no. 5 (2010): 1049–1062.

Damaceanu Mariana-Dana, Rusu Radu-Dan, Cristea Mariana, Musteata Valentina-Elena, Bruma Maria, and Wolinska-Grabczyk Aleksandra. Insights into the chain and local mobility of some aromatic polyamides and their influence on the physicochemical properties. *Macromol. Chem. Phys.* 215 no. 16 (2014): 1573–1587.

Da Silva N. Ana Lúcia, Rocha C. G. Marisa, Coutinho M. B. Fernanda, Bretas E. S. Rosário, and Scuracchio Carlos. Rheological and thermal properties of binary blends of polypropylene and poly(ethylene-co-1-octene). *J. Appl. Polym. Sci.* 79 no. 9 (2000): 1634–1639.

De Kee Daniel, and Wisbrun F. Kurt. Polymer rheology. *Phys. Today.* 51 no. 6 (1998): 24–29.

Dufor Bruno, Konynov Kaloian, Pakula Tadeusz, and Matyjaszewski Krzysztow. PMMA-PBA 3-arm star block copolymer thermoplastic elastomer. *Macromol. Chem. Phys.* 209 no. 16 (2008a): 1686–1693.

Dufor Bruno, Tang Chuanbing, Konynov Kaloian, Zhang Ying, Pakula Tadeusz, and Matyjaszewski Krzysztow. Polar three-arm star block copolymer thermoplastic elastomers based on polyacrylonitrile. *Macromolecules.* 41 no. 7 (2008b): 2451–2458.

Dufresne Alain. Dynamic mechanical analysis of the interphase in bacterial polyester/cellulose whiskers natural composites. *Compos. Interface.* 7 no. 1 (2000): 53–67.

Fannin C. Paul. Wideband measurement and analysis techniques for the determination of the frequency-dependent, complex susceptibility of magnetic fluids. *Adv. Chem. Phys.* 104 (1998): 181–292.

Fried L. Joel Sub-T_g Transitions. In *Physical properties of polymers handbook*, James E. Mark (ed.), 217–232, New York: Springer, 2007.

Gauthier Catherine, Reynaud Emmanuelle, Vassoille V. Rene, and Ladouce-Stelandre Laurence. Analysis of the non-linear viscoelastic behavior of silica filled styrene butadiene rubber. *Polymer.* 45 no. 8 (2004): 2761–2771.

Gradin Per, Howgate G. Paul, Selden Ragnar, and Brown A. Richard. Dynamic-mechanical properties. In *Comprehensive Polymer Science and Supplements, V. 2 Polymer Properties*, Sir Geoffrey Allen and John C. Bevington (eds.), 533–569, New York: Pergamon Press, 1989.

Han Chang Dae, Kim Jinhwan, Kim Jin Kon. Determination of the order-disorder transition temperature of the blockcopolymers. *Macromolecules*. 22 no. 1 (1989): 383–394.

Hara Masanori, and Sauer A. John. Synergism in mechanical properties of polymer/polymer blends. *J. Macromol. Sci. Pol. Rev.* 38 no. 2 (1998): 327–362.

Huang Menglong, Tunnicliffe W. Lewis, Thomas G. Alan, and Busfield J. C. James. The glass transition, segmental relaxations and viscoelastic behaviour of particulate-reinforced natural rubber. *Eur. Polym. J.* 67 (2015): 232–341.

Ionita Daniela, Cristea Mariana, Gaina Constantin, and Banabic Dorel. Tailoring the hard domain cohesiveness in polyurethanes by interplay between the functionality and the content of chain extender. *RSC Adv.* 5 no. 94 (2015): 76852–76861.

Jäckle Josef. Models of the glass transition. *Rep. Prog. Phys.* 49 no. 2 (1986): 171–231.

Jekle Mario, and Becker Thomas. Wheat dough microstructure: The relation between visual structure and mechanical properties. *CRC Crit. Rev. Food Sci.* 55 no. 3 (2015): 369–382.

Jones S. David, Tian Yiwei, Abu-Diak Osama, and Andrews P. Gawin. Pharmaceutical applications of dynamic mechanical thermal analysis. *Adv. Drug Deliv. Rev.* 64 no. 5 (2012): 383–478.

Jose Seno, Francis Bejoy, Thomas Sabu, and Karger-Kocsis Józef. Morphology and mechanical properties of polyamide 12/polypropylene. *Polymer.* 47 no. 11 (2006): 3874–3888.

Jose Seno, Thomas Sabu, Parameswaranpillai Jyotishkumar, Aprem Abi Santhosh, and Karger-Kocsis Józef. Dynamic mechanical properties of immiscible polymer systems with and without compatibilizer. *Polym. Test.* 44 (2015): 168–175.

Lacík Igor, Krupa Igor, Stach Marek, Kučma Anton, and Chodák Ivan. Thermal lag and its practical consequence in the dynamic mechanical analysis of polymers. *Polym. Test.* 19 no. 7 (2000): 755–771.

Lakowicz R. Joseph. *Principles of Fluorescence Spectroscopy*. 3rd ed. New York: Springer, 2006.

Lee Y-C. Charles, and Ivan J. Goldfarb. The glass transition temperature of partially cured polymers as a non-equilibrium parameter and its effect on observed dynamic mechanical behavior. *Polym. Eng. Sci.* 21 no. 14 (1981): 951–957.

Lipatov S. Yuri, and Alekseeva T. Tatiana. Phase-separated interpenetrating polymer networks. *Adv. Polym. Sci.* 208 (2007): 1–227.

Liu Chaoren, Pineda Eloi, and Crespo Daniel. Mechanical relaxation of metallic glasses. An overview of experimental data and theoretical model. *Metals.* 5 no. 2 (1978): 1073–1111.

Liu Wanjun, Mohanty K. Amar, Drzal T. Lawrence, et al. Injection molded glass fiber reinforced poly(trimethylene terephtalate) composites: Fabrication and properties evaluation. *Ind. Eng. Chem. Res.* 44 no. 4 (2005): 857–862.

Makrocka-Rydzyk Monika, Nowaczyk Grzegosz, Głowinkowski Stanisław, and Jurga Stefan. Dynamic mechanical study of molecular dynamics in ethylene-norbornene copolymers. *Polymer.* 51 no. 4 (2010): 908–912.

Mano F. João, Reis L. Rui, and Cunha M. António. Dynamic mechanical analysis in polymers for medical applications. In *Polymer based systems on tissue engineering, replacement and regeneration, NATO Science Series*, Dordrecht: Kluwer, Rui L. Reis and Daniel Cohn (eds.), 86, 139–164, 2002.

Menard P. Kevin. *Dynamic mechanical analysis: A practical introduction*. Boca Raton, FL: CRC Press, 1999.

Menard P. Kevin, and Menard R. Noah. Dynamic mechanical analysis in the analysis of polymers and rubbers. In *Encyclopedia of polymer science and technology*, 1–33, 2015. http://onlinelibrary.wiley.com/doi/10.1002/0471440264.pst102.pub2/full.

Mezger Thomas G. *The rheology handbook*. Hannover, Germany: Vincentz Network, 2006.

Monemian Seyedali, and Korley T. J. LaShanda. Exploring the role of supramolecular associations in the mechanical toughening of the interpenetrating polymer networks. *Macromolecules*. 48 no. 19 (2015): 7146–7155.

Musto Pellegrino, Abbate Mario, Lavorgna Marino, Ragosta Giuseppe, and Scarinzi Gennaro. Microstructural features, diffusion and molecular relaxations in polyimide/silica hybrids. *Polymer*. 47 no. 17 (2006): 6172–6186.

Nicotera Isabella, Coppola Luigi, Oliviero Cesare, and Ranieri A. Giuseppe. Spectromechanical properties of polymeric gel electrolytes and blends. *Macromol. Symp.* 247 no. 1 (2007): 282–294.

Pakula, Tadeusz, Koynov Kaloian, Boerner Hans, et al. Effect of chain topology on the self-organization and the mechanical properties of poly(n-butyl acrylate)-b-polystyrene blockcopolymers. *Polymer*. 52 no. 12 (2011): 2576–2583.

Paramesvaranpillai Jyotishkumar, Joseph George, Chellappan V. Resmi, Zahakariah K. Ajesh, and Hameed Nishar. The effect of propylene-graft-maleic anhydride on the morphology and dynamic mechanical properties of polypropylene/polystyrene blends. *J. Polym. Res.* 22 no. 2 (2015): 2.

Picciochi Ricardo, Wang Yaming, Alves M. Natália, and Mano F. João. Glass transition of semi-crystalline PLLA with different morphologies as studied by dynamic mechanical analysis. *Coll. Polym. Sci.* 285 no. 5 (2007): 575–580.

Pinheiro Ana and Mano F. João. Study of the glass transition on the viscous-forming and powder materials using dynamic mechanical analysis. *Polym. Test.* 28 no. 1 (2009): 89–95.

Racles Carmen, Cristea Mariana, Doroftei Florica, and Alexandru Mihaela. All-polymer composites from two incompatible polymers. *Soft Mater.* 11 no. 4 (2013): 421–429.

Razzaq Muhammad Yasar, Anhalt Mathias, Frormann Lars, and Weidenfeller Bernd. Mechanical spectroscopy of magnetite filled polyurethane shape memory polymers. *Mater. Sci. Eng. A*. 471 no. 1–2 (2007): 57–62.

Robertson Christopher G. Dynamic mechanical properties. In *Encyclopedia of polymeric nanomaterials*, Shiro Kobayashi and Klaus Mullen (eds.), 1–9. Berlin: Springer Verlag, 2014.

Roling Bernhard. Mechanical loss spectroscopy on inorganic glasses and glass ceramics. *Curr. Opin. Solid St. M.* 5 no. 2–3 (2001): 203–210.

Runt P. James, and Fitzgerald J. John. *Dielectric spectroscopy of polymeric materials. Fundamentals and applications*. Washington, DC: American Chemical Society, 1997.

Sava Ion. FeS_2/polyimide composite films. Synthesis and characterization. *Rev. Roum. Chim.* 54 no. 11–12 (2009): 957–962.

Sava Ion, Burescu Ada, Stoica Iuliana, et al. Properties of some azo-copolyimide thin films used in the formation of photoinduced surface relief gratings. *RSC Adv.* 5 no. 14 (2015): 10125–10133.

Schlesing Willi, Buhk Maren, and Osterhold Michael. Dynamic mechanical analysis in coatings industry. *Prog. Org. Coating*. 49 no. 3 (2004): 197–208.

Sepe P. Michael. *Dynamic mechanical analysis for plastics engineering*. Norwich: Plastics Design Library, 1998.

Sikorska Wanda, Adamus Grażyna, Dobrzyński Piotr, et al. Forensic engineering of advanced polymeric materials—Part II: The effect of the solvent-free non-woven fabrics formation method on the release rate of lactic and glycolic acids from the tin-free poly(lactide-co-glycolide) nonwovens. *Polym. Degrad. Stabil.* 110 (2014): 518–528.

Smola Anna, Dobrzyński Piotr, Cristea Mariana, et al. Bioresorbable terpolymers based on L-lactide, glycolide and trimethylene carbonate with shape memory behaviour. *Polym. Chem.* 5 no. 7 (2014): 2442–2452.

Sperling H. Leslie. *Introduction to physical polymer science*. Hoboken, NJ: Wiley-Interscience, 2006.

Sperling H. Leslie, and Hu R. Maria. Interpenetrating polymer networks. In *Polymer blends handbook*, Leszek A. Utracki and Charles Wilkie (eds.), 677–724. New York: Springer, 2014.

Spitalsky Zdenko, Tasis Dimitrios, Papagelis Konstantinos, and Galiotis Costas. Carbon nanotube-polymer composites: Chemistry, processing, mechanical and electrical properties. *Prog. Polym. Sci.* 35 no. 3 (2010): 357–401.

Stark Wolfgang. Investigation of the curing behavior of carbon fibre epoxy prepreg by dynamic mechanical analysis DMA. *Polym. Test.* 32 no. 2 (2013): 231–239.

Struik C. E. Leen. The mechanical behavior and physical ageing of semicrystalline polymers: 2. *Polymer.* 28 no. 9 (1987): 1534–1542.

Suresh I. Kattimuttathu, Sitaramam S. Bhamidipalli, and Raju V. S. N. Kothapally. Effect of copolymer composition on the dynamic mechanical and thermal behaviour of butyl acrylate-acrylonitrile copolymer. *Macromol. Mater. Eng.* 288 no. 12 (2003): 980–983.

Thomas Sabu, Boudenne Abderrahim, Ibos Laurent, and Candeau Yves. Physical, thermophysical and interfacial properties of multiphase polymer systems: State of art, new challenges and opportunities. In *Handbook of multiphase polymer systems*, Abderrahim Boudenne, Laurent Ibos, Yves Candau, and Sabu Thomas (eds.), 1–12. Chichester, UK: Wiley, 2011.

Vélez-Ruiz F. Jorge, Barbosa Cánovas V. Gustavo, and Peleg Micha. Rheological properties of selected dairy products. *CRC Crit. Rev. Food Sci.* 37 no. 4 (1997): 311–359.

Warren J. Frederick, Royall G. Paul, Butterworth J. Peter, and Ellis L. Peter. Immersion mode material pocket dynamic mechanical analysis (IMP-DMA): A novel tool to study gelatinization of purified starches and starch containing-plant materials. *Carbohydr. Polym.* 90 no. 1 (2012): 628–636.

Wen Jianye. Some mechanical properties of typical polymer-based composites. In *Physical properties of polymers handbook*, James E. Mark (ed.) 487–496, New York: Springer, 2007.

Wetton E. Ray, and Corish J. Patrick. DMTA studies of polymer blends and compatibility. *Polym. Test.* 8 no. 5 (1989): 303–312.

Willbourn H. Anthony. The glass transition in polymers with the $(CH_2)_n$ group. *Trans. Faraday Soc.* 54 (1958): 717–729.

Williams L. Malcom, Landel F. Robert, and Ferry D. John. The temperature dependence of relaxations mechanisms in amorphous polymers and other glass-forming liquids. *J. Am. Chem. Soc.* 77 no. 14 (1955): 3701–3707.

Wunderlich Bernhard. Effect of decoupling of molecular segments, microscopic stress-transfer and confinement of the nanophases in semicrystalline polymers. *Macromol. Rapid Commun.* 26 no. 19 (2005): 1521–1531.

Xavier S. Francis. Properties and performance of polymer blends. In *Polymer blends handbook*, Leszek A. Utracki, Charles A. Wilkie (eds.), 1031–1202, New York: Springer, 2014.

Xie Fengwei, Long Yu, Chen Ling, and Li Lin. A new study of starch gelation under shear using dynamic mechanical analysis. *Carbohydr. Polym.* 72 no. 2 (2008): 229–234.

Yoon N. Hyun, and Eisenberg Adi. Dynamic mechanical properties of metallic glasses. *J. Non-Cryst. Solids.* 29 no. 3 (1978): 357–364.

11 Designing Thermophysical Behavior of Polymers with Nanometric Inclusions for Heat Dissipation in Electronic Devices

Andreea Irina Barzic

CONTENTS

11.1 INTRODUCTION

The development of modern technologies demands electronic devices with superior performance reflected in multiplication of their functionality (Tong 2011). One of the challenges in electronic packaging is to overcome barriers concerning thermal management, which is a common problem that can reduce a system's reliability (Oechsner and Murch 2011). The current requirements for smaller, more capable, and more efficient electronic power devices involve components that can dissipate heat efficiently and materials that exhibit an enhanced capability for dissipating heat and maintaining compatibility with the die and electronic packaging. In response to these critical needs, many efforts have been made to revolutionize thermal management materials and technologies for active and passive cooling in order to solve the vexing problem at the device and package level (Wang et al. 2011). Heat release from electronic devices by excellent thermal conduction implicates the utilization of a heat sink to dissipate heat from an electronic package and various industrial processes based on heating and cooling. For efficient thermal conduction, thermal interface or thermally conductive materials must present enhanced heat transfer ability with low contact resistance (Yang et al. 2008). If the latter aspect is not accomplished, the use of a thermal conductive material is too cost-consuming. Among the materials that promise integrable and cost-effective thermal management solutions, reinforced polymers represent a good economic alternative to improve heat dissipation from the heat source (Tong 2011). In comparison with traditional thermally conductive metals, polymer composites have better potential in terms of resistance to corrosion, light weight, and processability for their purposes (Moniruzzaman and Winey 2006; Han and Fina 2011; Chauhan et al. 2012; Choi and Kim 2013; Shtein et al. 2015).

Keeping these factors in view, this chapter is focused on describing the thermal conduction processes in reinforced polymers starting from the matrix itself and continuing to the multiplication of the conduction path by inserting various types of nanofillers. The most important theoretical models and measurement techniques used for thermal conductivity of filled polymers are reviewed. Current trends in manufacturing, such as producing multiphase materials with enhanced thermal conduction by controlling the nanofiller–matrix interactions and composition, are presented in relation to heat dissipation purposes in electronics.

11.2 BASIC CONCEPTS OF HEAT TRANSFER AND THERMAL CONDUCTION

The aim of thermal management in electronic packaging is to efficiently dissipate heat from the semiconductor junction to the ambient environment. This process includes heat transfer within the semiconductor component package, then dissipation from it to a heat dissipater, and finally dissipation to the environment through other cooling systems. To obtain an efficient thermal connection along these paths demands a good understanding of basic concepts of heat transfer as well as knowledge of available interface materials and how their key physical characteristics influence the heat transfer process.

11.2.1 Polymer Matrix

Thermal conduction is the phenomenon by which heat is transported from high- to low-temperature regions of media that are in intimate contact. The ability of a material to transfer heat is described by thermal conductivity, which is a microstructure sensitive property. For polymers, the internal heat flow is much more complicated when compared with metals, glasses, and ceramics owing to their large number of motion units (side groups, branched chains, chain units, etc.). Polymers are known to have intrinsic thermal conductivity much lower than that of metals or ceramic materials; therefore, they are good thermal insulators. Further enhancement of their conduction features can be accomplished by reinforcement with inorganic or organic fillers (Ebadi-Dehaghani and Nazempour 2012).

Generally, the processing of these materials in a solid phase is performed at elevated temperatures, often above their melting point. This may require a long period of time and high costs due to the low thermal conductivity of polymers. Therefore, the cooling process (or annealing) can also be determined by heat transport characteristics of polymers, which eventually influence the physical properties of the materials. Careful consideration in designing polymer processing is essential for obtaining the desired features in the final product.

For one-dimensional and rectilinear heat flow, the steady-state heat transfer in polymeric materials can be described by Fourier's law of heat conduction:

$$q = -k_m \frac{dT}{dx} \tag{11.1}$$

where q is the heat flux along x, dT/dx is the temperature gradient per unit length, and k_m is the thermal conductivity of the polymer medium in the direction of heat flow.

Heat transfer involves the transport of energy from one place to another by energy carriers. In liquids, energy can be transported by diffusion and advection (an overall drift in a certain direction) of molecules. In solids, the main carriers of energy are represented by phonons, electrons, or photons. For conjugated polymers, the valence electrons gain energy, travel toward the colder areas of the material, and transfer their energy to other atoms. The amount of energy transported is determined by the number of excited electrons and their mobility (i.e., lattice imperfections and temperature). The quantized modes of vibration taking place in a rigid crystal lattice, known as phonons, are the primary mechanism of heat conduction in most polymers because free movement of electrons is limited (Majumdar 1998). The thermal energy associated with phonons is transported in the direction of their motion.

In theoretical prediction, the Debye equation is usually used to calculate the thermal conductivity of polymers (Han and Fina 2011).

$$k_m = \frac{C_p \cdot v \cdot l}{3} \tag{11.2}$$

where C_p is the specific heat capacity per unit volume, v is the average phonon velocity.

In the case of amorphous polymers, the phonon mean free path is an extremely small constant (a few angstroms) due to phonon scattering from numerous defects, leading to a very low thermal conductivity of polymers (Agari et al. 1997). The thermal conductivity depends on the degree of crystallinity because vibration and movement/rotation of macromolecular chains transfer heat energy.

Thus, a polymer with a highly crystalline and ordered structure will have higher conductivity than an amorphous polymer.

Another theoretical approach—proposed by Bicerano—allows evaluation of thermal conductivity of polymers starting from connectivity index formalism (Bicerano 1996):

$$k_m = 0.135614 + 0.126611 \cdot {}^1\chi^{BB}/N + 0.108563(N_N + N_O - 0.125N_H)/N \quad (11.3)$$

where ${}^1\chi^{BB}$ is the fraction of first-order connectivity index that contributes to the bonding among the backbone atoms; N_N, N_O, and N_H are the number of nitrogen, oxygen, and hydrogen atoms in the structural unit, respectively; and N is the number of non-hydrogen atoms.

Different types of heat transfer processes are called modes. The main modes of heat transfer are convection, radiation, and conduction. For a temperature gradient that exists between a surface and a moving fluid, one should use the term convection. The radiation mode of heat transfer is driven by electromagnetic waves emitted from all surfaces of finite temperature, so there is a net heat transfer by radiation between two surfaces at different temperatures. When a temperature gradient exists in a stationary medium, heat flows under the law of conduction heat transfer. In the case of solid materials, such as polymers, conduction is the dominant mechanism for heat transfer, involving mainly lattice vibrations and, in few cases, the transfer of kinetic thermal energy from one electron to another.

The polymer matrices in advanced thermally conductive composites include thermosetting and thermoplastic types. Figure 11.1 shows thermal conduction properties of some typical polymer matrices. The most used polymers for composite

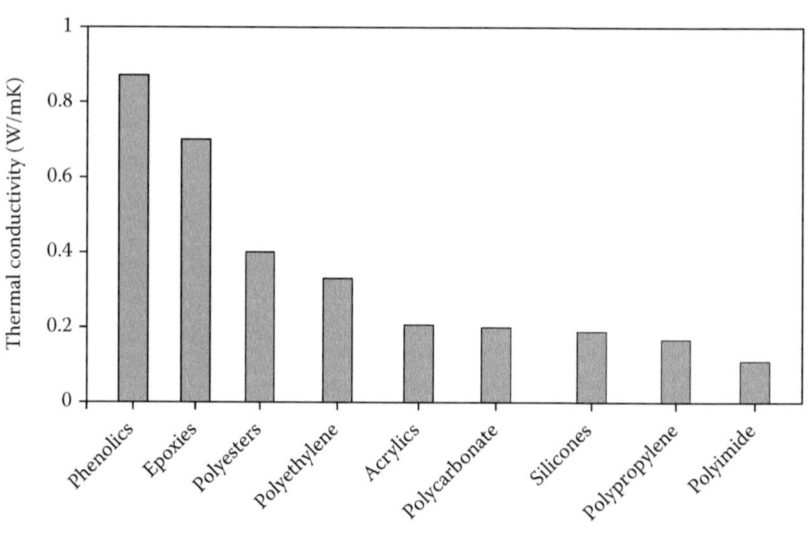

FIGURE 11.1 Thermal conductivity of typical polymer matrices used in preparation of polymer composites. (Adapted from Tong Xingcun Colin. *Advanced materials for thermal management of electronics packaging*. New York: Springer, 2011.)

preparation are phenolics, epoxies, polyesters, polyethylene, acrylics, polycarbonate, silicones, polypropylene, and polyimides.

11.2.2 Reinforced Polymers

11.2.2.1 Theoretical Modeling

Various types of fillers with high thermal conductivities (such as graphite, metals, and ceramics) have been added to polymers to create composite materials with improved properties. Many modeling efforts have been made to predict the effective thermal conductivity of polymer composite materials. For small amounts of reinforcement, the effective thermal conductivity of the polymer composite generally increases linearly with an increasing volumetric fraction. However, at a high percent of reinforcement, the thermal conductivity behaves nonlinearly with the volumetric fraction. Several analytical and numerical models have been proposed for the dispersed phase; at small volume content, the filler seems to be evenly dispersed in the homogeneous phase of the matrix. Hence, in this case, there is no contact among reinforcement particles. At a higher volume content, it is very probable that fillers are brought in contact with others, thus forming conductive paths. If these paths are aligned along the direction of the temperature gradient, they will create a preferable path for heat transfer. These kinds of conductive chains may behave differently with percolation of electrically conductive connection (Tong 2011; Barzic et al. 2014). However, this arrangement of thermally conductive filler inside the polymer subsequently generates a drastic increase in the effective thermal conductivity of the composite material. The Maxwell–Eucken (Tong 2011) model allows evaluation of the effective thermal conductivity of composite material at low reinforcement amounts (equation [11.4]). They considered spherical shape fillers that are randomly distributed inside the polymer. The results provided by this model are suitable only for low amounts of filler.

$$k_c = k_m \frac{2k_m + k_p + 2\phi_p\left(k_p - k_m\right)}{2k_m + k_p - 2\phi_p\left(k_p - k_m\right)} \tag{11.4}$$

where the subscript "m" is for matrix, "c" is for composite, and "p" is for the particles introduced in the matrix; ϕ_p is the volume fraction of filler.

The Agari–Uno model (Agari and Uno 1986) shows that the thermal conductivity data are deviating from the Maxwell–Eucken equation at about a 10% volumetric fraction. At a high filler volume, the measured value was much higher than the prediction using existing correlations. This approach is based on series and parallel models and considers the crystallinity of both phases and interfacial factor A, as shown in equation (11.5):

$$\lg k_c = \lg A + \phi_p C_p \lg k_p + (1 - \phi_p) \lg C_m k_m \tag{11.5}$$

Semi-empirical correlation has been developed to fit the data obtained from experiments. The resulting analytical and numerical models work well for the disperse phase and are reliable as long as the filler is homogeneously dispersed. However, if the filler is in the attached phase, only a few models give promising results. Rayleigh extended the Maxwell approach (Tavman 2003) and developed the effective medium approximation model, assuming an isotropic distribution of spherical fillers embedded in the intersections of a cubic lattice (at a high volume fraction):

$$\frac{k_c}{k_p} = 1 + \frac{3\phi_p}{\left(\frac{k_m + 2k_p}{k_m - k_p}\right) - \phi_p + 1.569\left(\frac{k_m - k_p}{3k_m - 4k_p}\right)\phi_p^{10/3}} \quad (11.6)$$

Since then, many modifications have been added to this model, making it increasingly sophisticated.

Other models take into consideration the effects of shape, size, and interfacial resistance on thermal conductivity. However, these models are unable to predict the effective thermal conductivity accurately if contact among filler particles exists. The Cheng and Vachon (Tavman 2003) model assumes a parabolic distribution of disperse phase (spheres or fibers) in a solid matrix. When $k_m > k_p$, thermal conductivity of the polymer composite is given by equation (11.7):

$$\frac{1}{k_c} = \left[\frac{1}{\sqrt{C(k_p - k_m)(k_m + B(k_p - k_m))}} \cdot \ln \frac{\sqrt{k_m + B(k_p - k_m)} + \frac{B}{2}\sqrt{C(k_p - k_m)}}{\sqrt{k_m + B(k_p - k_m)} - \frac{B}{2}\sqrt{C(k_p - k_m)}} + \frac{1 - B}{k_m}\right] \quad (11.7)$$

where $B = \sqrt{\frac{3\phi_p}{2}}$ and $C = -4\sqrt{\frac{2}{3\phi_p}}$.

The Lewis–Nielsen model considers the effect of the shape of the filler and the orientation or type of packing for a two-phase system or single-phase reinforcement composite, resulting in equation (11.8) for effective thermal conductivity (Tavman 2003):

$$k_c = k_m\left[\frac{1 + AB\phi_p}{1 - B\psi\phi_p}\right] \quad (11.8)$$

where $A = K_E - 1$; K_E is the Einstein coefficient; $B = (k_p/k_m - 1)/(k_p/k_m + A)$; and $\psi = 1 + \phi_p \cdot \left[(1 - \varphi_m)/\varphi_m^2\right]$. The constant A depends on the shape and orientation of dispersed filler; φ_m is the maximum packing fraction of the dispersed particles. The data for parameters A and φ_m are given in literature for different geometrical shapes and orientations of the filler.

Contact resistance between matrix and particles (known as Kapitza resistance) was investigated by Hasselman (1987), who developed formulas for spherical, cylindrical, and flat-plate geometry to estimate the effective thermal conductivity of polymer composites with interfacial thermal barrier resistance—equation (11.9).

$$k_c = \left[\frac{k_p(1+2\kappa) + 2k_m + 2\phi_p\left(k_p(1-\kappa) - k_m\right)}{k_p(1+2\kappa) + 2k_m - \phi_p\left(k_p(1-\kappa) - k_m\right)} \right] k_m \qquad (11.9)$$

where $\kappa = a_K/a_p$ and $a_K = R_K k_m$ in which R_K is the Kapitza resistance.

This approach is suitable for a much-diluted system. The particle size impact on the effective thermal conductivity is minimum if the interfacial resistance was zero or infinite. In case of a finite resistance, decreasing the particle size usually reduces the effective thermal conductivity for a given volume amount.

For the polymers containing filler that touch each other, the percolation theory has been developed. This assumes a sharp increase in the effective conductivity of the disordered media, polymer matrix composite, at a critical volume fraction of the reinforcement known as the percolation threshold (ϕ_{percol}) at which long-range connectivity of the system appears. The model that best expresses these aspects is the one created by Vysotsky (Vysotsky and Roldughin 1999), which presumes a percolation network of nanofiller particles inside the polymer matrix as shown in equation (11.10):

$$k_c = k_p\left(k_{percol}/k_p\right)^{\left(\frac{1-\phi_p}{1-\phi_{percol}}\right)^n} \qquad (11.10)$$

where n is the percolation network exponent, which depends on several factors such as shape, size, and distribution of the filler.

11.2.2.2 Computational Modeling

The computing methods, especially those based on meshing, are able to describe the heat conduction in polymers with various reinforcement degrees. The analysis procedure involves (Rightley et al. 2007) the following steps: (1) generating a computational geometric model representing the structure of the composite; (2) simulating the heat transfer across the multiphase material for a set of boundary conditions using a finite element thermal analysis tool; and (3) calculating and evaluating the results.

For solid materials, one can use a two-dimensional (2D) or three-dimensional (3D) finite element mesh generation toolkit—CUBIT (Rightley et al. 2007). Usually, a geometry file can be ported into CUBIT to create a computation mesh system for analysis on a PC platform. Among the meshing techniques provided, one can mention are 3D mapping, multiple sweeping, paving, plastering, and whisker weaving. In addition to mesh generation, CUBIT presents complex solid modeling capabilities, such as advanced geometry creation and modification, Boolean operations, web cutting, healing, defeaturing, auto decomposition, and nonmanifold topology (Tong 2011).

Three-dimensional heat transfer analysis can be performed using a computational program such as Calore. It is built upon the SIERRA finite element framework to run on a computer in a manner that relates mesh management, external field parameters, and other specific algorithms (Rightley et al. 2007). Advanced thermal analysis capabilities consist of anisotropic conduction, enclosure radiation, thermal contact, and chemical reaction. Calore also exhibits multiple computational features, including mesh adaptivity, element death, and automatic selection of the time step size. For a given degree of reinforcement, the measured thermal conductivity of the composite is repeatable, even though the filler is randomly distributed in the polymer. These measured values generally follow a distribution with a narrow standard deviation, showing that the samples are prepared from a well-controlled and consistent manufacturing process. The computational model is constructed by comparing its predictions against measurement. For a low volume content of filler that is randomly distributed, the numerical algorithm searches and puts a reinforcement agent into an open space within a defined volume. However, when the volume amount of filler gets too large, the "search-and-place" algorithm becomes inefficient and time-consuming. To solve this problem, a new numerical algorithm was created, which has filler initially distributed in a crystallographic structure. Then for the next 100 iterations, it allows the filler to move randomly. This approach functions very well for constructing a computational model and for calculating the effective thermal conductivity of a reinforced polymer. Nevertheless, randomly distributed filler particles have a drawback because they produce significant variations in the calculated thermal conductivity. A large number of simulations are required to reduce the noises and calculate an average value. Obtaining an average is important for recognizing the pattern of thermal conductivity behavior as a function of volume content of filler.

When chains of attached filler are generated in the direction of a temperature gradient, a preferable path for heat conduction will be formed, determining a major reduction of temperature at the hotspot and, hence, a higher effective thermal conductivity. The effect of nonzero contact resistance between adhesive polymer and filler on the overall effective thermal conductivity of the composite was examined. When the reinforcement particles interact with each other, the effective thermal conductivity of the composite is not significantly changed if the contact resistance between polymer and filler increases significantly. For those cases in which the filler particles are dispersed evenly without any contact with any other particle, the effective thermal conductivity of the composite does change drastically.

11.3 MEASUREMENT TECHNIQUES OF THERMOPHYSICAL PROPERTIES

The selection of the thermal management materials for electronic packaging purposes demands close examination of thermophysical characteristics, such as thermal conductivity and diffusivity, specific heat capacity, coefficient of thermal expansion, and thermal shock resistance. A variety of measurement techniques have been developed to evaluate these properties, but this chapter focuses on thermal conductivity and diffusivity evaluation methods. Each of them is suitable for a limited range of materials, depending on the thermal properties and the medium temperature. The precise determination of the thermal properties of bulk composite materials is challenging. For instance, loss terms for the heat input intended to flow through the sample usually exist and can be difficult to quantify.

Measurement methods can be divided into two classes: steady-state and non-steady-state (or transient) methods. The steady-state techniques involve experiments in which the temperature of the studied material does not change with time. The major drawback arises from the fact that the method involves a well-engineered experimental setup. Classical steady-state methods allow evaluation of the temperature difference across the specimens in response to an applied heating power, either as an absolute value or with regard to a reference material put in series or in parallel to the composite sample. The problem is that these methods are often time-consuming and demand samples of a large thickness (Tong 2011).

On the other hand, transient techniques involve a measurement during the heating up process. The advantage is that experiments can be performed relatively fast because there is no need to wait for a steady state. The disadvantage is that the mathematical analysis of the data is more difficult. Among the non–steady-state techniques, the most important are hot wire and hot plate, temperature wave, and laser flash methods (dos Santos 2007; Nikolaev 2007). Among these, laser–flash thermal diffusivity measurement is widely used; it is a relatively fast method that uses small composite samples (Gaal et al. 2004). Such experiments involve irradiation of the reinforced polymer surface with a very short laser pulse, and the temperature rise is evaluated on the opposite side of the specimen—allowing estimation of the thermal diffusivity of the material—after proper mathematical elaboration. For conductivity measurement, steady-state methods (up to 1200 K) and pulse methods (over 1500 K in particular) are usually utilized. Transient methods are preferred for diffusivity measurement. This demands an accurate recording of the time dependence of temperature following a transient or periodic temperature perturbation at a sample boundary. As a result of the wide domains of thermal property, there is no single technique that can be used for measurement of either property (thermal conductivity, in particular). To achieve acceptable values for the evaluated property, the material type and its range of property values above its operational temperature range determine the type of method used as well as the dimension and conjunction of the tested composite and apparatus. In order to choose an adequate technique, the following criteria must be taken into consideration: (1) appropriate sample size and shape; (2) specific temperature range for each technique; and (3) the magnitude of thermal conductivity (because low-conductivity materials require different experimental approaches than high-conductivity multiphase materials). The main techniques used for measurement of thermal conductivity and diffusivity of polymer composites are depicted in Figure 11.2. Two of the most important are further described here: the guarded plate and laser flash techniques.

The guarded hot plate is traditionally known to be the only absolute method for thermal conductivity determination in a steady state of homogeneous single or multiphase materials. It allows a global measurement of uncertainty under 2% (Nikolaev 2007; Tong 2011). This method reproduces a uniform, unidirectional, and constant thermal flux density existing through an infinite homogeneous sample (with plate shape) which is placed between infinite planes of constant temperature. The examined specimen is sandwiched between an electrically heated hot plate kept at a higher temperature and a cold plate maintained at a lower temperature. The heat dissipated by the Joule effect in the hot plate travels to the cold plate through the sample but also backward and laterally on the edges of the hot plate. Back and lateral guard zones are

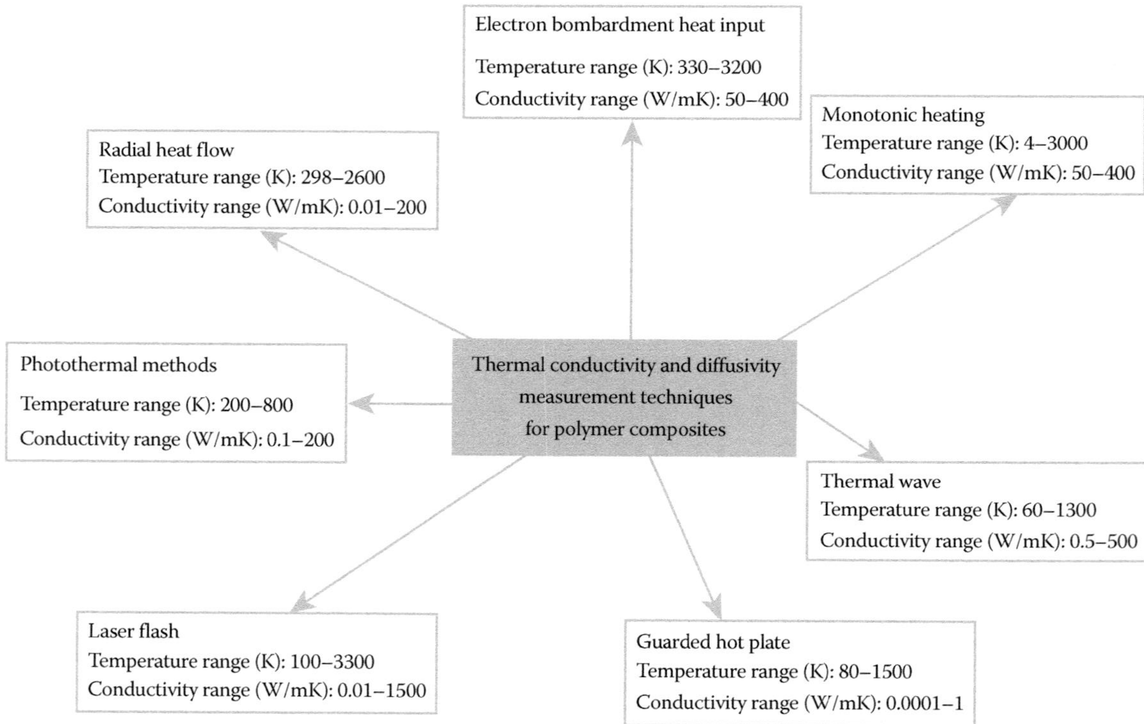

FIGURE 11.2 Schematic representation of the main techniques used in the determination of thermal conductivity and diffusivity of reinforced polymers.

then mandatory to neutralize these thermal leaks. In this case, all unwanted thermal transfers are canceled by keeping the different guard zones at the same temperature with precise control. Thus, the hot plate is divided into two parts. The first one is represented by the heater or measurement zone, which presents a known area, and is heated electrically. The second zone is an adjacent ring heated with a different circuit. These parts of the experimental setup are separated by a small gap, and the temperature difference between them is controlled by a thermopile. An additional guard plate is placed at the back face of the hot plate to provide protection from backward thermal fluxes. This back guard plate is subjected to high temperature from a separate system and is associated with appropriate temperature differential control as in lateral gap. If all three elements are found to be at exactly the same temperature, the heat generated in the measurement plate will flow unidirectionally toward the cold plate. All the other heat fluxes are supposed to be removed. Using a guarded hot plate device, represented schematically in Figure 11.3, the thermal conductivity is obtained by the one-dimensional Fourier equation.

The laser flash method is a relatively newer technique used to evaluate thermal diffusivity for a variety of different materials (Gaal et al. 2004). An energy pulse heats one side of a plane-parallel sample. The temperature increase on the backside caused by the energy input is time-dependently detected. The energy reaches the backside in a shorter time if the specimen's thermal diffusivity is higher. In a vertical setup, a flash lamp heats the sample from the bottom side, and the augmentation of the temperature is monitored by a detector placed on top (see Figure 11.4). For measuring the thermal diffusivity, which is strongly temperature-dependent, the sample can be positioned in a furnace at a elevated temperature. The best results are obtained when experimenting with homogenous material and input energy, but also when the laser pulse has time dependence similar to Dirac delta function. The obtaining of accurate data involve utilization of a reference sample with given density, for which is easy to determine specific heat and implicitly thermal conductivity.

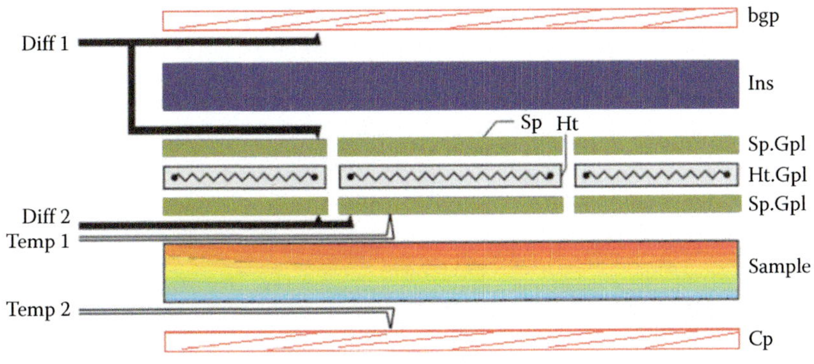

FIGURE 11.3 Schematic representation of a hot guarded plate device, where Cp = cold plate, Diff 1 = imbalance sensing element 1, Diff 2 = imbalance sensing element 2, bgp = back guard plate, Ht = measurement zone heating element, H.Gpl = lateral guard heating element, Ins = thermal insulation, Sp = measurement zone surface plate, Sp.Gpl = lateral guard surface plate, Temp 1 = temperature sensor 1, and Temp 2 = temperature sensor 2.

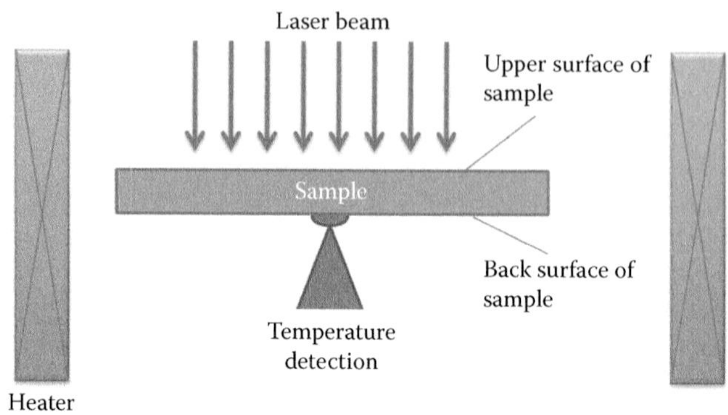

FIGURE 11.4 Schematic representation of a laser device.

11.4 DESIGN OF THERMALLY CONDUCTIVE MULTIPHASE MATERIALS

The heat conduction ability of polymers is considerably enhanced by adding fillers with appropriate thermal features. Various types of fillers or reinforcements currently are used to produce multiphase polymer materials with high-performance thermal characteristics. Their classification and thermal conductivity values are presented in Figure 11.5.

The thermophysical properties of multiphase systems are affected by matrix and filler characteristics. In the case of the polymer phase, the microstructure is the most important feature that influences thermal conduction ability. When discussing the filler, one must take into consideration filler physicochemical properties but also several microstructural parameters, such as the diameter, length, shape, distribution, volume fraction, the alignment, and the packing arrangement. Fillers may be in the form of fibers or particles uniformly or randomly placed in the polymer matrix material. Therefore, thermal conduction of particle-filled polymers is isotropic, although

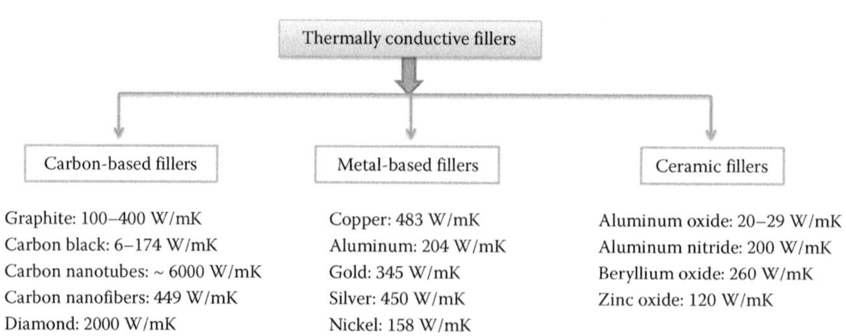

FIGURE 11.5 Classification of the thermally conductive fillers primarily used for multiphase polymer preparation.

for fibers or tubular-filled composites, the thermophysical properties are anisotropic. Exceptions from/to latter case are noticed for very short, randomly distributed fibers (Tavman 2003). Among the properties that can be affected by particle size and shape are electrical conductivity, thermal conductivity, tensile strength, viscosity, weight loss, and rheology. Starting from these aspects, one may divide thermally conductive multiphase materials into two categories: isotropic and anisotropic.

11.4.1 Materials with Isotropic Thermal Conductivity

Reinforcement of polymers with spherically shaped particles has been widely investigated (Tekce et al. 2007; Chauhan et al. 2012). The use of such reinforcement agents decreases the viscosity, although flakes, needles, and fibers offer ideal overlapping conditions for isotropic adhesive-conducting materials. The manufacturer therefore can optimize the composite properties to a specific use by mixing filler particles with different shapes. Another factor important in processing such composites is the solvent in which the filler is dispersed. It should have a high flash point and low vapor pressure because of heat generation. In order to ensure a homogeneous dispersion of the filler inside the polymer matrix, one should select an appropriate surfactant (Ghorabi et al. 2012). Its suitability is dictated by its solubility in the utilized solvent for providing even distribution. The surfactant has major roles in preventing the particle agglomeration (fluid energy milling) and in avoiding cold welding (ball milling) in both milling methods. In most cases, fatty acids are used as surfactants. For thermal interface materials, it is essential for the resulting conductivity of the final product that the surfactant exhibits compatibility with the polymer matrix. Moreover, the size of the reinforcement particles is important. The effect of the particle size of alumina on the thermal conductivity of silicone rubber was reported (Zhou et al. 2007). Thermal conductivity of such a composite was improved with alumina particles with a smaller diameter at the same volume fraction.

The polymer performance and production efficiency can be enhanced as a function of the basic features of the reinforcement fillers. In the attempt to achieve fillers with increased performance, the following features must be monitored: density, flame retardancy, mechanical resistance, thermal conductivity, and magnetic properties. Nanoparticles of carbides, nitrides, and carbonitrides can be used to reinforce polymer matrix nanocomposites with desirable thermal conductivity. However, current trends in the design of these materials reveal that is not enough to choose a well-performing material for each component of the heat dissipation path. In addition, careful attention must be paid to the manner in which these materials interact with each other. A filler that conducts heat well but does not wet the matrix may lead to poor results compared to a lower conductivity filler that does wet the matrix. In other words, a major fact that leads to interfacial resistance is faulty physical contact between filler and matrix, which primarily depends on surface wettability (Han and Fina 2011).

From a theoretical point of view, the scattering of phonons in multiphase systems is determined by the existence of an interfacial thermal barrier from acoustic mismatch. In a simplified model, the transmission of a phonon between two phases is affected by the presence of common vibration frequencies for the two phases. Thus, it can be assumed that only low frequency phonon modes of fillers are effective when they interact with a polymer only through weak dispersion forces (Shenogin et al. 2004). Phonon scattering at the interfaces and at plating/surface finish interfaces should be optimized in a thermal design. The investigations made in this direction led to four approaches that help in designing products/devices with lower total thermal resistance (Tong 2011):

- Enhance thermal conductivity of the interface materials and bulk heat spreading and dissipating materials.
- Increase wetting or bonding to reduce contact resistance at the surface.
- Enlarge flatness of heat conductive component and reduce the interfacial thickness with adjacent component.
- Diminish the number of interfaces in the thermal management package.

11.4.2 Materials with Anisotropic Thermal Conductivity

Reinforcement agents shaped like fibers or tubes exhibit strong anisotropic thermal conduction abilities. For instance, carbon nanotubes (CNTs) present high conductivity along the tube axis and reduced thermal conduction abilities perpendicularly to it. In the case of nanotubes with multiple walls, the thermal conductivity at room temperature increased with decreasing of the diameter and the number of walls. In the latter case, thermal conductivity ranged from about 500 W/mK for an outer diameter of 28 nm to 2069 W/mK for a 10-nm diameter (Han and Fina 2011). Moreover, when the tube length is much longer than the mean free path of the energy-carrying phonons, the thermal conductivity becomes constant. On the other hand, carbon nanofibers (CNFs) present a thermal conductivity in the range of 12–1950 W/mK (measured along their long axis). This large difference can be attributed to various factors such as sample diameter, synthesis method, structural defects, and heat treatment (Mayhew and Prakash 2013). However, CNFs present poor thermal conductivity along their radial direction—lower by three orders of magnitude than those along their axial direction.

Localized aggregation of fillers can be achieved on the basis of colloidal-physics methods using polymer emulsions or latexes. When combined with nano sized filler, the polymer macromolecules generate excluded volume that creates a segregated network of filler. It may be possible to form a segregated network of anisotropic shape fillers and enhance thermal conductivity at very low concentrations (Han and Fina 2011). For example, semicrystalline polyurethane dispersions were used as a latex host to accommodate CNTs following colloidal-physics concepts. The continuous nanotube-rich phase was formed in the interstitial space among the latex particles. This structural arrangement determined an enhancement by ~210% in thermal conductivity, namely this parameter increased from 0.15W/mK to 0.47 W/mK at CNTs of 3 wt% (Cai and Song 2008). In this context, the colloidal-physics method appears to be useful in controlling the segregation of fillers (such as CNTs) and in forming an interconnecting network to obtain thermally conductive polymer/CNT composites at a low CNT amount.

There are also other methods that allow controlled segregation of conductive particles. In particular, co-continuous immiscible polymer blends have proven to be an effective method to build a percolation network for enhanced conductivity purposes (Pötschke and Paul 2003). Co-continuous structures are able to create conductive paths in the multiphase material, selectively dispersing reinforcement particles, such as CNTs in one of the phases. For instance, CNT master batches represent a good alternative for forming co-continuous structures, generating significant reduction in electrical resistivity due to the interconnecting network of fillers restricted by polymer chains. Another approach relies on introducing of preformed conductive networks leads to multiphase materials with precise control of conductive pathways through their volume. For example, Ji et al. (2009) obtained CNT/carbon/polystyrene composites by in situ polymerization of a monomer in preformed CNT/carbon foams, which were produced from a template and had a more continuous conductive structure than the CNT networks created by other methods in composites. The thermal conductivity of composites with 1 wt% amount of CNT/carbon foam led to augmentation of 37%, and this further increased to 155% when 20 wt% of CNT/carbon was inserted.

Thermal transport in such multiphase materials can be made more efficient by aligning the nanofiller in a certain direction. Alignment is understood as a preferred orientation of a fiber or tube (in its longitudinal axes) within a 3D sample. This may provide a great advantage when thermal conductivity enhancement is desired in a preferential direction, which is similar to the case of a pure filler network (Barzic and Barzic 2015). Fillers with anisotropic shape, such as CNTs or CNFs, can be oriented in polymer composites by processing, CNT arrays, or by external force.

Alignment of fillers with high aspect ratios often takes place during processing of the composite material. The degree of alignment increases with decreasing fiber/tube diameter (due to greater extensional flow) and decreases with increasing reinforcement amounts (due to agglomeration and restrictions in motion from neighboring particles). When multiphase systems are prepared by the melting method, CNTs or CNFs tend to orient parallel to the flow direction. The good dispersion and alignment of fillers with an anisotropic shape, combined with good adhesion with the matrix are essential for achieving mechanical and thermal properties' improvement. When thermal conductivity has been measured parallel to alignment direction of the filler, it was found to be higher in comparison with the same sample with random filler orientation. This suggests that heat conduction along aligned CNTs or CNFs is more successful. Moreover, addition of isotropic Al particles lowers the difference in conduction in the axial and transverse directions and thus acts as a connecting factor between aligned pathways (Ghose et al. 2008). The orientation of CNTs can be related to the deformation applied during the optimized milling process, which results in obvious improvements in the heat conduction ability of the composite (Han and Fina 2011).

Alignment of fillers with a high aspect ratio can also take place if they are aligned in the matrix using preformed CNT arrays. Huang et al. (2005) prepared silicone elastomer/CNT composites by using in situ injection of matrix into CNT arrays. They showed that thermal conductivity of aligned composites is 105% higher than that of pure elastomers. Similar enhancements in heat conduction were obtained for composites processed by infiltrating polymers in aligned multiwall CNTs. The overall performance of the filler array interface is affected by thermal resistance of the contacts between filler ends and opposing substrate surface or the opposing array. A good solution is functionalization by microwave curing that creates stronger bonding between the two phases (Brunetti et al. 2008).

Alignment of fillers, particularly CNTs, can also be realized by applying external fields, such as magnetic (Gonnet et al. 2006), electric (Park et al. 2006), or force fields (stretching process).

The electric field-assisted alignment of fillers was widely investigated for optimization of conduction features of composites. Several factors, such as magnitude, frequency, and application time of the applied electric field, were found to influence the degree of reinforcement particle alignment. One study (Park et al. 2006) made of CNTs revealed that, below a critical value of electric field, there is no orientation; after which, the degree of alignment starts to increase up to a plateau value. Increasing the voltage beyond this point resulted in more skewed alignment near the electrodes, possibly due to excessive thermal energy. Notice should be taken regarding Joule heating (higher than 100°C), which tends to locally disrupt or bend the aligned CNTs. At a fixed voltage, alignment began almost instantly and grew with increasing field application time until saturation (around 10 min). In the case of field frequency, the degree of orientation increased with increasing frequency, saturating at 10 Hz. Interestingly, the filler started to align to some degree at frequencies as low as 0.001 Hz, which is effectively a DC condition, although this requires a much longer field application time.

Gonnet et al. (2006) accomplished in-plane CNT alignment under a strong magnetic field. However, similar thermal conductivities were noticed in the aligned and perpendicular direction of the nanocomposites with epoxy matrix, even though the magnetically pre-aligned fillers with anisotropic shape exhibited greater thermal conductivity along the alignment direction.

Stretching the matrix represents another method of aligning CNTs or CNFs. This is limited to polymers that present elastomeric characteristics. The change in orientation on stretching can be reversible (i.e., equilibrium) with the orientational order parameter becoming null after the imposed strain is removed (Ahire et al. 2008).

11.5 TYPES OF MULTIPHASE POLYMERS FOR THERMAL MANAGEMENT APPLICATIONS

11.5.1 Polymer–Carbon Systems

Carbon fibers derived from polyacrylonitrile present low heat characteristics, so these type of composites are suited for applications where this aspect is mandatory. The investigation of polymers reinforced with pitch-based carbon fibers, vapor grown carbon fibers, carbon foam, and CNTs having high thermal conductivities opens up significant new markets based on multiple potential applications.

An essential aspect that should be considered in the analysis of competing thermal management materials is that convection often is the limiting factor, rather than material thermal conductivity. This is the reason for which, in many cases, heat dissipation is not improved by increasing thermal conductivity beyond a certain value.

So, in certain designs, the solution was to replace aluminum—and even copper (Cu)—with carbon-reinforced polymers having much lower thermal conductivities. The use of polymer–carbon composites is growing considerably. They have been used in thermal applications as heat conductors or insulators owing to their thermal conductivity and thermal expansivity that can be ranged and even oriented (as in the case of fillers with high aspect ratio) in a preferred direction. Structural features are by far the most studied and applied for this type of composite as reflected in the numerous studies of a variety of carbon-dispersed phase structures, from macrolaminates to the CNTs (single and multiwalled). When supplementary reinforcement with anisotropic fillers that can act as a local heating source is performed, some composites display self-healing of cracks without significant opening.

11.5.2 Polymer–Metal Systems

Polymers reinforced with metal fillers possess intrinsic advantages including low cost and ease of processing, high electric conductivity, paramagnetism, high thermal conductivity, as well as good mechanical properties. The attention toward this type of materials was attracted mainly by the fact that the electric and magnetic characteristics of such materials are close to those of metals. Insertion of aluminum particles in polyethylene oxide led to a considerable increase in the conductivity of a pure polymer by over one with an order of magnitude of 1–2 wt%. The advantage of using aluminum instead of a ceramic as the filler is that maximum conductivity is achieved for the latter at a concentration of around 40 wt%. This is reflected in the cost of the material's production.

Among the metals used as fillers for polymers are copper and nickel powders. For systems based on epoxy resin or poly(vinyl chloride), the investigation of concentration dependence on the electric conductivity of the resulting composites reveals that less than 1 vol% was enough to attain the percolation concentration and a big jump in electric conductivity (around 12 orders of magnitude). On the other hand, thermal conductivity increases linearly with the metal powders' concentration. The advantage of the fluidity of polymer melt and the ductile properties of Cu films were exploited to produce multiphase systems by the Cu metallization of polystyrene (PS) bead and the hot press molding of Cu-plated PS beads. The hexagonal-shaped 3D Cu shell networks in the PS matrix demonstrated isotropic conductive abilities at extremely low Cu amounts. In contrast to the conventional simple melt-mixed Cu beads/PS composites at the metallic filler concentration of 23.0 vol%, the PS composites with Cu shell networks indeed showed 60 times larger thermal conductivity (Yu et al. 2013). Electrically and thermally conductive composites were prepared from a high-density polyethylene (HDPE) matrix blended with a special grade of branch-structured nickel particles. The thermal conductivity obtained using this filler content was found to be 1.99 W/mK. The percolation concentration of the filler within the HDPE matrix, as revealed by electrical conductivity measurements, was determined to be 8 vol% (Krupa et al. 2013). Other polymers modified by reinforcement with metallic particles include polyethylene (Kumlutaş et al. 2003), polypropylene (Boudenne et al. 2005), polyamide (Tekce et al. 2007), polyvinylchloride, and epoxy resins (Mamunya et al. 2002), highlighting the fact that thermal conductivity performance is affected by the thermal conductivity of the metallic fillers, the particle shape and size, the volume fraction, and the spatial arrangement in the polymer matrix.

11.5.3 Polymer–Ceramic Systems

Polymer–ceramic composites are composed of an organic matrix and inorganic fillers, such as oxides or silicates. The thermal features of these composites are determined mainly by chemical cross linking among the organic components and among the organic components and functional groups on the surfaces of the filler particles. Polymer–ceramic composites combine typical properties of ceramics—such as high thermal stability, thermal conductivity, and dimensional stability—with typical characteristics of plastics, namely plastic formability and functional variability. This allows the processing of such materials by plastic-forming techniques, in particular, injection molding or solution casting. Fabrication of composites by solution blending of aluminum nitride with poly(4-vinylpyridine) led to materials with a thermal conductivity 69.15 times higher than with the pure polymer, owing to the percolation network formation being rheologically analyzed (Barzic et al. 2014).

This category of multiphase materials has been used for power electronics, where they must endure high thermal loads. Therefore, they often are encountered in the encapsulation of high-performance resistors or metallic contact bars in heating registers. Other practical purposes are focused on adhesives for attaching electrical components to printed circuit boards, high power components such as transistors for heat sinks, laminates, hybrid supports for electrical components, and encapsulating compositions for use in integrated circuits where dissipation of heat is the main requirement of the encapsulant. In this context, thermal interface features are investigated for producing heat-reducing grease from silicone oil (as a base material) and a zinc oxide or alumina powder, as well as aluminum nitride as a thickener to achieve improved thermal conductivity. Thixotropic thermally conductive dielectric materials have found particular application in the cooling of electronic components, especially in direct contact with integrated circuit chips (Tong 2011). The composition of the materials contains a liquid silicone carrier, a thermal filler powder chosen from the group consisting of lamellar aluminum nitride, dendritic zinc oxide, lamellar boron nitride, or mixtures thereof, in combination with silica fibers that function as a bleed inhibiting agent. Polymer ceramic composites have also been used in the energy storage, electric, electronic, biomedical, and structural fields. Alumina and zirconia are fillers often inserted in polymers to obtain high-performance electrolyte composites.

11.5.4 Polymer–Combined Filler Multiphase Systems

Polymers can be filled with hybrid particles or with a combination of fillers in order to better control their thermophysical behavior (Lee et al. 2006; Choi and Kim 2013). However, there are relatively few reports of polymers containing mixed fillers for heat conduction purposes. To improve thermal conductivity of an epoxy, silicon carbide microparticles were combined with graphite nanoplatelets. The bidimensional structure with the latter's high aspect ratio enables use as heat conductive bridges among tridimensional microparticles, thus contributing to the formation of a more efficient 3D percolating network for heat flow. This leads to higher thermal conductivity at relatively lower filler contents, which is important for decreasing the density and viscosity and for improving the processability of composites. Thermal conductivity was enhanced 26.1 times compared to the matrix for 7 wt% with graphite nanoplatelets + 53 wt% silicon carbide (Zhou et al. 2013).

Other combinations of fillers can be found in functionalized multiwalled CNTs within amino groups and micro–aluminum nitride particles (Teng et al. 2012). To form a covalent bond between the fillers and the epoxy matrix, poly(glycidyl methacrylate) (PGMA) was grafted onto the surface of CNTs. This led to good dispersion and interfacial adhesion among multiple phases, so the hybrid fillers provided an synergetic effect on heat conductive networks. The thermal conductivity of epoxy composites containing 25 vol.% modified aluminum nitride and 1 vol.% functionalized CNTs is 1.21 W/mK, comparable to that of epoxy composites containing 50 vol.% ceramic particles (1.25 W/mK), which can lower by half the quantity of aluminium-based filler used.

A bimodal distribution of polygonal aluminum nitride and planar boron nitride fillers was used for the examination of effects of the particle size and the relative composition in an attempt to enhance heat conduction of epoxy (Hong et al. 2012). The optimal thermal conductive path was strongly influenced by the packing efficiency and interfacial resistance of the particles in a sensitive way. Maximum thermal conductivity was achieved up to 8.0 W/mK in the 1:1 volume ratio of both types of particles. This methodology revealed that different particles should be packed closely to fill up the interstitial space and, simultaneously, the contact resistance and the contact area of the fillers should be balanced to maximum enhancement of the thermal conductivity.

Another study pointed out the effects of multiple carbon fillers on the electrical and thermal conductivity of polycarbonate-based resins (King et al. 2012). Three different carbon fillers (carbon black [CB], carbon nanotubes [CNTs], and exfoliated graphite nanoplatelets [GNPs]) were analyzed via three different combinations of two different fillers (CB/CNT, CB/GNP, and CNT/GNP). In the case of the single fillers, a statistically significant increase was noticed at the 95% confidence level in composite electrical and thermal conduction. But too many filler interactions statistically impacted the composite electrical and thermal conductivity.

11.6 OUTLOOK AND FUTURE CHALLENGES

Significant progress has been made during the last few years in the search for high-thermal-conductivity multiphase polymer materials. There is a large interest in obtaining composites with high thermal conductivity for a variety of practical purposes, such as printed circuit boards, insulated gate bipolar transistors, and rotating machine insulation. One of the major challenges in composites is producing good thermophysical properties and good dielectric properties. Overall, inorganic fillers with high thermal conductivity, low dielectric constant, high electrical resistivity, low coefficient of thermal expansion, and high aspect ratio have been shown to be ideal fillers for thermal conductivity enhancement of dielectric polymer composites. Among the reinforcement agents, boron nitride seems to be a good choice for heat conductive and concomitantly electrically insulating polymer composites. Reduction of the interfacial resistance between the filler particles and the host polymer in a composite remains a major challenge, but perhaps multiple fillers inserted in macromolecular matrices might be the solution.

REFERENCES

Agari Yasuyuki, Ueda Akira, Omura Yoshihiko, and Nagai Susumu. Thermal diffusivity and conductivity of PMMA–PC blends. *Polymer.* 38 no. 4 (1997): 801–807.

Agari Yasuyuki, and Uno Taiko. Estimation on thermal conductivities of filled polymers. *J. Appl. Polym. Sci.* 32 no. 7 (1986): 5705–5712.

Ahire Samit V., Huang Yan Yan, and Terentjev Eugene M. Polymers with aligned carbon nanotubes: Active composite materials. *Polymer.* 49 no. 18 (2008): 3841–3854.

Barzic Razvan Florin, and Barzic Andreea Irina. Thermal conduction in polystyrene/carbon nanotubes: Effects of nanofiller orientation and percolation process. *Rev. Roum. Chim.* 6 no. 7–8 (2015): 803–807.

Barzic Razvan Florin, Barzic Andreea Irina, and Dumitrascu Gheorghe. Percolation network formation in poly(4-vinylpyridine)/aluminum nitride nanocomposites: Rheological, dielectric, and thermal investigations. *Polym. Compos.* 35 no. 8 (2014): 1543–1552.

Bicerano Josef. Prediction of the properties of the polymers from their structures. *J. Macromol. Sci. Part C.* 36 no. 1 (1996): 161–196.

Boudenne Abderrahim, Ibos Laurent, Fois Maura, Majeste Jean Charles, and Géhin Eveline. Electrical and thermal behavior of polypropylene filled with copper particles. *Composites Part A.* 36 no. 11 (2005): 1545–1554.

Brunetti Fulvio G., Herrero M. Antonia, de Muñoz Juan M., et al. Microwave-induced multiple functionalization of carbon nanotubes. *J. Am. Chem. Soc.* 130 no. 25 (2008): 8094–8100.

Cai Dongyu, and Song Mo. Latex technology as a simple route to improve the thermal conductivity of a carbon nanotube/polymer composite. *Carbon.* 46 no. 15 (2008): 2107–2112.

Chauhan Deepti, Singhvi Nilima, and Singh Ramvir. Effect of geometry of filler particles on the effective thermal conductivity of two-phase systems. *Int. J. Modern Nonlin. Theor. Appl.* 1 (2012): 40–46.

Choi Seran and Kim Jooheon. Thermal conductivity of epoxy composites with a binary-particle system of aluminum oxide and aluminum nitride fillers. *Composites Part B.* 51 (2013): 140–147.

dos Santos Wilson Nunes. Thermal properties of polymers by non-steady-state techniques. *Polym. Test.* 26 no. 4 (2007): 556–566.

Ebadi-Dehaghani Hassan, and Nazempour Monireh. Thermal conductivity of nanoparticles filled polymers. In *Smart nanoparticles technology*, Ebadi-Dehaghani Hassan, Nazempour Monireh (eds.), pp. 519–540. Rijeka: InTech, 2012.

Gaal Peter S., Thermitus Marc Antoine, and Stroe Daniela E. Thermal conductivity measurements using the flash method. *J. Therm. Anal. Calorim.* 78 no. 1 (2004): 185–189.

Ghorabi Shima, Rajabi Laleh, Siavash Madaeni Sayed, Zinadini Sirus, and Ashraf Derakhshan Ali. Effects of three surfactant types of anionic, cationic and non-ionic on tensile properties and fracture surface morphology of epoxy/MWCNT nanocomposites. *Iran. Polym. J.* 21 no. 2 (2012):121–130.

Ghose Sayata, Watson Kent A., Working D.C., Connell John W., Smith Joseph G., Jr., and Sun Ya Ping. Thermal conductivity of ethylene vinyl acetate copolymer/nanofiller blends. *Compos. Sci. Technol.* 68 no. 7–8 (2008): 1843–1853.

Gonnet Philippe, Liang Zhiyong, Choi Eun Sang, et al. Thermal conductivity of magnetically aligned carbon nanotube buckypapers and nanocomposites. *Curr. Appl. Phys.* 6 no. 1 (2006): 119–122.

Han Zhidong and Fina Alberto. Thermal conductivity of carbon nanotubes and their polymer nanocomposites: A review. *Prog. Polym. Sci.* 36 no. 7 (2011): 914–944.

Hasselman Diedricus P.H., and Johnson Lloyd F. Effective thermal conductivity of composites with interfacial thermal barrier resistance. *J. Compos. Mater.* 21 no. 6 (1987): 508–515.

Huang H., Liu C., Wu Y., and Fan S. Aligned carbon nanotube composite films for thermal management. *Adv. Mater.* 17 no. 13 (2005): 1652–1656.

Hong Jung-Pyo, Yoon Sung-Woon, Hwang Taeseon, Oh Joon-Suk, Hong Seung-Chul, Lee Youngkwan, and Nam Jae-Do. High thermal conductivity epoxy composites with bimodal distribution of aluminum nitride and boron nitride fillers. *Thermochim. Acta.* 537 (2012): 70–75.

Ji Lijun, Stevens Molly M., Zhu Yuefeng, Gong Qianming, Wu Jianjun, and Liang Ji. Preparation and properties of multi-walled carbon nanotube/carbon/polystyrene composites. *Carbon.* 47 no. 11 (2009): 2733–2741.

King Julia A., Via Michael D., Mils Owens P., Alpers Daniel S., Sutherland John W., and Bogucki Gregg R. Effects of multiple carbon fillers on the electrical and thermal conductivity and tensile and flexural modulus of polycarbonate-based resins. *J. Compos. Mater.* 46 no. 3 (2012): 331–350.

Krupa Igor, Cecen Volkan, Boudenne Abderrahim, Prokeš Jan, and Novák Igor. The mechanical and adhesive properties of electrically and thermally conductive polymeric composites based on high density polyethylene filled with nickel powder. *Mater. Design.* 51 (2013): 620–628.

Kumlutaş Dilek, Tavman İsmail Hakki, and Çoban M. Turhan. Thermal conductivity of particle filled polyethylene composite materials. *Compos. Sci. Technol.* 63 no. 1 (2003): 113–117.

Lee Geon-Woong, Park Min, Kim Junkyung, Lee Jae Ik, and Yoon Ho Gyu. Enhanced thermal conductivity of polymer composites filled with hybrid filler. *Composites Part A.* 37 no. 5 (2006): 727–734.

Majumdar Arunava. Microscale transport phenomena. In *Handbook of heat transfer*, Rohsenow Warren M., Hartnett James P., Cho Young I., (eds.), pp. 8.1–8.8. New York: McGraw-Hill, 1998.

Mamunya Ye P., Davydenko V.V., Pissis P., and Lebedev E.V. Electrical and thermal conductivity of polymers filled with metal powders. *Eur. Polym. J.* 38 no. 9 (2002): 1887–1897.

Mayhew Eric, and Prakash Vikas. Thermal sonductivity of individual carbon nanofibers. *Carbon.* 62 (2013): 493–500.

Moniruzzaman Mohammad, and Winey Karen I. Polymer nanocomposites containing carbon nanotubes. *Macromolecules.* 39 no. 16 (2006): 5194–5205.

Nikolaev Iavn. An experimental study of soil thermal conductivity using a guarded hot plate apparatus. Dissertation at Ryerson University of Toronto Canada, 2007.

Oechsner Andreas and Murch Graeme E. *Heat transfer in multi-phase materials.* New York: Springer, 2011.

Park Cheol, Wilkinson John, Banda Sumanth, Ounaies Zoubeida, Wise Kristopher E., Sauti Godfrey, Lillehei Peter T., and Harrison Joycelyn S. Aligned single-wall carbon nanotube polymer composites using an electric field. *J. Polym. Sci. Part B.* 44 no. 12 (2006): 1751–1762.

Pötschke Petra, and Paul Donald R. Formation of co-continuous structures in melt-mixed immiscible polymer blends. *J. Macromol. Sci. Part C.* 43 no. 1 (2003): 87–141.

Rightley Michael J., Emerson John A., Wong C. Channy, Huber Dale L., and Jakaboski Blake. Advancement in thermal interface materials for future high-performance electronic applications: Part 1, 2007, http://prod.sandia.gov/techlib/access-control.cgi/2007/070417.pdf. Accessed on January 2016.

Shenogin Sergei, Xue Liping, Ozisik Rahmi, Keblinski Pawel, and Cahill David G. Role of thermal boundary resistance on the heat flow in carbon-nanotube composites. *J. Appl. Phys.* 95 (2004): 8136–8144.

Shtein Michael, Nadiv Roey, Buzaglo Matat, and Regev Oren. Graphene-based hybrid composites for efficient thermal management of electronic devices. *ACS Appl. Mater. Interf.* 7 no. 42 (2015): 23725–23730.

Tavman Ismail Hakki. Thermal conductivity of particle reinforced polymer composites. In *Nanoengineered nanofibrous materials book series NATO science series ii, mathematics, physics and chemistry*, Guceri Selcuk, Gogotsi Yuri G., Kuznetsov Vladimir (eds.), pp. 451–458. Dordrecht: Kluwer Academic Publishers, 2003.

Tekce H. Serkan, Kumlutas Dilek, and Tavman Ismail Hakki. Effect of particle shape on thermal conductivity of copper reinforced polymer composites. *J. Reinforced Plast. Compos.* 26 no. 1 (2007): 113–121.

Teng Chih-Chun, Ma Chen-Chi M., Chiou Kuo-Chan, and Lee Tzong-Ming. Synergetic effect of thermal conductive properties of epoxy composites containing functionalized multi-walled carbon nanotubes and aluminum nitride. *Composites Part B.* 43 no. 2 (2012): 265–271.

Tong Xingcun Colin. *Advanced materials for thermal management of electronics packaging.* New York: Springer, 2011.

Vysotsky V.V., and Roldughin V.I. Aggregate structure and percolation properties of metal-filled polymer films. *Coll. Surf. Part A.* 160 no. 2 (1999): 171–180.

Wang Ru-Min, Zheng Shui-Rong, and Zheng Ya-Ping. *Polymer matrix composites and technology.* Oxford: Woodhead, 2011.

Yang Qiang, Shuai Li, Zhou Jinjin, Lu Fachuang, and Pan Xuejun. Functionalization of multiwalled carbon nanotubes by pyrene-labeled hydroxypropyl cellulose. *J. Phys. Chem. B.* 112 no. 41 (2008): 12934–12939.

Yu Seunggun, Lee Jang-Woo, Han Tae Hee, et al. Copper shell networks in polymer composites for efficient thermal conduction. *ACS Appl. Mater. Interf.* 5 no. 22 (2013): 11618–11622.

Zhou Tianle, Wang Xin, Cheng Ping, Wang Tianchi, and Xiong Dang Sheng. Improving the thermal conductivity of epoxy resin by the addition of a mixture of graphite nanoplatelets and silicon carbide microparticles. *EXPRESS Polym. Lett.* 17 no. 7 (2013): 585–594.

Zhou Wenying, Qi Shuhua, Tu Chunchao, Zhao Hongzhen, Wang Caifeng, and Kou Jingli. Effect of the particle size of Al_2O_3 on the properties of filled heat-conductive silicone rubber. *J. Appl. Polym. Sci.* 104 no. 2 (2007): 1312–1318.

12 Polymer Systems Containing Electrically Conductive Single and Multiphases

Andreea Irina Barzic

CONTENTS

12.1 INTRODUCTION

Electrical properties of materials are described by their behavior in the presence of an electric field. Basically, the response of any material to an electric field can be separated into two main parts: dielectric response and electrical conduction. Polymer-based materials with good conduction properties have been the subject of fundamental scientific interest due to their tremendous potential in various applications (Gul 1996). Even though most polymeric materials are not conductors of electricity, they are easy to fabricate into complex shapes at a reduced expense. In particular, thermoplastics are moldable or extrudable into various shapes and sizes (Thomas et al. 2015). Besides this, the specific weight of industrial standard plastics is about one-third the specific weight of the corresponding metal materials. Hence, polymers that conduct electricity are in demand for various applications in which the conductivity of metal is not required. Macromolecular compounds can be formulated to have specific conductivity features (from very low to very high conductivity), making them desirable for various applications (Kim et al. 2004; Mao et al. 2012). If the level of electrical charge transfer is not enough, modifications to the polymer must be made to increase conductivity. This has resulted in plastics that can be used in four distinct application categories of increasing conductivity:

- Insulating (wire coating)
- Dissipative ("antistatic" polymers)
- Conductive (materials capable of conducting modest amounts of electrical current)
- Highly conductive or shielding (materials capable of conducting a significant amount of electrical current)

In the natural state, most thermoplastic polymers exhibit relatively high surface resistivity (of 10^{12}–10^{16} ohm/sq), which inhibits the conduction of electricity (Solymar and Walsh 2010).

Antistatic polymers allow electrons to travel across the surface from areas of high charge to areas of lower charge (reducing the tendency to create sparks). In general, antistatic materials have a surface resistivity of 10^9–10^{11} ohm/sq.

For conductive polymers, the surface resistivity is lower (about 10^1–10^7 ohm/sq). The continuous evolution of current technologies has stimulated researchers to develop new and low-cost materials with excellent electrical conductivity. To attain this level of conduction, the polymer is reinforced with electrically conductive additives (Grossiord et al. 2008; Jin et al. 2013). The additives introduced in conductive plastics significantly affect the properties of the polymers, but not all of these changes have been beneficial. The tensile properties and stiffness/flexural moduli often increase while the elongation and impact strength are reduced due to the addition of conductive particles.

The chapter describes some fundamental aspects concerning electrical conduction in polymers (in undoped and doped states) and their composites. The current characterization methodologies for electrical conductivity are presented. The performance of polymers containing single or multiple conductive phases is reviewed in accordance with the applications of these multiphase materials in electronics, optoelectronics, and in other fields.

12.2 FUNDAMENTALS OF ELECTRICAL CONDUCTIVITY OF POLYMERIC MATERIALS

12.2.1 Electrical Conduction in Polymers

Electrical conductivity is an intrinsic property that quantifies the ability of a material to conduct an electrical current. Application of a potential difference across a conductor determines its charges to flow, giving rise to an electric current.

The conductivity (σ) is defined by the ratio of the current density to the electric field and is the reciprocal of electrical resistivity (ρ). Generally, materials can be classified from the point of view of their conduction abilities into insulators, semiconductors, and conductors (Drobny 2012).

With regard to polymers, it is widely known that most of them are found in the category of insulators. However, there are a few conducting polymers that display behavior similar to metals and semiconductors. Intrinsically conductive polymers belong to one of five families: polyacetylene, polyaniline, polypyrrole, polythiophene, or polyphenylvinylenes. The common features of these polymers are as follows:

- Conjugated system along the polymer backbone.
- Generally planar structure.
- *Intra*chain conductivity is much larger than *inter*chain conductivity. The anisotropy ratio should be in the range of 10^3–10^4, providing information on the dimensionality of the system. In a certain anisotropy range of conductivity, these systems can be considered unidimensional.

Electrically conducting polymers, in their doped form, are mainly prepared by the following three methods:

- Chemical synthesis of the polymer followed by doping with doping agents
- Chemical synthesis of the polymer and its subsequent doping by an electrochemical method
- Electrochemical polymerization and simultaneous doping with desired dopant (in a single procedure)

The electrical conduction ability of these compounds arises from the delocalized bond structure, which can be enhanced through "doping" with an appropriate agent. Typically, "doping" the conductive polymers involves oxidizing (electron-accepting or p-doping) or reducing (electron-donating or n-doping) the compound. Thus, charge carriers (from the addition or removal of electrons) are inserted into the conduction or valence bands significantly increasing the electrical conductivity. Though there are polymers with many structural similarities, they have a wide range of conductivity depending upon the doping percent, the alignment of the polymer chain, and the purity of the sample. Also, various structural factors are shown to affect the conduction of polymers. Heterogeneity of the chains seems to yield carrier localization, leading to the decreased potential for holes or electrons (Boudenne et al. 2011). On the other hand, the homogeneous character of the backbone affects the width of the highest occupied π-band. The latter is related to the degree of delocalization of the π-systems from the main chain and also to the mobility of the charge carriers in these bands. When the ionization potential is lowered and the width of the highest occupied π-band increases, the band gap decreases, resulting in a better conductor. Thus, low ionization potential and large electron affinity are mandatory for good conduction properties (Drobny 2012).

The nature of processes inducing high conductivity in conducting polymers is distinct from other materials. This is due to two important structural features:

- *Polymers are molecular in character and lack long-range order:* Synergetic effects of molecularity and disorder determine profound differences between the fundamental physics phenomena that occur in traditional covalent network semiconductors and those that are characteristic of polymers.
- *Doping in polymers is a charge transfer reaction:* Doping reactions of polymers give rise to partial oxidation or reduction of polymers. Exposure to a doping agent leads to the formation of a positively (or negatively) charged polymeric complex and a counter ion, which is the reduced or the oxidized form of the oxidant or reductant.

In order to better understand the nature of charge carriers in polymeric materials, the basic notions of solid-state physics are applied. It is known that, for high conductivity, a periodic array of lattice atoms is mandatory for coherent scattering of electron waves. Further, the periodic arrangement of atoms in a crystal and the strong interactions among these atoms cause a widening of energy levels into energy bands. The highest occupied electronic levels constitute the valence band (VB) and the lowest unoccupied levels form the conduction band (CB). The energy difference between the top of the VB and the bottom of the CB is known as the band gap (Eg), and this band gap determines the intrinsic electrical properties of the polymers (Drobny 2012). Insulators present wide gaps between valence and conduction bands, although the energy gaps for semiconductors are narrow (Barzic et al. 2013a, 2013b). Conducting polymers have a conductivity temperature dependence similar to that of semiconductors—namely it increases (exponentially) with increasing temperature and decreasing gap energy (although this is opposite for metals). This suggests that certain aspects of semiconductor theory also may be used for polymers but with certain adaptations of the band model because these materials may exist in amorphous as well as in crystalline form or, more commonly, as a mixture of both. So, according to electron theory and the band structure concept, one needs to know the degree of order (periodicity of the atoms) because only ordered and strongly interacting atoms or molecules lead to distinct and wide electron bands.

For undoped conjugated polymers, the gap energy can be greater than 2 eV, which is too great for thermally activated conduction (Boudenne et al. 2011). Subsequent doping of the conducting polymers will result in a saturation of the conductivity (0.1–10 kS/cm). Initially, the increase in conductivity observed upon doping organic conjugated polymers was thought to result from the formation of unfilled electronic bands. Later, it was proven that conductivity is not related to unpaired electrons but rather to spinless charge carriers. It has been found that high conductivities obtained upon doping in polymers can be ascribed to formation of self-localized excitations, such as solitons, polarons, and bipolarons (Drobny 2012). These quasiparticles originate in strong interactions within the charge on the chain (electron or hole) acquired as a result of doping. These aspects of the molecular structure are the direct consequence of the strong electron–phonon interaction found in these quasi-one-dimensional polymers. Thus, charge-carrying species in doped organic conjugated polymers are not free electrons or holes as in the case of inorganic semiconductors, but they are quasiparticles that may travel relatively freely through the material.

In order to have a clearer understanding of the nature of charge carriers and the mechanisms of their conduction, the conducting polymers may be divided into two categories (Thomas et al. 2015; Bakhshi 2007):

- *Polymers with a degenerate ground state:* The degeneracy of the ground state of a polymer introduces the possibility of structural defects (kinks) in chains that are viewed as a change in the bond alternation. A localized electronic state with energy at mid-gap can be attributed to the structural kink. Optical absorption occurs for transitions across the band gap to the mid-gap state. The mid-gap optical absorption gives direct evidence for charged soliton states in a doped polymer.
- *Polymers with a nondegenerate ground state:* These types of polymers do not support solitons because they have an energetically nondegenerate ground state; therefore, the structures on either side of the charged defect would have different energy levels. In this case, the charge exchange will lead to formation of pairs of neutral soliton (S^0)–negative soliton (S^-) of zero spin or S^0–positively charged soliton (S^+) of zero spin. These pairs will be strongly localized to form a polaron, which gives rise to two levels in the band gap, and several optical–absorption bands may be noticed. For such structures, formation of one bipolaron is thermodynamically more stable than that of two separated polarons.

Despite intensive research, the relationship among morphology, chain structure, and conductivity is still not entirely clarified. Generally, it is supposed that conductivity should be higher for a higher degree of crystallinity and better alignment of the chains. However, this could not be confirmed for all amorphous polymers. Moreover, there are some issues regarding the solubility and processability of doped polymers, which are restricted to a great extent because of the insolubility and infusibility of these structures. The possible elimination of doping in obtaining conducting polymers while still achieving high conductivity is one of the greatest motivations for the need for small band-gap polymers. The reinforcement of polymers with conductive particles has provided new perspectives for development of flexible components for specific applications in the electronics industry.

12.2.2 Single and Multiphase Conductive Polymer Materials

The electrical properties of polymers can be fine-tuned using organic synthesis methods (Thomas et al. 2015) and by advanced techniques of filler dispersion inside the continuous polymer phase. Therefore, polymers are frequently blended with different kinds of fillers, resulting multiphase materials with an electrical conductivity comparable to that of doped semiconductors. The most frequently used reinforcement agents for the improvement of electrical conductivity in polymers are as follows:

- *Metal powders:* These include fillers such as iron (good magnetic properties), copper (reduced conduction due to oxidization tendency), aluminum (fibers and flakes with treated surface to avoid oxidation), nickel (spheres and flakes particles can be coated by silver to improve their resistivity against high humidity), silver (spherical or flake silver particles with excellent conducting features, unlike less costly metals; its oxide, sulfate, and carbonate are also good conductors), and noble metals (gold or palladium—the tendency of agglomeration of spherical particles is higher for gold, which has a lower percolation threshold) (Jiang et al. 2006; Li and Sundararaj 2015).
- *Metalized fillers:* Metal-coated inorganic fibers (especially those coated with silver) have the advantage of a lower density of the final filler and a lower price, as well as variability in the shape of the filler. Metallization can be undertaken using an electroless metallization method (Boudenne et al. 2011).
- *Carbon black:* This is an elemental carbon in a form different from diamond, coke, charcoal, and graphite. It consists of spherical-like particles linked with each other in the form of aggregates. The conductivity of this filler is much lower than that of metals or graphite, but it has the advantage of low cost, low density, and, especially, a specific structure that enables the formation of a conductive network within a matrix at a low filler concentration (Zois et al. 2001).
- *Graphite:* This is a soft, crystalline form of carbon (in addition to diamond and fullerenes). It exhibits perfect basal cleavage. There are two general types of graphite: natural and synthetic.
- *Graphene:* Graphene consists of single planar layers of hexagonally arranged sp^2 carbons and has unusual electronic properties. One of its derivatives is graphene oxide, which is hydrophilic and has a larger interlayer distance than graphite (Mittal and Chaudhry 2015).
- *Carbon nanotubes:* These are perhaps the most important filler for designing electrically conductive composites. They have two distinct forms—single-walled carbon nanotubes (SWCNTs), which are composed of a graphene sheet rolled into a cylinder, and multiwalled carbon nanotubes (MWCNTs), which consist of multiple concentric equiaxed graphene cylinders. SWCNTs can either be metallic or can be semiconductors as a function of their chirality and diameter (Moniruzzaman and Winey 2006).
- *Conductive polymers:* Conductive polymers may also be used as the filler in nonconductive polymers. The resulting material will exhibit unusual electronic properties, such as low ionization potentials, low energy optical transitions, and high electron affinities. Instead of the doping level, the conductivity in conductive polymers (density of localized states) also can be controlled by modifications of the polymer backbone, using

copolymerization of the conducting polymer with its derivative method (Boudenne et al. 2011).

- *Fillers coated by conducting polymers:* Inorganic–organic particles and fibers can be coated with conducting polymers such polypyrrole (Ouyang and Chance 1998).

The electrical conductivity of a reinforced polymer depends on the volume content of the filler. For very low filler fractions, the mean distance between conducting particles is large, and the conductance is limited by the polymer matrix. When a certain amount of filler is introduced, the conductive particles get closer and form linkages, which result in an initial conducting path through the whole composite. The corresponding filler content is called the percolation threshold. In this concentration domain, the conductivity can rise drastically by several orders of magnitude for small volumes of filler content. Finally, at high loadings, the increasing number of conducting paths forms a three-dimensional network, which results in high conductivity and lower sensitivity to small changes in volume fraction.

The concept of connectivity allows classifying different types of bi- or multiphase materials considering that each phase in a composite may be self-connected in zero, one, two, or three dimensions. Percolation theory gives a phenomenological description of the conductivity of a system near a metal–insulator transition. This approach assumes a large, preferably infinite, regular periodic lattice; in a random site, this lattice has a probability, P. All sites are bonded to adjacent ones by a conductive bond. At a critical probability, Pc, an infinite conducting cluster is first contoured in the lattice. At an initial percolating path through the material, the conductivity of the lattice changes from insulating to an infinite value. Above Pc, the conductivity (σ) is found to increase as shown in Equation 12.1:

$$\sigma(P) \propto (P - Pc)^n \qquad (12.1)$$

where n is a parameter with values between 1.65 and 2 for three-dimensional lattices, and Pc denotes the percolation threshold.

The percolation behavior is determined by the shape of the filler and also by geometry and the dimension of the regarded composite sample (Boudenne et al. 2011; Barzic et al. 2014).

In an effective media theory of a composite, a spherical or ellipsoidal grain is considered to be surrounded by a mixture, which has the effective conductivity of the composite medium. It is mainly used for composite materials with well-separated subphases for the prediction and explanation of large volume average values of electrical properties. An excellent overview has been provided by Landauer (1977).

The process of charge carrier transport may occur in two steps: (1) the injection of charge carriers into the material and (2) the movement of charge carriers through the material via hopping, tunneling, ballistic transport, diffusion, or metallic conduction. For composites with at least two phases, several aspects have to be considered: the conduction in the polymer matrix, in the filler material, between adjacent filler particles, and from the filler into the matrix and vice versa. Three different regimes of charge carrier transport are possible, which are affected by the morphology of the compounded composite (Landauer 1977):

- *Regime one:* The fillers are so far away one from one another that conductivity is determined by transport processes within the polymer host matrix.
- *Regime two:* The fillers are well separated, but their mean distance is below a certain threshold so that electrical field-assisted tunneling can occur between neighboring reinforcement particles. For composite systems near the percolation limit, the conductivity is influenced by the properties of the polymer, the filler material, the interface, and by the dispersion of the filler in the matrix.
- *Regime three:* At high reinforcement amounts, the fillers are in close contact so the conduction of charge carriers occurs through the continuous structure of the chain of filler particles in the matrix. The conductivity is mainly determined by the filler and its microscopic contacts to adjacent particles.

The conductivity models are made for simplified materials that, in practice, exhibit special properties or morphologies that have a severe influence on the overall conduction mechanism.

12.3 CHARACTERIZATION METHODOLOGIES FOR ELECTRICAL CONDUCTIVITY

Electrical conductivity or resistivity is a key physical property of all materials of interest in the electronics industry. The manner in which a material opposes the flow of electric current can be determined through several measuring methods, such as eddy current testing, the two-point technique, the four-point technique, and the van der Pauw technique (Boudenne et al. 2011).

12.3.1 Eddy Current Testing

Eddy current is based on electromagnetic induction for conducting analysis. When an alternating current travels through a conductor, a testing material, or sensor, a magnetic field is formed in and around that material. This magnetic field expands as the electric conduction reaches a superior limit and suddenly diminishes as the current is reduced to zero. If another testing sample is brought in the vicinity of this changing magnetic field, the current will be induced in this second conductor, which is determined and correlated with the conductivity of the tested material. The measurements for electrical conductivity are usually to null an absolute probe in air (in contact with the sample surface). Because conductivity depends on temperature, the experiments must be performed at a constant temperature, and adjustments must be made for temperature variations when necessary. The thickness of the analyzed material should generally be greater than three standard depths of penetration. In this manner, the eddy currents at the back side of the specimen are sufficiently weaker when compared with the variations in the sample thickness that are not observed during the measurements.

Eddy current testing also serves as a nondestructive test tool for crack detection, material thickness measurements, and coating thickness evaluation, as well as conductivity calculations for sample recognition, heat damage detection, case depth determination, and heat treatment monitoring (Boudenne et al. 2011). The main advantage of this technique is that it can provide data for materials with various defects, complex shapes, and wide size ranges. On the other hand, this testing method presents some limitations, including (1) only conductive materials can be examined, (2) the sample surface must be homogeneous for probe scans since surface roughness may interfere with testing results, (3) the depth of signal penetration is limited, and (4) flaws (such as delamination) that are collateral to probe spire and also to the direction of the sample scanning, are not detected. Another drawback of this technique is its limitations for nonmagnetic materials because the change in impedance of the coil probe can be related to the conductivity of the material.

12.3.2 Two-Point Technique

The two-point technique primarily is based on the determination of the two-point resistivity of the probe material. If the resistance and physical dimensions of a bar of material are determined, one can evaluate the electrical resistivity and, implicitly, the conductivity. The specimen used in such experiments must be cut into the shape of a rectangular bar of length l, height h, and width w. Copper wires are attached to both ends of the bar across which is applied a voltage, causing a current that travels through the bar (Tong 2011). An ammeter is connected in series with the bar and voltage source to evaluate the current I that flows through the bar. The voltage drop read on the ammeter should be negligible. The resistance R of the bar is given by the ratio between the voltage and the electrical current. The two-point resistivity of the material can be obtained by (Heaney 1999)

$$\rho = \frac{Rwh}{l} \tag{12.2}$$

where ρ is the resistivity in $\Omega \cdot m$ and R is the measured resistance in Ω.

In practice, measuring resistivity with a two-point technique is often not reliable because of some resistance between the contact wires and the material or in the measuring equipment itself. This additional resistance makes the resistivity of the material measure higher than it really is.

12.3.3 Four-Point Technique

Using the four-point measurement technique on a material (similar to a bar with known dimensions) requires four wires attached to the specimen. A current source forces a constant current through the ends of the sample bar. The amount of current passing through the bar is monitored by a separate ammeter. The voltage V produced across the inner part of the bar is determined by a voltmeter. The four-point resistivity of the material can then be expressed as (Heaney 1999; Tong 2011)

$$\rho = \frac{Vwh}{Is} \tag{12.3}$$

where ρ is the resistivity in Ω · m, V is voltage measured by the voltmeter in volts, w is width of the sample bar measured in meters, h represents the height of the sample bar measured in meters, I is current the ammeter measures flowing through the sample in amperes, and s is distance between the two points where the voltmeter wires make contact to the bar, measured in meters.

12.3.4 Van der Pauw Technique

The four-point measurement technique starts with the assumption that the sample has the shape of a rectangular thin film or a bar. The van der Pauw technique is a more general four-point resistivity measurement technique that allows examination of samples that present arbitrary shapes, with no requirement concerning the physical dimensions of the sample. There are four conditions that must be satisfied to use this experimental method (Heaney 1999):

- The sample must have a flat shape of uniform thickness.
- The sample must not have any isolated holes.
- The sample must be homogeneous and isotropic.
- All four contacts must be located at the edges of the sample.

Another aspect concerns the area of each contact, which must be lower with an order of magnitude than that of analyzed material. For samples with reduced dimensions, this might not be possible or practical. If sufficiently small contacts are not achievable, it is still possible to perform accurate van der Pauw resistivity analysis, using geometric correction factors to account for the finite size of the contacts.

In the described techniques, the contact resistance-related issues usually are more typical in the through-plane measurements than in the in-plane measurements. However, surface conductivity measurements are also affected by the contact resistance. These types of experiments are sensitive to undesired electric noise (Tong 2011). Also, the qualities of all electric connections are of great importance. Preconditions for reliable results are careful construction of the measuring unit and sample preparation and quality management of testing procedures with reference probes. The environment has effects on the conductivity values and temperature, but humidity in particular may cause inaccurate results (Boudenne et al. 2011).

Plastic composites are more or less inhomogeneous particularly due to the anisotropy of filler concentration. Plastic processing methods involving shear flow can lead to separation of the fillers and polymer matrix. For instance, a skin-core structure can be formed where the outer surface has a lower filler concentration than the inner part of the specimen. This phenomenon generates samples with a nonconductive surface (although the conductivity of the core may be high). Typically, the conductivity may range point by point in specimens. On the surface, the conductivity may change depending on the direction on the surface. All these aspects show the need for good statistical reliability. This can be obtained by repeating measurements and by using parallel samples plus reference samples.

12.4 ELECTRICALLY CONDUCTIVE SINGLE/MULTIPHASE SYSTEMS

The electrical conductivity of polymer composites is affected by several factors, namely polymer characteristics, filler features, the interactions among the phases, the processing technique, and the multiphase morphology (Strumpler and Glatz-Reichenbach 1999; Boudenne et al. 2011; Mittal 2012). All these aspects are further reviewed here.

12.4.1 Effect of Polymer Matrix Properties

The selection of polymeric matrix depends on the final application of electrically conductive composites. For a specific purpose, an optimal balance among mechanical, optical, thermal, ecological, and other properties must be maintained. Various classes of polymers have been used for composite preparation, such as soft elastomers and rubbers, linear or branched thermoplastics, or hard duromers as epoxies. All these materials cover a wide range of properties. The main physical and structural parameters that influence the final electrical behavior of multiphase materials are the polymer surface chemistry, degree of crystallinity, and the thermal and mechanical behavior.

The *surface chemistry* of a matrix is important, particularly when considering adverse environmental factors. For instance, some plastics absorb water from the atmosphere during use, and this affects the electrical and mechanical properties.

The *crystalline* character of the matrix portion has a strong influence on the percolation concentration of the filler within the polymer especially considering that the filler particles can be distributed only in the amorphous part of the matrix.

The *operating temperature* also should be carefully considered in evaluating the electrical performance of a polymer because this parameter is reflected in the thermomechanical behavior of the matrix. For polymers, three reversible structural transitions are observed under the influence of the temperature: crystallization and melting in a semicrystalline phase and a glass transition in the amorphous phase. During all these processes, a relatively large volume change, or a pronounced change, occurs in the thermal expansion. The latter has a considerable influence on the electrical conductivity. When approaching the melting temperature, a strong positive temperature coefficient of resistance in conducting polymers was observed (Strumpler and Glatz-Reichenbach 1999). The resistivity and relative linear thermal expansion range with changes in the temperature. For example, a composite of high-density polyethylene containing 50 vol% TiB_2 has a melting point at about 133°C. Far below the melting temperature (30°C), the electrical resistivity is 0.01 $\Omega \cdot$ cm, which increases gradually with rising temperatures up to 80°C (Strumpler and Glatz-Reichenbach 1999). Near the melting point, a huge step-like increase of more than eight orders of magnitude is observed. This may be explained because below the melting temperature, the filler is closely packed, forming a conducting network in the matrix; however, at elevated temperatures, the polymer expands more than the filler, thereby reducing the contact pressure among the adjacent particles and interrupting the conducting paths. Similar results were reported for carbon black dispersed in polyethylene (Heaney 1996). The transition from low resistivity to high resistivity can be utilized for current limitation in small elements for the protection of circuit boards.

Regarding the *mechanical performance* of the polymer, particular attention should be paid to internal stresses caused by the following characteristics that directly affect the conductivity of composites:

- *Internal elastic properties:* The shrinkage of the polymer during processing might generate high internal stress (Strumpler and Glatz-Reichenbach 1999). This lowers the interparticle resistance but may also induce cracks and mechanical failure. The origin of the shrinkage stress differs for duromers and elastomers on the one hand and thermoplastics on the other. The curing of duromers and elastomers generates the reduction of free volume, creating internal stresses. Thermoplastics are processed above the melting temperature, and the thermal shrinkage during solidification builds up internal stresses. Several studies on composites prepared from an epoxy matrix reinforced with Ag, Ni, or TiB_2 (Miller 1966; Gul 1996) revealed that hardening or shearing of the polymer decreases the resistivity, and even larger changes in this property are noticed during curing.
- *External mechanical actuation:* Application of an external stress modifies the distribution of fillers, particularly at low loadings in soft polymers, due to their low Young's moduli. At higher reinforcement percents, the composite becomes more rigid and is more likely to change the number and efficiency of the electrical contacts by the application of external forces (Yoshikawa et al. 1990). Application of hydrostatic or uniaxial pressure (10–100 MPa) increases the conductivity of epoxy reinforced with high amounts (>20 vol%) of various fillers such as carbon back, Fe_3O_4, or SnO_2 (Yoshikawa et al. 1990; Pramanik et al. 1993).

Macromolecular compounds with good mechanical properties are used in regular cable compounds; after reinforcement, they gain a volume resistivity in the range of 1–100 $\Omega \cdot$ cm.

12.4.2 Effect of Filler Properties

There are several filler characteristics that strongly affect the conduction abilities of polymer composites. The most studied aspects include the size, hardness, shape, distribution, and wetting with regard to composite conductivity (Strumpler and Glatz-Reichenbach 1999; Boudenne et al. 2011).

The reinforcement of polymers with particles of higher radius leads to a decrease in constriction resistance. If the protrusions that form the contacts are assumed to be larger in radius with increasing particle size, larger filler particles should result in a higher composite conductivity (McLeod et al. 1984; Ruschau et al. 1992).

The filler size also has an impact on the heating of a composite by an electrical current. Smaller particles determine a larger number of contacts and lower volumes of matrix material embedded within the network of metallic particles. The heating of a polymer reinforced with smaller particles would occur faster owing to a reduced mean distance of the matrix material to the next neighboring particle contacts, where the energy is generated (Strumpler et al. 1997). Aside of matrix characteristics, this aspect is essential for understanding the nonlinear dependence of the resistivity versus temperature.

Filler hardness interferes in analysis of the contact resistance, which can be correlated with Young's modulus. Particles with surface hardness (ceramic silicides, borides, or carbides) lead to small point-like contact spot areas. In the absence of surface oxide layers, even these small contacts can provide relatively high conductivity. A positive aspect of hard particles is that they do not stick together, and the contact is easy to release by reducing the contact pressure (Strumpler et al. 1995). On the other hand, particles of ductile metals (silver) present a larger contact area with good adhesion to neighbors and therefore exhibiting a lower contact resistance. Relatively soft fillers tend to stick together and are difficult to separate using external forces.

The filler shape influences the percolation threshold, which can be drastically reduced for particles with an aspect ratio larger than one. Fiber-like fillers tend to connect more easily, thus forming conduction paths within the composite. Therefore, insertion of TiB_2 particles, Al fibers, and carbon black in high-density polyethylene gives a percolation threshold at 22%, 12%, and 8%, respectively (Bigg and Stutz 1983). Carbon black-filled composites are utilized for antistatic transport boxes for electronic devices and antistatic chip carriers, while metal-reinforced polymers are used for electromagnetic interference-shielded housings of electronic equipment such as computers, monitors, or cellular telephones.

The distribution of reinforcement particles in a composite largely depends on the processing technique that is used. Fibers, flakes, or carbon black can be aligned along the deformation direction during extrusion or injection molding processes. Anisotropic conductivity in filled polymers can be achieved if the filler is oriented by applying electric or magnetic fields during the processing (Park et al. 2006; Camponeschi et al. 2007).

Polymers wet the fillers and penetrate into their void spaces, removing the air bubbles within filler agglomerates from the polymeric phase. The matrix is presumed to penetrate not only the void between the aggregates but also within them.

12.4.3 Effect of Matrix–Filler Interactions

The interfacial compatibility between polymer and filler is determined by the wettability and quality of dispersion of the filler. Due to their completely different chemical and structural characters, fillers present higher polarity than polymers. This is reflected in a lower total surface free energy for the macromolecular compounds (a few tens mJ/m^2) with regard to common inorganic particles (a few hundreds mJ/m^2). The big difference in surface tension between polymers and fillers diminishes the compatibility among the composite phases as well as lowering the quality of dispersion (Boudenne et al. 2011). Hence, to enhance compatibility and dispersability of the composite components, surface treatment of fillers, polymers, or of both is frequently performed. Surface modification of a filler involves a simple treatment of surfaces by surfactants (fatty acids) or various covalent or noncovalent surface modifications using low molecular compounds or even certain polymers (Tong 2011). Composites with medium conductivity (10^2–10^7 Ω/sq) are useful for electrostatic discharge purposes. In addition, if wetting characteristics indicate good adhesion abilities, the multiphase material is well suited as a conductive adhesive.

12.4.4 Effect of Multiphase Morphology

The distribution of the reinforcement agent within the matrix can be very effectively impacted by the use of multiphase polymer blends. The microstructure morphology is determined by the polymer structure in the blend, the system composition and miscibility, which can lead to immiscible or co-continuous distribution of phases. (Utracki 1989). The incorporation of the filler into one of the polymer phases or into the phase interface has a strong influence on the electrical conductivity of composites, concomitantly reducing the percolation concentration.

Conductive multiphase composites have been obtained by dilution of polyoxymethylene/carbon black in the polyethylene-based phase (Lipatov et al. 1983). In this manner, the filler is localized at the interface between the polymeric components. Though a reinforcement agent was excluded to the interface of the polyoxymethylene phase as it crystallized, a small percent remained dispersed in the component. The conductive network at the polyethylene/polyoxymethylene interface enhanced the conductivity level at lower filler amounts.

Composites prepared by filling with carbon black natural rubber/polyethylene or polystyrene/ethylene–propylene random copolymer blends have an electrical conductivity much higher than that for the individual components at the same loading level (Geuskens et al. 1987). The addition of carbon black to these immiscible polymers lowered the size of the dispersed phase.

Depending on the blended polymers, one can achieve two types of carbon black distribution—one where the filler is homogeneously incorporated in one phase of the blend and another where the filler is preferably localized at the interface between the two polymers. This leads to so-called double percolation, which is explained by the fact that electrical conductivity was determined both by the filler concentration in the filler-rich phase and also by the structural continuity of this phase (Sumita et al. 1991, 1992).

The electrical conductivity in thermoplastics containing carbon black can be also influenced by using core-shell graft copolymers. The core can be made of a crosslinked polyethylene-vinyl acetate copolymer and crosslinked polybutylacrylate with a polyvinylchloride. It has been found that less than half the amount of carbon black was needed in the core-shell copolymer system. In order to reach a given level of conductivity, half the amount of carbon black was needed in the multiphase matrix (when compared with conventionally formulated composites with a single polymer matrix) without a loss of flexibility in the final material (Bueppelmann and Weber 1994). The morphological aspects of such multiphase systems are important in fuel cell applications, where such conductive materials make up the polar plates.

12.4.5 Effect of Stability and Endurance

In addition to the long-term properties of reinforced polymers, conducting composites are very sensitive to changes in inter-particle resistance due to environmental influences. When metal-filled polymers work at high temperatures, the conductive phase undergoes oxidization thus significantly limiting conductivity. (Gul 1996). For instance, in the case of carbon black, the resistivity strongly increases (Meyer 1974) in the presence of oxygen at elevated temperatures. Also, the mechanical properties of the polymer are modified upon continuous heating. The conductivity of multiphase materials should be maintained in a certain range in order to use them in resistance measurements.

12.5 CONCLUDING REMARKS AND FUTURE PERSPECTIVES

Current technologies are in continuous development requiring new materials that are able to perform a greater number of functions. The investigations in the area of conductive polymeric materials will extend the idea of multiphase systems that provide the desired properties through control of micro- and nanostructures. As noted earlier, selection of the appropriate materials and working conditions must be considered very carefully for any concrete application. However, the preparation procedure and the processing routes also have a great impact on the features of the final product. Therefore, research will be focused on these aspects—but not exclusively. Perhaps, in the future, scientists will focus more on the intelligent behavior of this type of material.

REFERENCES

Bakhshi A.K. Fundamentals of electrically conducting polymer. Course of Department of Chemistry, University of Delhi, 2007.

Barzic Andreea Irina, Stoica Iuliana, Fifere Nicusor, Dobromire Marius, Hulubei Camelia, Dorohoi Dana-Ortansa, and Harabagiu Valeria. Transparency and absorption edges of disiloxane modified copolyimides. *J. Molec. Struct.* 1044 (2013b): 206–214.

Barzic Andreea Irina, Stoica Iuliana, Fifere Nicusor, Vlad Cristina Doina, and Hulubei Camelia. Morphological effects on transparency and absorption edges of some semi-alicyclic polyimides. *J. Polym. Res.* 20 (2013a): 130–137.

Barzic Razvan Florin, Barzic Andreea Irina, and Dumitrascu Gheorghe. Percolation network formation in poly(4-vinylpyridine)/aluminum nitride nanocomposites: Rheological, dielectric, and thermal investigations. *Polym. Compos.* 35 no. 8 (2014): 1543–1552.

Bigg D.M., and Stutz D.E. Plastic composites for electromagnetic interference shielding applications. *Polym. Compos.* 4 no. 1 (1983): 40–46.

Boudenne Abderrahim, Ibos Laurent, Candau Yves, and Thomas Sabu. *Handbook of multiphase polymer systems.* West Sussex: Wiley, 2011.

Bueppelmann K., and Weber K. Electrically conductive materials based on graft copolymers. *Plast. Rubber Compos. Process. Appl.* 22 (1994): 47–50.

Camponeschi Erin, Vance Richard, Al-Haikc Marwan, Garmestani Hamid, and Tannenbaum Rina. Properties of carbon nanotube–polymer composites aligned in a magnetic field. *Carbon.* 45 no. 10 (2007): 2037–2046.

Drobny Jiri George. *Polymers for electricity and electronics: Materials, properties, and applications.* Hoboken, NJ: Wiley, 2012.

Geuskens G., Gielens J.L., Geshef D., and Deltour R. The electrical conductivity of polymer blends filled with carbon-black. *Eur. Polym. J.* 23 no. 12 (1987): 993–995.

Grossiord Nadia, Loos Joachim, van Laake Lucas, et al. High-conductivity polymer nanocomposites obtained by tailoring the characteristics of carbon nanotube fillers. *Adv. Funct. Mater.* 18 (2008): 3226–3234.

Gul V.E. *Structure and properties of conducting polymer composites (new concepts in polymer science)*. Boca Raton, FL: CRC Press, 1996.

Heaney Michael B. Resistance-expansion-temperature behavior of a disordered conductor–insulator composite. *Appl. Phys. Lett.* 69 no. 17 (1996): 2602–2604.

Heaney Michael B. Electrical conductivity and resistivity. In *The measurement, instrumentation, and sensors: Handbook*, 43-1-43-14, Boca Raton, FL: CRC Press LLC, 1999.

Jiang Hongjin, Moon Kyoung-Sik, Li Yi, and Wong C. P. Surface functionalized silver nanoparticles for ultrahigh conductive polymer composites. *Chem. Mater.* 18 no. 13 (2006): 2969–2973.

Jin Jie, Lin Yue, Song Mo, Gui Chenxi, and Leesirisan Siriwan. Enhancing the electrical conductivity of polymer composites. *Eur. Polym. J.* 49 no. 5 (2013): 1066–1072.

Kim Bohwon, Koncar Vladan, and Devaux Eric. Electrical properties of conductive polymers: PET –nanocomposites' fibres. *AUTEX Res. J.* 4 no. 1 (2004): 9–13.

Landauer Rolf. Electrical conductivity in inhomogeneous media. *American Institute of Physics Conference Proceedings: Electrical Transport and Optical Properties of Inhomogeneous Media.* 40 (1977): 2–45.

Lipatov Yu. S., Mamunya Ye.V., Gladyreva N.A., and Lebedev Ye.V. Influence of the mode of carbon black distribution on the electrical conductivity of the binary blend of polymers. *Polym. Sci. U.S.S.R.* 25 no. 7 (1983): 1714–1722.

Li Yan, and Sundararaj Uttandaraman. Comparative study on electrical properties of copper nanowire/polypropylene and carbon nanotube/polypropylene composites. *AIChE J.* 61 no. 1 (2015): 296–303.

Mao Cui, Zhu Yutian, and Jiang Wei. Design of electrical conductive composites: Tuning the morphology to improve the electrical properties of graphene filled immiscible polymer blends. *ACS Appl. Mater. Interf.* 4 no. 10 (2012): 5281–5286.

McLeod A.D., Haggerty J.S., and Sadoway D.R. Electrical Resistivities of monocrystalline and polycrystalline TiB2. *J. Am. Ceram. Soc.* 67 no. 11 (1984): 705–708.

Meyer J. Stability of polymer composites as positive-temperature-coefficient resistors. *Polym. Eng. Sci.* 14 no. 10 (1974): 706–716.

Miller B. Polymerization behavior of silver-filled epoxy resins by resistivity measurements. *J. Appl. Polym. Sci.* 10 no. 2 (1966): 217–228.

Mittal Vikas. *Polymer-graphene nanocomposites*. Dorchester: Royal Society of Chemistry, 2012.

Mittal Vikas, and Chaudhry Ali Usman. Polymer – graphene nanocomposites: Effect of polymer matrix and filler amount on properties. *Macromol. Mater. Eng.* 300 no. 5 (2015): 510–521.

Moniruzzaman Mohammad, and Winey Karen I. Polymer nanocomposites containing carbon nanotubes. *Macromolecules.* 39 no. 16 (2006): 5194–5205.

Ouyang Meng, and Chance Chi Ming. Conductive polymer composites prepared by polypyrrole-coated poly(vinyl chloride) powder: Relationship between conductivity and surface morphology. *Polymer.* 39 no. 10 (1998): 1857–1862.

Park Cheol, Wilkinson John, Banda Sumanth, Ounaies Zoubeida, Wise Kristopher E., Sauti Godfrey, Lillehei Peter T., and Harrison Joycelyn S. Aligned single-wall carbon nanotube polymer composites using an electric field. *J. Polym. Sci. Part B.* 44 no. 12 (2006): 1751–1762.

Pramanik P. K., Khastagir D., and Saha T. N. Effect of extensional strain on the resistivity of electrically conductive nitrile-rubber composites filled with carbon filler. *J. Mater. Sci.* 28 no. 13 (1993): 3539–3546.

Ruschau G. R., Yoshikawa S., and Newnham R. E. Resistivities of conductive composites. *J. Appl. Phys.* 72 no. 3 (1992): 953–959.

Solymar Laszlo, and Walsh Donald. *Electrical properties of materials.* Oxford: Oxford University Press, 2010.

Strumpler R., and Glatz-Reichenbach J. Conducting polymer composites. *J. Electroceram.* 3 no. 4 (1999): 329–346.

Strumpler R., Loitzl R., and Ritzer L. European Patent 0 696 036 A1, 1995.

Strumpler R., Maidorn G., and Rhyner J. Fast current limitation by conducting polymer composites. *J. Appl. Phys.* 81 no. 10 (1997): 6786–6794.

Sumita Masao, Sakata Kazuya, Asai Shigeo, Miyasaka Keizo, and Nakagawa Hideaki. Dispersion of fillers and the electrical conductivity of polymer blends filled with carbon black. *Polym. Bull.* 25 no. 2 (1991): 265–271.

Sumita Masao, Sakata Kazuya, Hayakawa Y., Asai Shigeo, Miyasaka Keizo, and Tanemura M. Double percolation effect on the electrical conductivity of conductive particles filled polymer blends. *Colloid Polym. Sci.* 270 no. 2 (1992): 134–139.

Thomas Sabu, Shanks Robert, and Chandrasekharakurup Sarathchandran. *Design and applications of nanostructured polymer blends and nanocomposite systems.* Amsterdam: Elsevier, 2015.

Tong Xingcun Colin. *Advanced materials for thermal management of electronics packaging.* New York: Springer, 2011.

Utracki L.A. *Polymer alloys and blends.* Toronto: Hanser, 1989.

Yoshikawa Shoko, Ota Toshitaka, Newnham Robert, and Amin Ahmed. Piezoresistivity in polymer-ceramic composites. *J. Am. Ceram. Soc.* 73 no. 2 (1990): 263–267.

Zois Haralampos, Apekis Lazaros, and Omastová Mária. Electrical properties of carbon black-filled polymer composites. *Macromol. Symp.* 170 no. 1 (2001): 249–256.

13 Polymer Composites with Maleimide Compounds

Viorica Gaina, Oana Ursache, Fulga Tanasa, and Constantin Gaina

CONTENTS

13.1 INTRODUCTION

Advanced functional materials are among the key elements in the development of science and technology because they can achieve high-performance properties while contributing to the reduction of production costs and negative environmental impact. Given the molecular approaches in the chemical synthesis and progress in nanoscience, bismaleimide (BMI) resins can now be formulated as multiphase polymer systems able to respond to a wider variety of demands than classic BMIs did, ranging from engine parts for aerospace and the defense industry to cardiac valves and fuel cells.

In order to improve BMI resins' processability while retaining other important characteristics (excellent physical property retention at elevated temperatures and in wet environments, almost constant electrical properties over a wide range of temperatures, nonflammability, etc.), several modern approaches were utilized.

The reactive modification by using comonomers with wisely selected structure and characteristics takes into account changes that may occur in structure during and after the cure stage, at the molecular and supramolecular levels. The control of the molecular weight, use of low-melting low-viscosity reactive modifiers, and the design of bulky oligomers able to entail structural free volume must be considered. Kinetic and thermodynamic studies have proven that all factors affecting the process must act synergistically in order to tune the final properties of the BMI resins according to the envisaged purpose.

Formulating BMIs as matrices for composites and hybrid materials is another effective approach. The use of micro- and nanometer scale fillers allowed materials with new or improved properties. Studies of interactions at the interface among nanometric particles and a multicomponent polymer matrix have indicated that the interface itself can be equally important as the individual components regarding the overall effects because not all the principles from macro- and microscale can be used to explain the properties and behavior of nanocomposites. Combining these methods provides the ability to tailor and control the overall composition of these new materials, their structure (nanostructure, as well as supramolecular architecture), and also allows tunable properties by tunable structure–property relationship.

Given the importance of BMI resins, for a better understanding of different approaches and their effects, data presented herein were structured using the formulation as categorizing criterion. In addition, because the domain is developing at a fast pace, only information from the most recent decade was selected because benchmark references are considered already known. An assessment of current BMI resin applications showed how their features have been used and may effectively direct future research toward expanding this field.

13.2 BISMALEIMIDES MODIFIED WITH DIALLYLPHENOLS AND ALLYLATED NOVOLAC RESINS

BMIs are some of the most successful thermoset resins designed for high-temperature applications due to their outstanding oxidative and thermal stability, flame retardation, low propensity toward water sorption, and the fact that they are readily synthesized and, moreover, cost-effective. Still, given their aromatic nature and high cross-linking density, BMIs are brittle and display rather poor adhesion properties. Various approaches have been proposed in order to improve BMIs' elasticity and, subsequently, processability. Thus, the linear chain extension of comonomers prior to the cross-linking reaction via the Michael addition (MA) was possible for BMIs with aromatic diamines and aminobenzoic hydrazides, the incorporation of some thermoplastics (polyimides, polyethersulfones, etc.), and for the reaction of BMI with allyl-containing compounds yielding resins having high fracture toughness.

Given a resin based on BMI and *o,o′*-diallyl bisphenol A (DBA) (Guo et al. 2006), it was possible to study the effect of the cure on the resin properties in a continuous range of temperatures. The results suggested the cure evolved following a single mechanism, the curing degree increased along with the curing temperature, and the reaction rate decreased due to the diffusion-controlled reactions.

When modified novolac resins are used to modify BMIs properties, the typical procedure involves the allylation of novolac and, subsequently, its "ene" reaction with the selected BMI. The morphological transitions of such materials highly depend on the degree of allylation that thereafter influences the cross-linking density and hence the processability window (thermal and mechanical properties). For a particular series of BMI-allylated novolac resins having different amounts of BMI (Lu et al. 2006b), it was possible to demonstrate that the morphology varied with the degree of allylation: the single-phase system turned into a phase-separated structure for allylation degrees in the range 48%–59%. The higher the allylation, the lower the thermal and mechanical properties.

In the case of some resin-transfer molding (RTM) materials based on modified novolac and BMI, studies indicated that the nature of the BMI-allylated novolac (Yao et al. 2005) yielded materials with very high T_g.

As a general feature, those resins with improved processability have the ability to maintain their newly acquired properties even when used as matrices in composite formulations, no matter whether the reinforcement is glass fibers/fabrics, carbon fibers, or nanoparticles (silica, clays, metals and metallic oxides, etc.).

Kinetic models correlate the thermal and rheological behavior patterns of various composites with thermoset matrices. The curing temperature and cross-linking degree are generally influenced by the system geometry, thermal properties of the composite, and by the reaction rate. Studies performed on model T700 carbon fiber–BMI resin composites (Guo et al. 2005; Guo 2008) proved that even the conditions and time of storage of the uncured prepregs affect their reactivity and glass transition temperature, as well as the processability window. Even if the T_g is not severely affected, it was shown that the curing rate decreased along with the viscous flow at the curing temperature.

Carbon fibers and BMI-based composites having exceptional properties can be achieved by wisely selecting the matrix–filler pairs. Thus, liquid oxygen (LOX) compatible materials with remarkable high in-plane thermal conductivity were obtained using commercially available toughened Cytec BMI Rigidite 5250-4 RTM, as matrix, and GRANOC thermally conductive carbon fibers YSH50A and YSH60A (Gerzeski 2008). Their level of performance is within the accepted limits for such materials primarily used in the aerospace industry for shuttle boosters.

The BMI resins modified with diallylphenols and allylated novolac acquired improved processability but, upon combination with glass fibers or fabrics, they became fit for high-performance applications (automotive, aerospace, defense industry, etc.) due to their enhanced mechanical properties (mainly fatigue and fracture characteristics). The behavior of a composite made of a commercial resin based on 4,4′-bismaleimidodiphenylmethane (BDM) and DBA reinforced with glass fibers was investigated for virgin and aged prepregs and also in the presence of the glass fibers (Frigione and Kenny 2005). The model was validated by the experimental data and correlated well with other reports in the literature, suggesting that the presence of glass fibers influenced the crystallinity of the resin. Further studies (Bhuvana and Saroja Devi 2007; Ali et al. 2013) confirmed the beneficial effect of glass fibers, which gave composites mechanical properties superior to those of corresponding epoxies, even after water uptake; upon aging, the composites showed increased modulus values and significant strength retention.

Compounding BMIs modified with diallylphenols and allylated novolac resins with nanofillers such as silsesquioxanes, silica, silicon carbide, clays, metallic oxides, and carbon nanotubes aimed to obtain composites with reduced friction coefficients and increased wear resistance. Thus, it was demonstrated that an addition of 2% SiC to a BMI resin based on BDM and DBA results in increased bending and impact strength and a significantly reduced friction coefficient and wear rate during dry friction (Yan et al. 2005).

When surface-modified SiO_2 nanoparticles (the coupling agent was N-[2-aminoethyl]-γ-aminopropylmethyldimethoxy silane) were utilized for the aforementioned BMI resin (Yan et al. 2008), their effect was superior to those that were unmodified in terms of wear resistance and friction coefficient as substantiated by the tribological study.

Metal oxide nanopowders/nanoparticles were used in such composites for the same purpose. It was proven that the addition of potassium titanate (Liang et al. 2005), aluminum borate (Gu and Liang 2007), or zirconia (Yan et al. 2007) nanoparticles improved the wear resistance and decreased the friction coefficient of nanocomposites. When the nanoparticles were previously modified (in zirconia-based composites, with *N*-[2-aminoethyl]-γ-aminopropylmethyldimethoxysilane, and Si- and P-based compounds in potassium titanate containing composites), the tribological behavior of the composites was superior due to the improved dispersion of the filler.

Multiwalled carbon nanotubes (MWCNTs) were proven to also contribute to the increase in the wear resistance and to the reduction of the friction coefficient in modified BMI resins (Liu et al. 2007a), especially when MWCNTs are functionalized with carboxylic groups. This is because they induce a change in the main wear mechanism—from adhesive wear in neat resin to abrasive attrition—by changing the self-lubricating property of the worn surface, the dispersion of filler in the matrix, and the interfacial adhesion between filler particles and matrix.

Various silsesquioxanes were used for composite formulations with BMI resins in order to achieve high-performance materials with improved specific properties. Thus, composites with hydrogen silsesquioxane and BMI resin modified with DBA were prepared and studied to demonstrate that the T_g and thermal stability of the cured materials increased with the amount of silsesquioxane (Huang et al. 2010). A novel, incompletely condensed, polyhedral oligomeric silsesquioxane containing allyl groups was used in combination with a BMI resin based on DBA (Zeng et al. 2012) to obtain new hybrids with simultaneously improved thermal stability and dielectric properties. The entire system cured following the same mechanism as the prepreg, but the cross-linking density is higher for the composite and increased along with the amount of silsesquioxane.

Flame retardancy was supplementarily achieved when phosphorus-containing polyhedral oligomeric silsesquioxane and BMI/DBA were used (Wang et al. 2015), and the compact char resulted upon burning acted as a barrier, preventing the heat and mass transfer between the flame and the underlying material.

When an allylated novolac/BMI resin was compounded with polyhedral oligomeric silsesquioxane (POSS), in the presence of octamaleimidophenyl polyhedral silsesquioxane (OMPS) as a co-curing reagent, materials with POSS cages uniformly distributed in the network resulted (Li et al. 2007). Their enhanced thermal stability is the result of the higher cross-link density and the inorganic nature of POSS.

Interesting studies were conducted on BMI resins modified with clays in order to evaluate their effects as compared to the neat resins. Experimental data recorded for composites made of BMI/DBA resin and organically modified montmorillonites (Hu and Meng 2005) indicated all reactions that occur during the cure were affected by the presence of clays: the "ene" reaction was accelerated due to the acidity of the organic modifier; the curing took place as in the neat resin when the clays are highly exfoliated but at lower temperatures. For the same resin, it was demonstrated that adding up to 2% organically modified rectorite may increase the viscosity and enhance the reactivity as well as the flexural and impact strength (Yuan et al. 2009). Cracking and spalling phenomena were limited due to the uniform dispersion of clays, although the resistance toward hot water, thermal degradation temperature, and the initial storage modulus increased without a significant decrease of the T_g of cured resin.

Composites with increased dielectric properties were obtained when barium titanate nanoparticles were used in a BMI resin matrix (Chao et al. 2008). The dielectric constants and dielectric losses were enhanced by the increasing amount of nanoparticles (up to 60%).

BMI resins were modified with a wide range of other polymers in order to achieve improved properties. Toughened BMIs may be obtained by using polyetherketones (Han et al. 2009), while enhanced processing characteristics and low dielectric losses may be achieved for BMI by modification with allyl phenyl compounds, allyl epoxy resins, and epoxy acrylate resins (Liang et al. 2007). Flame retardancy was accomplished by modifying BMIs with fully end-capped hyperbranched polysiloxane (Zhuo et al. 2011a).

Higher toughness, excellent dielectric properties, and good processability were simultaneously achieved when a BMI/DBA system was compounded with poly(urea-formaldehyde) microcapsules filled with epoxydic resins (Yuan et al. 2011). A content of 2%–10% epoxy-filled microcapsules with various diameters caused enhanced mechanical properties (flexural strength, interlaminar shear strength, impact strength), due to the plastic deformation of capsules that also contributed to prevent the cracking progress.

Recently, interesting results were obtained for BMIs modified with eugenol (bio-based and synthetic) and eugenol-modified novolac (Shibata et al. 2013). These new resins underwent the "ene" reaction and subsequent thermal addition copolymerization in the same way as the curing reaction of DBA and BMI, and no phase separation was noticed for the cured resins.

13.3 BISMALEIMIDES MODIFIED WITH NOVOLAC/OTHER ALLYL COMPONENTS/OTHER COMPONENTS

To improve the toughness and reduce the viscosity of BMI, allyl compounds, diamine, or other compounds have been used as BMI's modifier. High-performance RTM bismaleimide resin with low cure temperature and post-cure temperature (≤200°C) based on BMI and *N*-allyl diaminodiphenylether (DDO) was prepared (Gu 2005). The effect of DDO and its stoichiometry on the performance of the modified resin showed that the resulting resin has good processing characteristics, meeting the requirements of the RTM technique; moreover, after cured and post-cured at 200°C, the resin has desirable thermal and mechanical properties, equaling the typical commercial BMI resin. An optimum molar ratio of BMI and DDO (1:0.55) leads to the injection temperature ranging between 70°C and 82°C and the pot life at 80°C is 300 min. Another modifier for BMI was a mixture of the diallyoxy azomethines containing different mesogens such as benzenoid, biphenyl, and tolyl structures, in which case a cured blend with high T_g and heat resistance was obtained (Yu et al. 2006). A microfiber-like liquid crystal-rich phase was detected in the samples with appropriate content of LC modifiers. Further addition of DBA in the blends shows little effect on phase structures of cured blends; however, the mechanical properties increased with a decrease in thermal properties. Lu et al. (2006a) reported the synthesis and properties of BMI-modified novolac resin/titania nanocomposites prepared by a sol–gel process of tetrabutyl titanate with allyl phenol novolac–BMI resin. The average original particle size of the dispersed phase in the nanocomposites was less than 100 nm, but particle aggregates existed in bigger sizes. The introduction of the nanosized TiO_2 particles led to increases in the T_g of the polymer and modulus of the material at lower temperatures but decreased the thermal resistance of the material and the modulus at higher temperatures.

To increase processability and thermal stability of BMI resins, Huang et al. (2008) prepared organic/inorganic hybrid BMI resin from octa(aminophenyl)silsesquioxane (OAPS), BMI, and dipropargyl ether of bisphenol A, lowering the viscosity and increasing the T_g (350°C) of hybrid resin with 5 wt% OAPS loading, but the increase of OAPS loading led to a decrease of the T_g and T_{d5} in air of the cured resins. Very good results, in terms of mechanical properties, T_g (242°C), thermal stability, flexibility, and flame retardancy, were obtained by a Diels–Alder (DA) reaction between BMI and bisfuran compounds containing reactive benzoxazine groups in the main chain, followed by their cross-linking under heat treatment by means of the ring-opening reaction of benzoxazine and the addition reaction of maleimide groups (Chou and Liu 2008). The proposed reactions of polybenzoxazine in the curing process are shown in Scheme 13.1.

Recent research has focused on the use of other allyl components based on renewable resources, such as tung oil (TO) or eugenol (Shibata et al. 2011a and 2011b), leading to the bio-based thermosetting resins. The bio-based thermosetting resins based on TO (Shibata et al. 2011a), with C=C ratio from 1/1 to ¼, presented an increase of T_g and T_{d5} with increasing BMI content from 111 to 150°C and 435 to 468°C, respectively, but maximal tensile strength (38.1 MPa) and modulus (2.6 GPa) were obtained for the cured products with the C=C ratios of ½ and 1/3, respectively. Because the partially cross-linked TO/BMI prepolymer was insoluble in general organic solvents, the model reaction based on TO and N-phenylmaleimide (PMI) with the C=C ratio 2/1, 1/1, ½, 1/3, and ¼ treated at 150°C for 24 h was used to evaluate the reaction mechanism between TO and BMI by NMR spectroscopy. As a result, it was revealed that the DA reaction preferentially occurred at 2/1 and that ene reaction and other reactions such as radical homo- and copolymerization gradually increased with decreasing C=C ratio of TO/PMI. The other bio-based thermosetting resins starting from eugenol with a molar ratio of allyl/BMI components of 1/1, ½, or 1/3 at 200°C, showed a similar behavior as in the case of TO-based resin regarding the T_g and T_{d5}, while the mechanical properties were superior to the resin

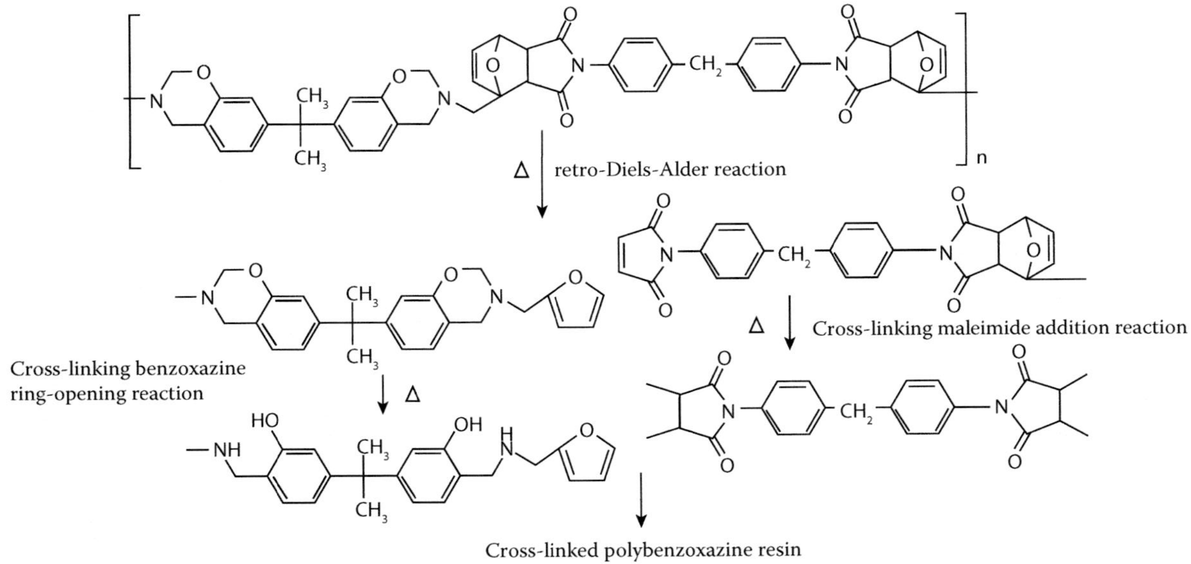

SCHEME 13.1 Preparation of cross-linked polybenzoxazine resin through possible reactions of the retro-DA reaction, the benzoxazine ring-opening reaction, and the maleimide addition reaction. (Adapted from Chou Chou I., and Liu Ying-Ling. *J. Polym. Sci. Part A*. 46 no. 19 (2008): 6509–6517.)

based on DBA and BMI (Shibata et al. 2011b). The FTIR study of the cured materials revealed that chain polymerization of the allyl and maleimide groups occurs for this resin, whereas a stepwise ene reaction and subsequent chain polymerization in addition to etherification occur for DBA/BMI.

New kinds of flame-retarding BMI resins with simultaneously good processing characteristics, high toughness, and outstanding thermal stability were prepared by copolymerizing BMI with a monomer containing three allyl groups (allyl triphenylborate) (Zhu et al. 2015). When compared with BMI/DBA resins, these resins had higher impact and flexural strengths that increased with the increase of allyl triphenylborate content (up to 137 MPa) and presented the same curing mechanism; however, their cross-linked networks have different chemical and aggregate structures, including the cross-linking density, free volume, the proportion of homopolymer of BMI and copolymer of BMI, and allyl triphenylborate.

BMI can be used for improving toughness and thermal properties of other resins such as formaldehyde resins epoxidated and modified with allyl maleate and thermally cured in the presence of BMI, diglycidyl ether of bisphenol A (DGEBA), and diaminodiphenylmethane (DDM) (Mustata and Bicu 2008); allylated/methacrylated novolac resins with epoxy BMI resins (Mustata 2011); epoxy modified novolac resin (Ye et al. 2012); novolac epoxy resin in the presence of 5% CTBN (Tian and Yang 2011); and bisphthalonitrile containing allyl groups (Zou et al. 2015). In the last case, the resulting copolymers have high T_g (>350°C), better processability, and shorter curing time, and the polymerization with BMI presented two stages of polymerization: the addition polymerization of allyl moieties and the ene reaction between allyl moiety and maleimide moiety at low temperature (around 270°C); the cyano moieties polymerized at high temperature (around 330°C). Their laminates with glass fiber exhibited better mechanical properties (>440 Mpa) and satisfactory thermal properties (>440°C).

13.4 BLENDS

BMI was used as a constituent of blends either as a modifier of an initial system of resins or as a component of the modified blend/system. So, BMIs were blended with polyethersulfone, and the effects of polyethersulfone molecular weight and ratio and curing temperature were studied (Liu et al. 2007b). After the phase separation took place, all the blends presented an increase of the complex viscosities. Benzoxazines were also used for blending with BMI, and the BMI content was found to influence the softening temperature of the alloy: an increase of BMI ratio leads to a higher softening temperature. The T_g of the blends followed the same trend as the softening temperature when increasing BMI content and so did the thermal stability (Takeichi et al. 2008). Resin networks with improved thermal, mechanical, and dielectric properties were obtained by curing BMI/aromatic diamine prepolymers with the epoxy resin DGEBA. It was noted that T_g and the flexural and shear storage modulus increased when adding BMI to the epoxy resin (Rajabi et al. 2006).

Novolac-based systems were modified with BMI in order to improve either T_g or thermal stability (Ambika et al. 2007; Zhou et al. 2009). For improving the toughness of allylated novolac/BDM resin, polydimethylsiloxane (PDMS) was added, and the goal was successfully achieved when using an optimum quantity of PDMS of a certain molecular weight (Li et al. 2006). PDMS incorporated in diglycidylether was used in developing an epoxy matrix system. In addition, a BMI was added, and the properties of the two systems were compared. The results led to the conclusion that incorporation of diglycidylether–PDMS into the epoxy resin decreases the tensile and flexural strength, while the BMI has an opposite effect (Suresh Kumar and Alagar 2006). Epoxy hybrid nanocomposites were obtained by in situ polymerization of DGEBA and DDM with OMPS. The modification of the epoxy resin resulted in homogeneous structures with better thermal stability and lower T_g (due to the increase of free volume and decrease of cross-link density) (Jothibasu et al. 2008). Epoxy resin was also modified with phthalide-containing BMI in various ratios. Equilibrium water content and T_g decreased with the percent of BMI, while glassy modulus and thermal stability increased (Xiong et al. 2011). For another epoxy resin/BMI system, modulus and char yield increased when increasing the BMI content, while the T_g values were higher than the curing temperature for all the samples; the cure process took place in two stages (Rao et al. 2005).

Resins with improved flame-retarding properties were obtained from BMI containing phosphorus cured with different kinds of agents (Tsai et al. 2009) or by blending a commercial phenolic epoxy resin containing phosphorous and melamine phenol formaldehyde novolac with various amounts of BMI (Chiang et al. 2013).

Resins with improved processability were obtained in distinct studies by adding different quantities of a second BMI in the BDM/DBA system. Increased BMI proportions provided increased elastic moduli, slightly decreased fracture toughness, and decreased T_g and viscosities in one case (Evsyukov et al. 2015), and improved toughness, fracture toughness, and bend strength in the other case (Qu et al. 2014).

m-Phenyl bismaleimide was utilized in the waste recovery domain for improvement of the mechanical properties either as a modifier of natural rubber/scrap rubber powder blends (Du et al. 2005) or as a reactive compatibilizer for post-consumer polyolefin waste using $Mg(OH)_2$ as filler (Khunová et al. 2009).

The influence of a commercially available additive on BMI resins was investigated and a reduction of the cure shrinkage rate was observed without changes in the cure mechanism. The quality of a laminate of BMI with carbon fiber was improved with the additive (Zhang et al. 2007).

13.5 SEMI- AND INTERPENETRATING POLYMER NETWORKS

Interpenetrating polymer networks (IPNs) have been the subject of extensive study since their advent in the 1960s. The IPN is a combination of two polymers in network form, at least one of which is synthesized and/or cross-linked in the immediate presence of the other. In this section, we will refer to the research in recent years on the IPN or semi-IPN based on BMI. A special class that is part of the IPN network is made up of the bismaleimide-triazine (BT) resins based on BMI and dicyanate ester (CE), which has many attractive properties such as very good thermal stability, moisture resistance, low dielectric constant, and low dielectric loss. The cure mechanism of their blends was highly controversial and sometimes conflicting, and all research done on this subject until the year 2006 led to the conclusion that CE and BMI cured independently in the blends and formed IPN. The cure mechanism may be affected by the use (or not) of different catalysts in the blends of CE and BMI. So, if no catalyst is added, the two components react independently and form two kinds of networks: polycyanurate and polybismaleimide; when a nonylphenol catalyst is involved, the only effect is the acceleration of the cure of CE, but when using *p*-toluene sulfonic acid, a homogeneous network is formed by the co-reaction between CE and BMI (Liu et al. 2006; Yu et al. 2006).

Some BT resins based on 2,6-dimethylphenol-dicyclopentadiene or 2,6-dimethylphenol-dipentene CE were prepared and compared with the BMI/bisphenol A dicyanate (BADCy) ester system (Hwang et al. 2007). The first two had a lower dissipation factor, dielectric constant, and moister absorption than the last one. The BMI/CE systems had very good thermal stability and T_g, which decreased when increasing the CE molar ratio. In order to obtain resins with special features, the novel BMI and CE, containing naphthalene ring and ether linkages, were synthesized and blended in the catalyzed system (iron acetylacetonate and nonylphenol) to form a novel IPNs BT resin with good thermal stability (448°C) (Yan et al. 2009).

Recently, a Chinese research group successfully synthesized new hyperbranched polysiloxanes (HBPSi), which were expected to retain remarkable thermal resistance, toughness, and dielectric properties, similar to traditional polysiloxane, as well as good processing characteristics for copolymerization with BT resins, suggesting a great potential for fabricating advanced structural and functional composites via RTM technique (Guan et al. 2011a). For the curing mechanism of DBA modified BMI (BD)/CE, BD and CE cross-link to form two discrete networks that interpenetrate with each other in the cured resin; consequently, it is an IPN structure. In the case of HBPSi/BD/CE, besides the aforementioned IPN structure in BD/CE, a complex "IPN-coupling structure" is proposed because HBPSi performs as a coupling or bridge role to connect both CE and BD resin networks, but this still needs further study. Compared with BD/CE, the HBPSi/BD/CE system has a decreased initial curing temperature and apparent activation energy (E_a). A series of novel hyperbranched polysiloxanes with polar aliphatic chains were synthesized and used to modify CE resin. Results show that these hyperbranched polysiloxanes have very good compatibility with CE resin, and the modified CE resins have increased toughness and better moisture resistance (Ji et al. 2010; Guan et al. 2011b; Zhou et al. 2011); however, the addition of these hyperbranched polysiloxanes tends to cause deterioration of the dielectric properties and flame retardancy because of the existence of a large amount of polar aliphatic chains in their molecules. Consequently, an HBPSi with abundant amounts of phenyl and silanol groups was synthesized through the hydrolysis of phenyltrimethoxysilane and was used as a new modifier for BCE resin; then, a series of HBPSi modified BCE resins with different mixing ratios were prepared (Zhuo et al. 2011b).

The focus of this chapter is not only on evaluating the effect of HBPSi and its content on flame retardancy and dielectric properties but also on investigating the flame-retarding mechanism; consequently, there also is a focus on developing a new way to prepare high-performance resins simultaneously with outstanding flame retardancy and dielectric properties for next-generation microelectric industries.

In order to further improve the toughness of CEs, the 4,5-epoxycyclohexane 1,2-dicarboxylic acid dilycidyl (TDE-85) epoxy resin, which possesses good processability and high resistance to heat (Wu et al. 2012), or bisphenol A benzoxazine (BOZ), which possesses excellent glass transition temperature, high module, low water absorption values, and good dielectric properties, was incorporated into the BT resin system to form IPNs (Wu et al. 2013). The introduction of TDE-85 into BMI/CE resin system or CE into BOZ/BMI resin system changed the chemical cross-linked network, which exerted a tremendous influence on the mechanical and thermal properties. The results showed that the modified resins with suitable TDE-85 or CE content had better comprehensive performances.

Semi-IPNs were prepared by the sol–gel technique through in situ polymerization of BMI in thermoplastic polyetherimide (PEI) as well as in polysulfone (PSF) (Kurdi and Kumar 2006). Structure and properties of semi-IPN membranes could be varied by controlling the thermoset BMI microdomain size interpenetrating within a thermoplastic polymer network. The size of this microdomain depended on the polymerization time of BMI and on the degree of thermoset/thermoplastic phase separation. These semi-IPN membranes showed an improved T_g but a decrease in their thermal stability and could be used for O_2-enriched-air applications where high selectivity is not required. At ambient temperature, it was possible to increase the permeance of semi-IPN membranes.

Many attempts have been made to modify epoxy resins by the incorporation of BMI, which enhances the mechanical and thermal properties according to its percentage of content. Thus, over the years, there have been a number of articles on the incorporation of BMI or BMI/DBA into CE with Shiff base linkages (Anuradha and Sarojadevi 2006); the polyurethane-modified epoxy system (Mahesh et al. 2006); the epoxy resin with diallyl phthalate (Shenoy et al. 2007); allyl functionalized dicyanate ester resin-bearing sulfoxide linkage (Anuradha et al. 2008); polyethersulfone-toughened epoxy systems (Rajasekaran and Alagar 2008); vegetable oil-based resin (Sithique et al. 2010), resulting in resins with significantly enhanced flame retardancy; tetrafunctional epoxy resin, resulting in a new class of IPN resin system for rigid riser applications at high temperatures of more than 280°C (Jena et al. 2015); and epoxy resin and 4,4′-diaminodiphenyl sulfone leading to synthesis of a new resin containing a phthalide cardo structure whose cure reaction has been studied (Xiong et al. 2013). The cure reaction between epoxy and amine and the anionic polymerization of the double bond occurred at lower temperature and then the MA of BMI and diamine, the thermally initiated polymerization of double bond and epoxy resin homopolymerization at higher temperature, as illustrated in Scheme 13.2.

13.6 COMPOSITES

13.6.1 Composites Based on Carbon-Fiber-Reinforced Bismaleimides

Thermal oxidative degradation of carbon-fiber-reinforced polymer composites has been intensively investigated because these materials are fit for aerospace applications. A comparative kinetic study has been undertaken on the gravimetric curves

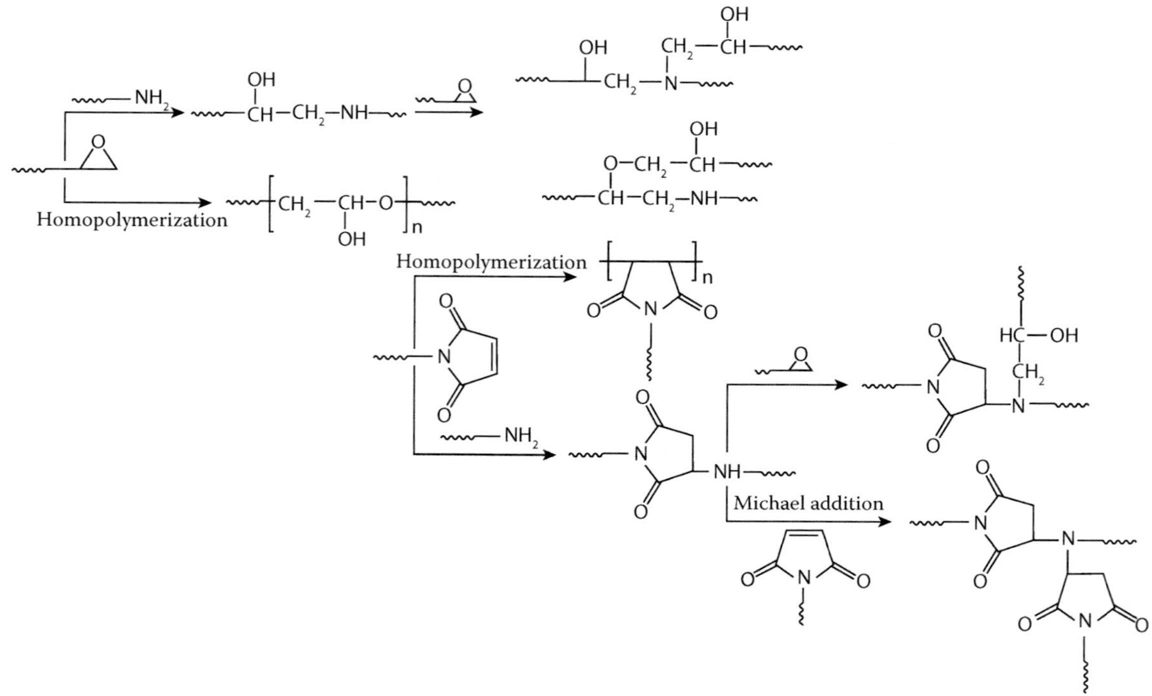

SCHEME 13.2 Proposed curing pathways of BMI/DDS/epoxy resin. (Adapted from Xiong Xuhai et al. *Temochim. Acta*. 559 (2013): 52–58.)

obtained for carbon fiber/BMI (T800H/BMI) and carbon fiber/amine cross-linked epoxy (IM7/ACE) composites and the corresponding unreinforced matrices (Colin et al. 2005b). It was found that cracks propagate differently toward the core according to both fiber orientation and nature of the composite system, and the composite behavior cannot be predicted using the previously established kinetic model for the unreinforced matrices. The mass loss is lowered in the presence of carbon fibers; a stabilization reaction, linked to the scavenging of peroxyl radicals by carbon, has been added to the kinetic model, and a corresponding parameter has been determined. This new model correctly predicts the gravimetric behavior of T800H/BMI and IM7/ACE composites.

Cryogenic studies were also reported as they were undertaken to assess the behavior of these materials. One approach was limiting the permeability of composites for cryogenic applications by limiting their microcracking (Bechel et al. 2006) using carbon-fiber-reinforced commercial BMIs. It was concluded that, by selecting the correct BMI and unloading the residual stresses, it is possible to reduce transverse cracking in composite cryogenic pressure-vessel applications that sustain thermal fatigue (including cryogenic temperatures) from processing. Another study (Wang et al. 2014), on the influence of cryogenic environment on mechanical properties of a unidirectional carbon-fiber-reinforced bismaleimide composite (T700/6421BMI composite), suggested that, after steeping in liquid nitrogen, the mechanical properties might be dominated by interfacial phenomena—the longer the steeping time, the higher the stress concentration at the matrix–fiber interface induced by the different values of the coefficient of thermal expansion (CTE), which might cause interfacial debonding.

Resin injection of bisphenol E cyanate ester, a low-viscosity resin that cures into a high-temperature thermoset polymer, has been investigated as a reliable repair method for restoring strength and stiffness in delaminated carbon fiber/BMI composites used in aircraft panels (Thunga et al. 2014). Evaluation of mechanical repair efficiency, using both bending stiffness and in-plain compressive strength of the composite panels as the repair metrics, showed values exceeding 100%. The IM7-G/5250-4 carbon/BMI unidirectional prepreg has been used in an increasing number of applications in the aerospace industry due to its superior availability for resin bonding and microfracture mending (Rider et al. 2012).

Cracking and oxidation processes were studied in carbon fiber/BMI (Hexel T800H/F655-2) and compared with carbon fiber/amine-cross-linked epoxy (Cytec Fiberite IM7/977-2) unidirectional composites (Colin et al. 2005a). Oxidation leads to shrinkage of the oxidized layer. Cracking basically results from the stress state induced by this shrinkage, with the critical conditions for crack initiation depending most likely on the structural changes induced by oxidation and, eventually, physical aging.

The static and fatigue strengths of a G40-800/5260 carbon fiber/BMI composite at room temperature (RT) and 150°C were investigated in order to evaluate its practicality (Shimokawa et al. 2008). The static and fatigue strengths showed relatively small reduction from RT to 150°C due to the use of a BMI resin as matrix.

Composites made of commercial BMIs reinforced with carbon fibers (HTM552–T_g after cure is approximately 195°C, and after postcure is 282°C) are often used in certain supersonic aircraft structures for use at high temperatures (≥200°C) due to their low-longitudinal thermal expansion and very low through-thickness strength (da Silva and Adams 2008).

Some studies on reducing the brittleness of BMI–carbon fiber composites are linked to the introduction of a thermoplastic component, such as the acrylonitrile–butadiene–styrene (ABS) terpolymer (Salini et al. 2010). Tensile and impact strengths at 6 wt% ABS reached maximum values while maintaining high thermal stability; therefore, the incorporation of ABS is an effective way of toughening the material with minimal reduction in high-temperature performance while maintaining mechanical strength. Another example is an amorphous polyetherketone with a phenolphthalein side group (PEK) used as a thermoplastic component in BMI 6421/T700SC composites processed as laminates (Cheng et al. 2006). Where there was evidence that a phase separation had occurred, the T_g of the toughened BMI laminates was slightly reduced due to the lower glass transition temperature of the thermoplast and the co-continuous, but not overlapped, the morphology of the granular structure may grant greater toughness and higher impact damage resistance. In the same trend of improving the interfacial interaction between BMI matrix and carbon fibers, the silane surface modification method was used (Fu et al. 2012) and the results revealed the largely reduced friction and wear of these composites because they displayed the strongest interfacial adhesion.

New developments in the field were achieved using nanoscale reinforcement. Thus, carbon nanofiber (CNF)-filled BMI composites were fabricated by using a commercial BMI prepolymer and an innovative melt compounding technique, that is, high shear thermokinetic mixing (Faraz et al. 2010). The poor results indicated that further chemical modification of fibers may improve interfacial adhesion and dispersion and, hence, mechanical properties. In another approach, sheets of millimeter-long epoxy-functionalized MWCNTs with stretch alignment were used to reinforce commercial BMI 5250-4 (Cheng et al. 2010). The resulting composites displayed unprecedented high tensile strength (3081 MPa) and modulus (350 GPa), much higher than that of unidirectional carbon-fiber-reinforced composites.

13.6.2 Composites with Glass Fibers

Poly(amido-imide)s (PAIs) were prepared by the intermolecular DA reaction of *N,N′*-1,4-phenylene bis[(2-furanyl methylamino)acetamide] (PBFA) with various BMIs, and glass fibers were used to produce composites of these (Patel and Patel 2008). These were proven to have good resistance to organic solvents and mineral acids and moderate resistance to heat; the PAIs produced in situ showed good adhesion to glass fibers; void-free composites could be prepared with good mechanical and electrical properties. Different novel polyimides and BMIs were designed and subsequently used as matrices for woven glass fabrics or as unidirectional glass fiber composites (Bhuvana and Saroja Devi 2007); they were subjected to thermal aging and moisture absorption, with their effects on tensile and flexural properties being studied. Mechanical properties were found to be superior to the corresponding epoxy composites, even after the retention of a significant amount of moisture.

In addition, the modulus values increased upon thermal aging, and the strength retention appeared appreciable.

Novel BMI resins have been developed, such as a thermosetting network with furan functionality and tailorable T_g as well as a glass sizing with maleimide functionality (Peterson et al. 2011) that reacts covalently with the furan-functionalized polymer matrix. Because the bonding at the glass–polymer interface is reversible, debonding of the glass fibers can be healed and mechanical properties can be recovered. Increasing the concentration of maleimide groups on the glass fiber yielded an increased healing efficiency.

13.6.3 Composites with Nanoclays and Ceramic Nanoparticles

The uses of organically modified clays for BMI-based composites were taken into consideration as a possible way to tune the processability window of these materials. The presence of the organic modifier in the structure of the organo-clay results in better compatibility between polymer and clay, given their different character.

Interactions between 2,2-bis[4-(4-maleimidophenoxy phenyl)]propane and Cloisite 15A added in various amounts were studied in order to assess their influence on melting and curing behavior (Surender et al. 2013). It was revealed that low amounts of clay (1%–4%) yielded increased values of the E_a, which is an indication of interactions between clay and BMI resin. At higher content (7%–9%), the chain mobility was restricted, and reactive sites became less available, which was expressed in the high value of E_a. The optimum was reached for 5% clay, which also resulted in a good dispersion as evidenced by SEM.

An interesting hybrid monomer of 4-(*N*-maleimido)phenylglycidylether (MPGE), which contains both oxirane and maleimide reactive groups, has been synthesized for high-performance thermosets. The corresponding cured resins exhibited good thermal stability, high glass transition temperatures, and superlative flame retardance. These properties were further enhanced by using three different organically modified clays in various resin formulations (Liu et al. 2005). The organic modifiers used for clays contributed to their exfoliations even by reacting with some oxirane groups in the resin.

Ceramic nanoparticles, namely barium titanate ($BaTiO_3$), were used in BMI-based composite formulation in order to obtain materials with excellent dielectric and mechanic properties appropriate for electronics. Thus, embedded capacitor materials with improved thermal stability were prepared by using organically modified $BaTiO_3$ to reinforce a mixed matrix made of an aromatic polyamide and an aromatic BMI (Kakimoto et al. 2006). The modifier was a metallophthallocyanine oligomer that granted polymer–ceramic composites increased dielectric constant. In addition, the film had a good compatibility with matrix polymers and a good affinity with electrode metals.

13.6.4 Composites with Fabric Reinforcements

A special group of BMI-based composites are those that use woven fabrics as reinforcement. Most of these materials are used in the aerospace industry due to their ability to be specifically shaped as structural parts and to their mechanical and thermal properties. Various types of fabrics may be used in such materials: carbon fabric (Cheng et al. 2008), eight-harness satin carbon fabric (Costa et al. 2005), eight-harness satin weave quartz fabric (Grace and Altan 2014), silica fabric (Luo et al. 2007), woven glass fibers (da Silva and Adams 2005), and even tridimensional braided carbon and Kevlar fabric (Wan et al. 2007) or three-dimensional braided graphite fibers (Takatoya and Susuki 2005).

All these types of composites showed remarkable mechanical properties over a wide range of temperatures due to the adhesion forces at the polymer–fabric interface. Some of them proved to be an excellent choice as joint adhesive (da Silva and Adams 2005) because they have reduced porosity, good ductility over a large interval of temperatures, and a superior rate of load transfer. Limiting microcracking, as a means of preventing damage and failure in the aerospace industry, was studied in three-dimensional composites in comparison with two-dimensional laminates (Takatoya and Susuki 2005). The results showed the superiority of three-dimensional composites due to their enhanced elasticity. The same conclusions were drawn when three-dimensional carbon and Kevlar braided fibers were used in composites; when compounded with BMI, significantly high flexural strength and modulus, as well as impact strength, were achieved (Wan et al. 2007). Good results were obtained when modified fibers were used in composite formulations, namely allylated silica fibers (Luo et al. 2007), and the level of performance increased along with the degree of allylation. Moisture absorption was also investigated in these BMI-based composites (Grace and Altan 2014); it was proven that these composites did not follow Fick's model of diffusion. Moreover, this factor, which is associated with the presence of voids in the composite structure, negatively affected the fracture parameters (Costa et al. 2005).

13.6.5 Miscellaneous

The research for new materials with improved properties included novel reactive unsaturated interfacial modifiers for polyolefin-based composites. Some functional maleimides, such as 1,3-phenylene dimaleimide and different linear alkyl dimaleimides (Liauw et al. 2006), were investigated. Most of them were very effective interfacial modifiers in polypropylene and poly(ethylene-*co*-vinylacetate) (18% w/w vinylacetate)-based composites.

BT resins are widely used in electronics and, when filled with hexagonal boron nitride, exhibited significantly improved thermal conductivity (Zeng et al. 2013).

It was also demonstrated that poly(p-phenylene benzobisoxazole) (PBO) fibers displayed improved interfacial adhesion in PBO-BMI composites when dielectric barrier discharge (DBD) plasma was used to modify them (Liu et al. 2014); the overall effect on the composites properties strongly depended on the degree of ablation.

BMI-based composites with exceptional mechanical properties were obtained using glass hollow microspheres that were previously submitted to a silane treatment (Koopman et al. 2006). All data indicated that interface adhesion between BMI and filler is strongly affected by the silylation reaction at the glass surface, which enhanced the compatibility.

An interesting study was recently conducted on BMI reinforced with silica particles obtained from renewable resources—namely rice husk ash that contains 90%–98% silica (Rajamani et al. 2013). For an improved adhesion at the polymer–silica interface, the particles were modified with stearic acid prior to their incorporation in the composite. It was concluded that well-dispersed silica particles enhanced the thermal stability of the resin, as was evidenced by the increased energy of activation.

13.7 CONCLUDING REMARKS

The ever-increasing societal demand for novel or upgraded, low cost, and environmentally friendly materials is the driving force of modern research. Because scientific approaches are mainly target-oriented, a better understanding of the inner processes and phenomena that occur in multiphase polymer systems based on BMI resins can provide solutions to current problems, new directions of development, as well as prospective applications. Data presented in this chapter offer an insight into preparing and processing BMI-based materials, optimizing formulations and the structure–properties–application relationship. New advances are expected in this field of research due to the evolution of these materials from micro- to nanostructured.

REFERENCES

Ali Nazakat, Wu Yin, Zhang Qiu Yu, and Chen Ying. Preparation and characterization glass fiber reinforced BMI/DABA/KH550 composite. *Adv. Mat. Res.* 650 no. 1 (2013): 220–225.

Ambika Devi K., Reghunadhan Nair C.P., and Ninan K.N. Diallyl bisphenol A-novolac epoxy system cocured with bisphenol-A-bismaleimide-cure and thermal properties. *J. Appl. Polym. Sci.* 106 no. 2 (2007): 1192–1200.

Anuradha Ganesan and Sarojadevi Muthusamy. Synthesis and characterization of Schiff base functionalized cyanate esters/BMI blends. *High Perform. Polym.* 18 no. 6 (2006): 1003–1018.

Anuradha Ganesan, Sarojadevi Muthusamy, and Sundararajan Pudupadi R. Synthesis, cure kinetics, and thermal properties of the bis (3-allyl-2-cyanatophenyl) sulphoxide/BMI blends. *J. Appl. Polym. Sci.* 110 no. 1 (2008): 517–522.

Bechel Vernon T., Negilski Mathew, and James Joshua. Limiting the permeability of composites for cryogenic applications. *Compos. Sci. Technol.* 66 no. 13 (2006): 2284–2295.

Bhuvana Sowrirajalu and Saroja Devi M. Bisphenol containing novel polyimides/glass fiber composites. *Polym.Compos.* 28 no. 3 (2007): 372–380.

Chao Fen, Liang Guozheng, Kong Weifeng, Zhang Zengping, and Wang Jinhe. Dielectric properties of polymer/ceramic composites based on thermosetting polymers. *Polym. Bull.* 60 no. 1 (2008): 129–136.

Cheng Qunfeng, Fang Zhengping, Xu Yahong, and Yi Xiao-Su. Improvement of the impact damage resistance of BMI/graphite laminates by the ex-situ method. *High Perform. Polym.* 18 no. 6 (2006): 907–917.

Cheng Qunfeng, Fang Zhengping, Yi Xiao-Su, An Xuefeng, Tang Bangming, and Xu Yahong. Ex-situ concept for toughening the RTMable BMI matrix composites. II. Improving the compression after impact. *J. Appl. Polym. Sci.* 108 no. 4 (2008): 2211–2217.

Cheng Qunfeng, Wang Ben, Zhang Chuck, and Liang Zhiyong. Functionalized carbon-nanotube sheet/bismaleimide nanocomposites: Mechanical and electrical performance beyond carbon-fiber composites. *Small.* 6 no. 6 (2010): 763–767.

Chiang Tzu Hsuan, Liu Chun Yu, and Dai Chih Yang. A study of the thermal, dielectric, and flame-retarding characteristics of various bismaleimide blended with halogen-free epoxy resin. *J. Polym. Res.* 20 no. 10 (2013): 274.

Chou Chou I., and Liu Ying-Ling. High performance thermosets from a curable Diels-Alder polymer possessing benzoxazine groups in the main chain. *J. Polym. Sci. Part A*. 46 no. 19 (2008): 6509–6517.

Colin Xavier, Mavel Anne, Marais Christian, and Verdu Jacques. Interaction between cracking and oxidation in organic matrix composites. *J. Compos. Mater.* 39 no. 15 (2005a):1371–1389.

Colin Xavier, Marais Christian, and Verdu Jacques. Kinetic modelling of the stabilizing effect of carbon fibres on thermal ageing of thermoset matrix composites. *Compos. Sci. Technol.* 65 no. 1 (2005b): 117–127.

Costa Michelle Leali, Rezende Mirabel Cerqueira, and De Almeida Sergio Frascino M. Strength of hygrothermally conditioned polymer composites with voids. *J. Compos. Mater.* 39 no. 21 (2005): 1943–1961.

Du Mingliang, Guo Baochun, and Jia Demin. Effects of thermal and UV-induced grafting of bismaleimide on mechanical performance of reclaimed rubber/natural rubber blends. *J. Polym. Res.* 12 no. 6 (2005): 473–482.

Evsyukov Sergey E., Pohlmann Tim, and Stenzenberger Horst D. m-Xylylene bismaleimide: A versatile building block for high-performance thermosets. *Polym. Adv. Technol.* 26 no. 6 (2015): 574–580.

Faraz M. Imran, Bhowmik Shantanu, De Ruijter Chrisliaan, et al. Thermal, morphological, and mechanical characterization of novel carbon nanofiber-filled bismaleimide composites. *J. Appl. Polym. Sci.* 117 no. 4 (2010): 2159–2167.

Frigione Mariaenrica, and Kenny Jose M. Effects of storage aging on the cure kinetics of bismaleimide prepregs. *Adv. Polym. Technol.* 24 no. 4 (2005): 253–265.

Fu Yuan-Fei, Xu Kan, Sun Zhao-Yao, Zhang Fu-Qiang, Chen De-Min, and Li Jing. The effect of silane surface treatment of carbon fiber on the tribological properties of bismaleimide (BMI) composite. *Polym. Plast. Technol. Eng.* 51 no. 10 (2012): 1068–1072.

Gerzeski Roger. LOX compatible toughened bismaleimide matrix thermally conductive fiber composites Part I: Composite fabrication viability and thermal conductivity measurement. *J. Test. Eval.* 36 no. 1 (2008): 1–14.

Grace Landon R., and Altan M. Cengiz. Three-dimensional anisotropic moisture absorption in quartz-reinforced bismaleimide laminates. *Polym. Eng. Sci.* 54 no. 1 (2014): 137–146.

Gu Aijuan. Novel high performance RTM bismaleimide resin with low cure temperature for advanced composites. *Polym. Adv. Technol.* 16 no. 7 (2005): 563–566.

Gu Aijuan, and Liang Guozheng. Novel high performance copper clad laminates based on bismaleimide/aluminium borate whisker hybrid matrix. *J. Appl. Polym. Sci.* 103 no. 2 (2007): 1325–1331.

Guan Qingbao, Gu Aijuan, Liang Guozheng, Yuan Li, Liao Fan, and Gong Yingwei. Curing kinetics and mechanism of novel high performance hyperbranched polysiloxane/bismaleimide/cyanate ester resins for resin transfer molding. *J. Appl. Polym. Sci.* 122 no. 1 (2011a): 304–312.

Guan Qingbao, Gu Aijuan, Liang Guozheng, Zhou Cheng, and Yuan Li. Preparation and properties of new high performance maleimide-triazine resins for resin transfer molding. *Polym. Adv. Technol.* 22 no. 12 (2011b): 1572–1580.

Guo Zhan-Sheng. Effects of storage aging on the cure kinetics of T700/BMI prepregs for advanced composites. *Polym. Compos.* 29 no. 11 (2008): 1269–1275.

Guo Zhansheng, Du Shanyi, Zhang Boming, and Wu Zhanjun. Cure kinetics of T700/BMI prepreg used for advanced thermoset composite. *J. Appl. Polym. Sci.* 97 no. 6 (2005): 2238–2241.

Guo Zhan-Sheng, Du Shan-Yi, Zhang Boming, and Wu Zhanjun. Cure characterization of a new bismaleimide resin using differential scanning calorimetry. *J. Macromol. Sci. Part A*. 43 no. 11 (2006): 1687–1693.

Han Yongjin, Liao Gongxiong, Xu Yajuan, Yu Guipeng, and Jian Xigao. Cure kinetics, phase behaviors, and fracture properties of bismaleimide resin toughened by poly (phthalazinone ether ketone). *Polym. Eng. Sci.* 49 no. 12 (2009): 2301–2308.

Hu Xiao, and Meng Jiru. Effect of organoclay on the curing reactions in bismaleimide/diallyl bisphenol A resin. *J. Polym. Sci. Part A*. 43 no. 5 (2005): 994–1006.

Huang Fuwei, Huang Farong, Zhou Yan, and Du Lei. Thermal properties of new bismaleimide resins containing hydrogen silsesquioxane and diallyl bisphenol A. *J. Appl. Polym. Sci.* 117 no. 5 (2010): 2932–2936.

Huang Fuwei, Rong Zhuxia, Shen Xuening, Huang Farong, Du Lei, and Li Zhongping. Organic/inorganic hybrid bismaleimide resin with octa (aminophenyl)silsesquioxane. *Polym. Eng. Sci.* 48 no. 5 (2008): 1022–1028.

Hwang Hann-Jang, Shieh Jeng-Yueh, Li Chun-Hung, and Wang Chun-Shan. Dielectric and thermal properties of polymer network based on bismaleimide resin and cyanate ester containing dicyclopentadiene or dipentene. III. *J. Appl. Polym. Sci.* 103 no. 3 (2007): 1942–1951.

Jena Rajeeb Kumar, Yue Chee Yoon, Sk Md Moniruzzaman, and Ghosh K. A novel high performance bismaleimide/diallyl bisphenol A (BMI/DBA)-epoxy interpenetrating network resin for rigid riser application. *RSC Adv.* 5 no. 97 (2015): 79888.

Ji Lifu, Gu Aijuan, Liang Guozheng, and Yuan Li. Novel modification of bismaleimide-triazine resin by reactive hyperbranched polysiloxane. *J. Mater. Sci.* 45 no. 7 (2010): 1859–1865.

Jothibasu Seetharaman, Premkumar Shanmugam, Alagar Muthukaruppan, and Hamerton Ian. Synthesis and characterization of a POSS-maleimide Precursor for hybrid nanocomposites. *High Perform. Polym.* 20 no. 1 (2008): 67–85.

Kakimoto Masa-aki, Takahashi Akio, Tsurumi Taka-aki, et al. Polymer-ceramic nanocomposites based on new concepts for embedded capacitor. *Mater. Sci. Eng. Part B.* 132 no. 1–2 (2006): 74–78.

Khunová Viera, Liauw Christopher M., Kelnar Ivan, and Snauko Marian. Elimination of separation processes for post-consumer polyolefin waste: Reactive blending using 1, 3-phenylene dimaleimide in presence of filler. *Macromol. Mater. Eng.* 294 no. 8 (2009): 502–509.

Koopman Mark, Chawla Krishan K., Carlisle Kipp B., and Gladyz Gary. Microstructural failure modes in three-phase glass syntactic foams. *J. Mater. Sci.* 41 no. 13 (2006): 4009–4014.

Kurdi Jamal, and Kumar Ashwani. Formation and thermal stability of BMI-based interpenetrating polymers for gas separation membranes. *J. Membr. Sci.* 280 no. 1–2 (2006): 234–244.

Li Wanwan, Liu Feng, Wei Liuhe, and Zhao Tong. Synthesis, morphology and properties of polydimethylsiloxane-modified allylated novolac/BMI. *Eur. Polym. J.* 42 no. 3 (2006): 580–592.

Li Wanwan, Liu Feng, Wei Liuhe, and Zhao Tong. Allylated novolac/4, 4′-bismaleimido-diphenylmethane resin containing polyhedral oligomeric silsesquioxane: Preparation, morphology and thermal stability. *J. Appl. Polym. Sci.* 104 no. 6 (2007): 3903–3908.

Liang Guozheng, Hu Xiaolan, and Lu Tingli. Inorganic whiskers reinforced bismaleimide composites Part II. *J. Mater. Sci.* 40 no. 7 (2005): 1743–1748.

Liang Guozheng, Zhang Zengping, and Yang Jieying. Study on properties of low dielectric loss resin matrix. *Polym. Bull.* 58 no. 5–6 (2007): 1021–1029.

Liauw Christopher Mark, Lees Graham Clayton, Rothon Roger Norman, et al. Evolution of reactive unsaturated interfacial modifiers for polyolefin based composites. *Compos. Interf.* 13 no. 8–9 (2006): 717–736.

Liu Lina, Gu Aijuan, Fang Zhengping, Tong Lifang, and Xu Zhongbin. The effects of the variations of carbon nanotubes on the micro-tribological behavior of carbon nanotubes/bismaleimide nanocomposite. *Composites Part A*. 38 no. 9 (2007a): 1957–1964.

Liu X., Yu Y., and Li S. Study on cure reaction of the blends of bismaleimide and dicyanate ester. *Polymer.* 47 no. 11 (2006): 3767–3773.

Liu Xiaoyun, Yu Yingfeng, Wang Minghai, Zhao Lin, Li Liang, and Li Shanjun. Study on the polyethersulfone/bismaleimide blends: Morphology and rheology during isothermal curing. *J. Mater. Sci.* 42 no. 6 (2007b): 2150–2156.

Liu Ying-Ling, Wang Yao-Hsuan, and Chen Hsueh-Shih. Novel thermosetting resins based on 4-(N-maleimido) phenylglycidylether, 4[a] preparation and properties of polymer-clay nanocomposites. *Macromol. Chem. Phys.* 206 no. 5 (2005): 600–606.

Liu Zhe, Chen Ping, Zhang Xiaoliang, and Yu Qi. Degradation of plasma-treated poly (p-phenylene benzobisoxazole) fiber and its adhesion with bismaleimide resin. *RSC Adv.* 4 no. 8 (2014): 3893–3899.

Lu Guotao, Huang Ying, Yan Yehai, Zhao Tong, and Yu Yunzhao. 4, 4′-Bismaleimidodiphenyl methane modified novolak resin/titania nanocomposites: Preparation and properties. *J. Appl. Polym. Sci.* 102 no. 1 (2006a): 52–57.

Lu Guotao, Huang Ying, Yan Yehai, Zhao Tong, and Yu Yunzhao. Morphology transition of bismaleimide-modified novolac resin. *J. Appl. Polym. Sci.* 102 no. 1 (2006b): 76–83.

Luo Zhen-hua, Liu Feng, Wei Liu-he, and Zhao Tong. Study on BMI modified allyl-functional novolac resin/silica cloth composites: Effect of allylation degree on thermal and mechanical properties. *Polym. Compos.* 28 no. 2 (2007): 180–185.

Mahesh K.P.O., Alagar Muthukaruppan, and Jothibasu S. A comparative study on the preparation and characterization of aromatic and aliphatic bismaleimides-modified polyurethane-epoxy interpenetrating polymer network matrices. *J. Appl. Polym. Sci.* 99 no. 6 (2006): 3592–3602.

Mustata Fanica. Thermosetting resin compositions based on novolac and modified novolac. *Adv. Polym. Technol.* 30 no. 3 (2011): 219–233.

Mustata Fanica, and Bicu Ioan. Synthesis and properties of multiallyl maleate resins from epoxy formaldehyde resins. *Ind. Eng. Chem. Res.* 47 no. 5 (2008): 1355–1364.

Patel Hashmukh S., and Patel Jigar A. Synthesis, characterization and glass reinforcement of poly(amido-imide)s. *Int. J. Polymer. Mater.* 57 no. 2(2008): 177–188.

Peterson Amy M., Jensen Robert E., and Palmese Giuseppe R. Thermoreversible and remendable glass-polymer interface for fiber-reinforced composites. *Compos. Sci. Technol.* 71 no. 5 (2011): 586–592.

Qu Chunyan, Zhao Liwei, Wang Dezhi, Li Hongfeng, Xiao Wanbao, and Yue Cheng'e. Bis[4-(4-maleimidephen-oxy)phenyl]propane/N,N-4,4′-bismaleimidodiphenylmethyene blend modified with diallyl bisphenol A. *J. Appl. Polym. Sci.* 131 no. 12 (2014): 40395.

Rajabi Laleh, and Malekzadeh Ghodratollah. Effects of bismaleimide resin on dielectric and dynamic mechanical properties of epoxy-based laminates. *Iran. Polym. J.* 15 no. 6 (2006): 447–455.

Rajamani Dharmaraj, Surender Rajendran, Mahendran Arunjunairaj, Muthusubramanian Shanmugam, and Vijayakumar Chinnaswamy Thangavel. Bismaleimide/rice husk silica reinforced composites synthesis and thermal degradation studies. *J. Therm. Anal. Calorim.* 114 no. 2 (2013): 883–893.

Rajasekaran Rajangam, and Alagar Muthukaruppan. Bismaleimides (1, 3-bismaleimidobenzene and 1, 1′-bis(4-maleimidophenyl) cyclohexane) modified polyethersulfone/epoxy matrices for engineering applications. *J. Compos. Mater.* 42 no. 10 (2008): 1031–1043.

Rao Butti S., Sireesha R., and Pasala A.R. Preparation and thermal properties of bismaleimide blends based on hydroxyphenyl maleimide. *Polym. Int.* 54 no. 8 (2005): 1103–1109.

Rider Andrew N., Brack Narelle, and Gopalakrishn J. Surface treatments and adhesives for bonded repairs to high temperature carbon-bismaleimide composite structure. *J. Adhes. Sci. Technol.* 26 no. 7 (2012): 911–937.

Salini K, Krishna Munishamaian, Narasimha Murthy H.N., Rai K.S., and Satheesh Chandran Maniyeri. The effect of the addition of acrylonitrile-butadiene styrene on the mechanical properties of bismaleimide/carbon composites. *Polym. Plast. Technol. Eng.* 49 no. 13 (2010): 1390–1394.

Shenoy M.A., Patil Mrinalini, and Shetty Aditya. Modification of epoxy resin by addition of bismaleimide and diallyl phthalate. *Polym. Eng. Sci.* 47 no. 11 (2007): 1881–1888.

Shibata Mitsuhiro, Teramoto Naozumi, Imada Ayumi, Neda Makiyo, and Sugimoto Shimon. Bio-based thermosetting bismaleimide resins using eugenol, bieugenol and eugenol novolac. *React. Funct. Polym.* 73 no. 8 (2013): 1086–1095.

Shibata Mitsuhiro, Teramoto Naozumi, and Nakamura Yu. High performance bio-based thermosetting resins composed of tung oil and bismaleimide. *J. Appl. Polym. Sci.* 119 no. 2 (2011a): 896–901.

Shibata Mitsuhiro, Teramoto Naozumi, Shimasaki Toshiaki, and Ogihara Megumi. High-performance bio-based bismaleimide resins using succinic acid and eugenol. *Polym. J.* 43 no. 11 (2011b): 916–922.

Shimokawa Toshiyuki, Kakuta Yoshiaki, Hamaguchi Yasumasa, and Aiyama Takenori. Static and fatigue strengths of a G40-800/5260 carbon fiber/bismaleimide composite material at room temperature and 150°C. *J. Compos. Mater.* 42 no. 7 (2008): 655–679.

da Silva Lucas F.M, and Adams Robert D. Measurement of the mechanical properties of structural adhesives in tension and shear over a wide range of temperatures. *J. Adhes. Sci. Technol.* 19 no. 2 (2005): 109–141.

da Silva Lucas F.M., and Adams Robert D. Effect of temperature on the mechanical and bonding properties of a carbon-reinforced bismaleimide. *Proc. IME. L J. Mater. Des. Appl.* 222 no. 1 (2008): 45–52.

Sithique Mohamed Aboobucker, Nagendiran Shanmugam, and Alagar Muthukaruppan. Synthesis and characterization of bismaleimide-modified, soy-based epoxy matrices for flame-retardant applications. *High Perform. Polym.* 22 no. 3 (2010): 328–344.

Surender Rajendran, Mahendran Arunjunai, Thamaraichelvan A., Alam Sarfaraz, and Vijayakumar Chinnaswamy Thangavel. Curing studies of bisphenol A based bismaleimide and Cloisite 15A nanoclay blends using differential scanning calorimetry and model-free kinetics. *J. Appl. Polym. Sci.* 128 no. 1 (2013): 712–724.

Suresh Kumar Ramasamy, and Alagar Muthukaruppan. Studies on mechanical, thermal, and morphology of diglycidylether-terminated polydimethylsiloxane-modified epoxy-bismaleimide matrices. *J. Appl. Polym. Sci.* 101 no. 1 (2006): 668–674.

Takatoya Takeshi, and Susuki Ippei. In-plane and out-of-plane characteristics of three-dimensional textile composites. *J. Compos. Mater.* 39 no 6 (2005): 543–556.

Takeichi Tsutomu, Saito Yuki, Agag Tarek, Muto Hiroyuki, and Kawauchi Takehiro. High-performance polymer alloys of polybenzoxazine and bismaleimide. *Polymer.* 49 no. 5 (2008): 1173–1179.

Thunga Mahendra, Bauer Amy, Obusek Kristine, Meilunas Ray, Akinc Mufit, and Kessler Michael R. Injection repair of carbon fiber/bismaleimide composite panels with bisphenol E cyanate ester resin. *Compos. Sci. Technol.* 100 (2014): 174–181.

Tian Xiao-Wei, and Yang Zhen-Guo. Effect of CTBN concentration on thermal and morphological properties of binary copolymers of o-cresol novolac epoxy resin and bismaleimide. *Polym. Plast. Technol. Eng.* 50 no. 5 (2011): 474–480.

Tsai Pai-Feng, Wu Chyong-Fen, Hsiao Yu-Che, and Shau Min-Da. Properties of novel polyimides containing bismaleimide and cyclic phosphine oxide. *J. Polym. Res.* 16 no 6 (2009): 673–680.

Wan Y.Z., Wang Y.L., He F., Huang Y., and Jiang H.J. Mechanical performance of hybrid bismaleimide composites reinforced with three-dimensional braided carbon and Kevlar fabrics. *Composites Part A*. 38 no 2 (2007): 495–504.

Wang Shaoquan, Dong Shangli, Gao Yu, Ye Zhuyu, and Lu Gang. Influence of liquid nitrogen steeping on mechanical properties of a unidirectional carbon fiber reinforced bismaleimide composite. *Polym. Polym. Compos.* 22 no. 2 (2014): 129–134.

Wang Zheng, Wu Wei, Zhong Yuhua, Ruan Mingzhu, and Hui Lin Lin. Flame-retardant materials based on phosphorus-containing polyhedral oligomeric silsesquioxane and bismaleimide/DBA with improved thermal resistance and dielectric properties. *J. Appl. Polym. Sci.* 132 no. 9 (2015): 41545–41555.

Wu Guanglei, Kou Kaichang, Chao Min, Zhuo Longhai, Zhang Jiaoqiang, and Li Ning. Preparation and characterization of bismaleimide-triazine/epoxy interpenetrating polymer networks. *Thermochim. Acta.* 537 (2012): 44–50.

Wu Guanglei, Kou Kaichang, Zhuo Longhai, Wang Yiqun, and Zhang Jiaoqiang. Preparation and characterization of novel dicyanate/benzoxazine/bismaleimide copolymer. *Thermochim. Acta.* 559 (2013): 86–91.

Xiong Xuhai, Chen Ping, Ren Rong, Lu Fang, and Yu Qi. Cure mechanism and thermal properties of the phthalide-containing bismaleimide/epoxy system. *Temochim. Acta.* 559 (2013): 52–58.

Xiong Xuhai, Chen Ping, Zhang Jinxiang, Yu Qi, and Wang Baichen. Preparation and properties of high performance phthalide-containing bismaleimide modified epoxy matrices. *J. Appl. Polym. Sci.* 121 no. 6 (2011): 3122–3130.

Yan Hong-qiang, Wang Hua-qing, and Cheng Jie. Interpenetrating polymer networks from the novel bismaleimide and cyanate containing naphthalene: Cure and thermal characteristics. *Eur. Polym. J.* 45 no. 8 (2009): 2383–2390.

Yan Hongxia, Li Pengbo, Ning Rongchang, Ma Xiaoyan, and Zhang Zengping. Tribological properties of bismaleimide composites with surface-modified SiO_2 nanoparticles. *J. Appl. Polym. Sci.* 110 no. 3 (2008): 1375–1381.

Yan Hongxia, Ning Rongchang, Liang Guozheng, Huang Ying, and Lu Tingli. The effect of silane coupling agent on the sliding wear behaviour of nanometer ZrO_2/bismaleimide composites. *J. Mater. Sci.* 42 no. 3 (2007): 958–965.

Yan Hongxia, Ning Rongchang, Liang Guozheng, and Ma Xiaoyan. The performances of BMI nanocomposites filled with nanometer SiC. *J. Appl. Polym. Sci.* 95 no. 4 (2005): 1246–1250.

Yao Yuan, Zhao Tong, and Yu Yunzhao. Novel thermosetting resin with a very high glass-transition temperature based on bismaleimide and allylated novolac. *J. Appl. Polym. Sci.* 97 no. 2 (2005): 443–448.

Ye Xiaochuan, Zeng Liming, and Liu Tao. Thermal stability and thermal degradation kinetic study of bismaleimideepoxy modified novolac resin. *Compos. Interf.* 19 no. 7 (2012): 461–473.

Yu Yingfeng, Wang Minghai, Liu Xiaoyun, Zhao Lin, Tang Xiaolin, and Li Shanjun. Synthesis and cure of liquid crystalline diallyl modifiers for bismaleimide resin. *J. Appl. Polym. Sci.* 101 no. 6 (2006): 4366–4371.

Yuan Li, Liang Guozheng, and Gu Aijuan. Novel fiber reinforced bismaleimide/diallyl bisphenol A/microcapsules composites. *Polym. Adv. Technol.* 22 no. 12 (2011): 2264–2272.

Yuan Li, Ma Xiaoyan, Gu Aijuan, et al. A novel organic rectorite modified bismaleimide/diallylbisphenol A system. *Polym. Adv. Technol.* 20 no. 11 (2009): 826–833.

Zeng Li, Liang Guozheng, Gu Aijuan, Yuan Li, Zhuo Dongxian, and Hu Jiang-Tao. High performance hybrids based on a novel incompletely condensed polyhedral oligomeric silsesquioxane and bismaleimide resin with improved thermal and dielectric properties. *J. Mater. Sci.* 47 no. 6 (2012): 2548–2558.

Zeng Xiaoliang, Yu Shuhui, and Sun Rong. Thermal behavior and dielectric property analysis of boron nitride-filled bismaleimide-triazine resin composites. *J. Appl. Polym. Sci.* 128 no. 3 (2013): 1353–1359.

Zhang B.Y., Li M., and Chen X.B. Influence of novel additive on BMI resin and BMI resin matrix composite. *J. Mater. Sci.* 42 no. 22 (2007): 9170–9175.

Zhou Bo Xuan, Huang Yi Jun, Zhang Xing Hong, Fu Zhi Sheng, and Qi Guo Rong. Thermal properties of an epoxy cresol-formaldehyde novolac/diaminodiphenyl sulfone system modified by bismaleimide containing tetramethylbiphenyl and aromatic ether structures. *Polym. Eng. Sci.* 49 no. 8 (2009): 1525–1532.

Zhou Cheng, Gu Aijuan, Liang Guozheng, and Yuan Li. Novel toughened cyanate ester resin with good dielectric properties and thermal stability by copolymerizing with hyperbranched polysiloxane and epoxy resin. *Polym. Adv. Technol.* 22 no. 5 (2011): 710–717.

Zhu Yanjing, Yuan Li, Liang Guozheng, and Gu Aijuan. Green flame retarding bismaleimide resin with simultaneously good processing characteristics, high toughness and outstanding thermal stability based on a multi-functional organic boron compound. *Polym. Degrad. Stab.* 118 (2015): 33–44.

Zhuo Dongxian, Gu Aijuan, Liang Guozheng, Hu Jiang-tao, Yuan Li, and Chen Xiangxiu. Flame retardancy materials based on a novel fully end-capped hyperbranched polysiloxane and bismaleimide/diallylbisphenol A resin with simultaneously improved integrated performance. *J. Mater. Chem.* 21 (2011a): 6584–6594.

Zhuo Dongxian, Gu Aijuan, Liang Guozheng, et al. Flame retardancy and flame retarding mechanism of high performance hyperbranched polysiloxane modified bismaleimide/cyanate ester resin. *Polym. Degrad. Stab.* 96 (2011b): 505–514.

Zou Xingqiang, Xu Mingzhen, Jia Kun, and Liu Xiaobo. Copolymerizing behavior and processability of allyl-functional bisphthalonitrile/bismaleimide system. *Polym. Compos.* (2015) doi: 10.1002/pc.

14 Fire-Resistant Epoxy Resins Containing Organophosphorus Compounds

Tachita Vlad-Bubulac, Corneliu Hamciuc, Diana Serbezeanu, Ionela-Daniela Carja, and Vicente J. Forrat Pérez

CONTENTS

14.1 INTRODUCTION

The specific chemistry of epoxy resins involves the existence, in the chemical structure of the raw material, of at least two reactive epoxide groups, besides aliphatic, aromatic, or heteroaromatic moieties (Pascault and Williams 2010). The higher the number of these oxirane cycles, the higher the functionality of the epoxy resin; and consequently, a higher cross-linked structure can be obtained (Kandola and Deli 2014). The easily accessible epoxide groups can undergo self-cross-linking in the presence of a catalyst (Levchik and Weil 2004) or can be cured using effective cross-linkers such as aliphatic (Wan et al. 2012), cycloaliphatic (Hamerton et al. 2014), and aromatic diamines (Braun et al. 2006), acids (Liu et al. 2010), anhydrides (Becker et al. 2001), phenols (Cai and Lin 2005), thiols (Luo et al. 2013), and so on. The curing process represents the final step of the polymerization, by which the raw material is irreversibly converted into an infusible three-dimensional polymer network. Such polymer networks are covalently linked and once converted into permanent, insoluble, and infusible thermosets they cannot be remelted or reshaped. These cross-linked structures possess some interesting and highly desirable characteristics such as increased adhesion to various substrates; resistance to moisture, solvents, and other chemicals; outstanding electrical insulation properties; high tensile strength and modulus combined with durability; low shrinkage upon curing and no volatiles released during curing; excellent dimensional stability; compatibility with a great number of materials; and so forth.

The first commercial epoxy resins produced from a common reaction between epichlorohydrin and bisphenol A (DGEBA) were pioneered, almost simultaneously, by the Swiss inventor P. Castan (under the Ciba Company license) and by S. O. Greenlee (working with Devoe Reynolds Company) back in the 1940s (Pascault and Williams 2010). Since then, the versatility of epoxies and the innovation in formulation has resulted in plenty of epoxy resin-based materials, with uses across an enormously wide variety of industries (Petrie 2005). Nowadays, engineered nanostructured epoxy-based materials can be manufactured and potentially used as novel functional materials for a number of new applications in

- Aircraft, aerospace, and automobile industries (structural molds, adhesives, paints, coatings, sealers, encapsulants for control modules, etc.)
- Construction and civil engineering (industrial flooring, grouts for roads and bridges, adhesives, maintenance paints, etc.)
- Chemical industry (linings for storage tanks, chemical plants-including coatings, pipes and pipe linings, filters)
- Electrical and electronics (construction and insulation of switchgears, transformers, turbine alternators; electric motor insulation; cable jointing; coatings for domestic electrical appliances; printed-circuit boards; packaging of active and passive components; encapsulation of electronic modules; adhesives)
- Marine engineering (primers and protective coatings for ships and marine structures)
- Telecommunications and multimedia, energy storage and production, medical applications, membranes, packaging, and so on.

Although the appliance of the epoxy thermosets is under continuous development, there are still many aspects to be improved. Epoxy resins are not the lowest cost materials available for most of the aforementioned applications. Thus, epoxy resins must provide added value to justify their additional cost. This added value is usually realized by the incorporation of a special property or combination of properties into the final product, which can often be designed through multiphase polymer system mechanisms.

14.2 FLAME RETARDANCY OF EPOXY RESINS

Due to the ongoing miniaturization trend of electronic devices and for meeting some critical application requirements, several techniques and approaches for improving the thermal stability and flame retardancy of the epoxy resins have been employed. Actually, all the modern-day electronic devices contain at least one printed wired board to support and connect their electronic components. The insulating lattice of these fabrics consists of epoxy composites manufactured mainly from three components: epoxy resin, woven glass fibers, and a flame retardant. For a long period, halogen-containing flame retardants (e.g., tetrabromobisphenol A [TBBA]) have been widely used to induce flame retardancy in such composites (Chen et al. 2005; Lin et al. 2010). However, halogenated epoxy resins release hydrogen halide (Purser 2000), polyhalogenated dibenzo-*p*-dioxin, and dibenzofuran (Weber and Greim 1997), during combustion, causing severe toxicity, corrosion, and ecological pollution. Because of these environmental and end-of-life issues, there is a global tendency of several research groups and companies to replace halogenated flame retardant with halogen-free alternatives.

Successful approaches to improve flame retardancy of epoxy resins imply interrupting the complex combustion process at a given stage to reduce the rate of burning and/or to change the mechanism of combustion at that stage (Salamone 1996). From a practical point of view, this is achieved either through an additive pathway by mechanical blending of a suitable flame-retardant compound with the flammable epoxy resin or by following reactive strategies by chemical incorporation of the flame retardant into the polymer by simple copolymerization or chemical modification of a preformed polymer (Joseph and Tretsiakova-Mcnally 2011).

The use of phosphorus-containing compounds as promising, halogen-free, flame retardants for epoxy resins has been extensively discussed and reviewed (Lu and Hamerton 2002; Levchik and Weil 2004; Levchik et al. 2005). It has been generally demonstrated that organophosphorus compounds have high flame-retardant efficiency in epoxy resins (Levchik et al. 1995; Hörold 1999; Liu 2001; Weil and Levchik 2004; Hergenrother et al. 2005; Deng et al. 2006; Hoang and Kim 2008; Hoang et al. 2008; Kilinc et al. 2015) and have also been found to generate less toxic gases and smoke compared to halogenated compounds (Wang and Shieh 1999; Shieh and Wang 2000). Phosphorus-based flame retardants have been used either as an additive (Xiang et al. 2007; Chen et al. 2012; Aslzadeh et al. 2013; Carja et al. 2014b; Jian et al. 2014; Xu et al. 2015) or reactive moiety of a polymer system (Liu 2002b; Liu et al. 2007; Petreus et al. 2010a, 2010b; Hamciuc et al. 2013; Toldy et al. 2014; Zhang et al. 2015). They are believed to act mainly in the condensed phase, by altering the rate and/or the pathway of the pyrolytic decomposition mechanism of polymer materials with the formation of a phosphorus-rich carbonaceous layer (Cochez et al. 2000; Ebdon et al. 2000; Duquesne et al. 2004; Deng et al. 2006; Brehme et al. 2011; Petreus et al. 2011; Benin et al. 2012).

Due to its P–H functionality, 9,10-dihydro-9-oxa-10-phosphaphenanthrene-10-oxide, denoted as DOPO, was assumed to compete with TBBA, as a reactive flame retardant. Thus, among the few phosphorus-based flame retardants that have been successfully introduced into the printed wired board market, DOPO was by far the most relevant (Döring et al. 2012).

14.3 DOPO-CONTAINING REACTIVE FLAME RETARDANTS

Phosphorus-containing flame retardants can be added to epoxy as additives, or they can be incorporated in an epoxy network by pre-reacting with epoxy resin or by using a phosphorus-based curing agent (Levchik and Weil 2004).

Reactive flame retardants are preferred in epoxy because they show less negative effects on physical properties-mostly glass transition temperature and hydrolytic stability (Morgan and Wilkie 2014). A close literature survey easily reveals the extensive use of DOPO molecules as an efficient flame retardant incorporated not only in epoxies but also into various other polymers [polyacrylonitrile, aromatic polyamide, polyesters, poly(ester-imide)s, etc.], generally preserving the thermal stability of the neat backbone. It has been shown that in DOPO-containing polymers, the cleavage of the P–O group does not lead to the scission of the main chain, while the O=P–O bond induces higher stability in comparison with the same functional group but from phosphonate or phosphate groups, and thus the thermal stability is little disrupted. In the first attempts, DOPO (I) was pre-reacted with diglycidyl ether of bisphenol A, resulting in a new resin with decreased functionality II (Scheme 14.1). DOPO-containing glass-fiber laminates, resulting from proper formulations, were classified V-1 with 1.6% P and V-0 with 2.1% P while preserving the stress at failure, the electrical characteristics, and glass transition temperatures of the standard laminates (Levchik et al. 2005). The limiting oxygen index values (LOI) of resin II increased from 22 to 28 when the phosphorus content increased from 0% to 1.6%.

Hamciuc et al. reported preparation of epoxy resins containing phosphorus and/or silicon atoms by using compound II and a co-monomer having polydimethylsiloxane segments. The cured epoxy resins exhibited good thermal stability with initial decomposition temperatures being in the range of 311°C–366°C and glass transition temperatures in the domain of 111°C–162°C. By introduction of DOPO units in the chemical structure of epoxy resin, a decrease in the glass transition temperature appeared as a consequence of a decrease in the cross-linking density. The presence of phosphorus and silicon atoms enhanced the LOI and thermal stability of the char residue. SEM measurements showed

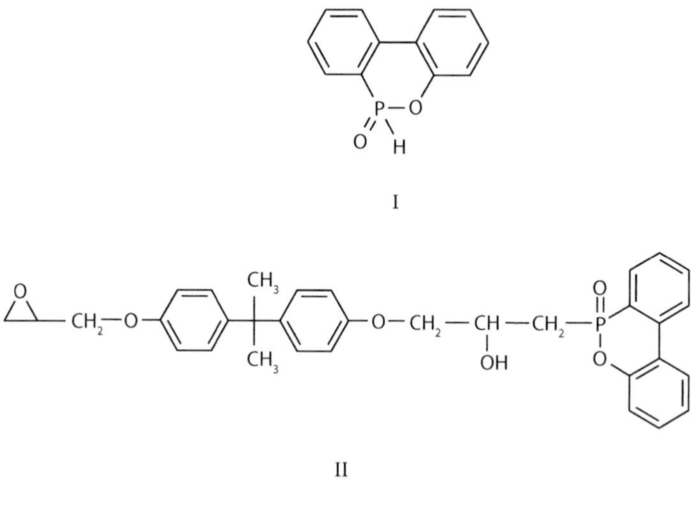

SCHEME 14.1 Chemical structure of DOPO (I) and adduct of DOPO with DGEBA (II).

different morphologies of the char residue, suggesting a more compact structure that increased the flame-retardant properties.

In the next step, due to its P–H versatility, DOPO was extensively used as a precursor for the preparation of a growing number of reactive diphenol, oxirane, or diamine curing agents. Thus, 2-(6-oxido-6*H*-dibenz[c,e][1,2]oxaphosphorin-6-yl)-1,4-benzenediol III and 2-(6-oxido-6*H*-dibenz[c,e][1,2]oxaphosphorin-6-yl)-1,4-naphthalenediol IV were prepared by nucleophilic addition reaction of P–H bond of DOPO with the carbonyl group of benzoquinone (Wang and Shieh 1998; Petreus et al. 2005) and naphthoquinone (Serbezeanu et al. 2010b), respectively. The oxirane analogue V was prepared by reaction of III with epichlorohydrin (Scheme 14.2).

Advanced epoxy resins with various epoxy equivalent weights were synthesized from both reactive diphenol III and its diglycidyl ether derivative V (Wang and Shieh 1999; Wang and Lin 2000). The char yield was found to be less than 1% in the non–flame-retardant system at 700°C under air, while it ranged from 10% to 18% in the phosphorus-containing epoxies. LOI increased from 22 to 28 when the phosphorus content increased from 0% to 1.4%. In epoxy formulation useful for encapsulation of electronic devices, the V-0 rating could be achieved with a phosphorus content of 1%, comparable to a bromine content of 7.2%. Other flame-retardant epoxy monomers were prepared by reacting diphenol III with DGEBA. Some studies have been performed to compare the performances of DOPO-based epoxies versus TBBA advanced epoxy resins (Wang and Lee 2000). Phosphorus was found to be more effective than bromine in reaching the flame retardancy because 2% P is enough to equal the efficiency of 20% Br. According to UL-94 tests, the V-0 rate could be achieved either with 1.4% P (wt) or 13.4% Br (wt) for highly cross-linked resins. Another well-established general rule is that the

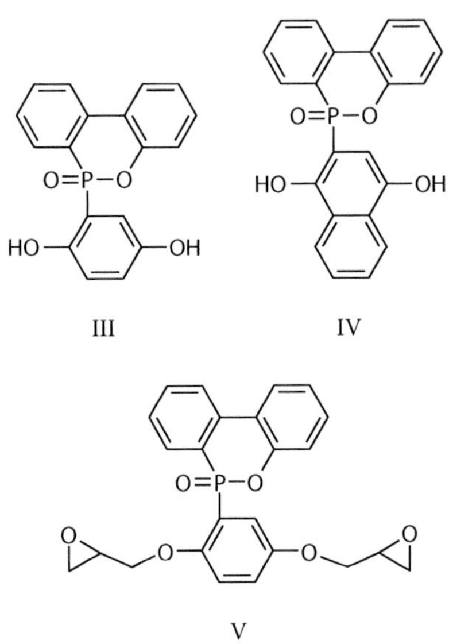

SCHEME 14.2 Chemical structures of compounds III, IV, and V.

higher the phosphorus content, the higher the LOI values (Lu and Hamerton 2002). The low solubility of diphenol III in acetone causes difficulties in the processing of printed-circuit laminates in the microelectronic industry. Thus, two DOPO-containing diacides, VI and VII were prepared from DOPO and maleic acid, or itaconic acid, respectively (Scheme 14.3).

Two series of advanced epoxy resins have been prepared from diacids VI and VII and DGEBA using 4,4′-diaminodiphenylsulphone (DDS) as a curing agent (Lin et al. 2000). The degradation temperatures were found to decrease with increasing P content, while the char yield increased with P content. The formulation based on diacid VII required slightly higher P content in order to achieve the passing rate in the UL-94 test. The DOPO-based diamine curing agent VIII (Scheme 14.4) was found to be a very efficient flame retardant (Wang and Lin 1999), with a reactivity toward DGEBA like that of 4,4′-diaminodiphenylmethane (DDM) and DDS. The char yield of such formulation at 700°C under nitrogen atmosphere was 32% compared with phosphorus-free epoxy resin (13%–15%).

Further attempts directed efforts to obtain DOPO-containing reactive flame retardants with a higher content of phosphorus. DOPO was reacted with a carbonyl group of 4,4′-dihydroxybenzophenone or 4,4′-diaminobenzophenone to give a double DOPO-containing bisphenol IX and diamine X, respectively (Scheme 14.5).

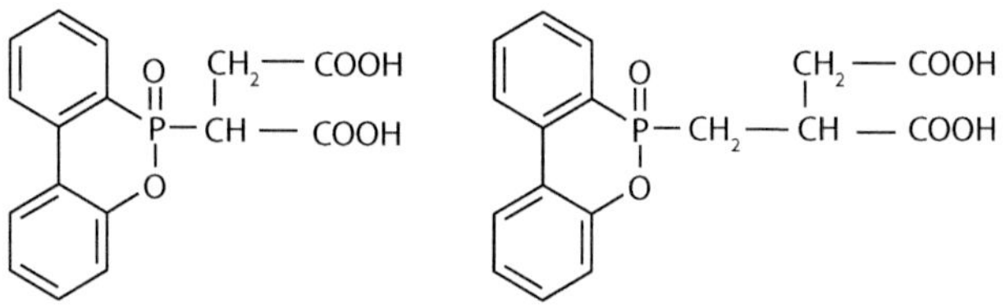

VI VII

SCHEME 14.3 Chemical structures of DOPO-containing diacids VI and VII.

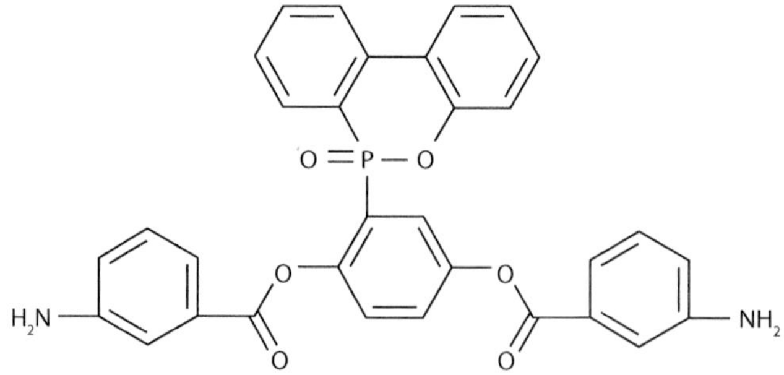

VIII

SCHEME 14.4 DOPO-based aromatic diamine VIII.

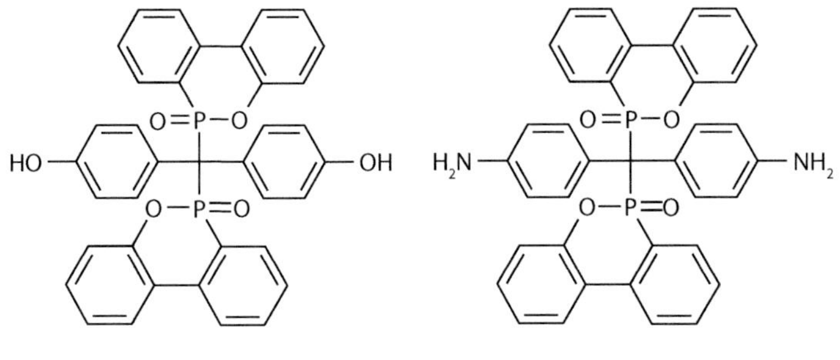

SCHEME 14.5 DOPO-containing bisphenol IX and diamine X.

The reaction of IX with epichlorohydrin or DGEBA resulted in DOPO-containing epoxy monomers that were further cured in proper conditions (Liu 2002b). These cured epoxy resins had high glass transition temperatures and high thermal stabilities. The char yields were found to increase with phosphorus content. Also, LOI values increased with phosphorus content. The authors investigated the influence of different curing agents on the LOI values. The DICY-cured epoxy resins showed higher LOI when compared to other diamines. This was attributed to a possible phosphorus-nitrogen synergism. When double DOPO-containing diamine X was used as a curing agent, extremely high LOI values, between 37 and 50, were obtained as a result of the high phosphorus content of the formulations.

Our group has also successfully developed the area of DOPO-containing reactive flame retardants. Besides the compounds I, III, and IV, already examined earlier, our laboratory synthesized DOPO-based dianhydrides, diols, diphenols, and so on. A few of them will be briefly discussed in context with the current work. Bisphenols III and IV were reacted with trimellitic anhydride chloride yielding two DOPO-containing dianhydrides, namely 1,4-[2-(6-oxido-6*H*-dibenz[c,e][1,2]oxaphosphorin-6-yl)]phenylene-bis(trimellitate) dianhydride XI (Serbezeanu et al. 2010a; Liang et al. 2013) and 1,4-[2-(6-oxido-6*H*-dibenz[c,e][1,2]oxaphosphorin-6-yl)]naphthalene-bis(trimellitate) dianhydride XII (Hamciuc et al. 2011a, 2011b). A bisphenol with preformed imide rings, 1,4-bis[*N*-(4-hydroxyphenyl)phthalimidyl-5-carboxylate]-2-(6-oxido-6*H*-dibenz [c,e][1,2]oxaphosphorin-6-yl)-naphtalene XIII, was obtained by reacting XII with 4-aminophenol (Serbezeanu et al. 2010d) (Scheme 14.6). The existence of both phosphorus and nitrogen atoms in the same structure is expected to induce improved fire resistance, providing a dual mechanism of degradation associated with the condensed and gas phases due to the synergistic effects (Lu and Hamerton 2002; Carja et al. 2016).

A new phosphorus-containing monomer, bis(N-hydroxyethyl-phthalimidyl)-bis(9,10-dihydro- 9-oxa-10-oxide-10-phosphaphenanthrene-10-yl)methane XIV, was prepared by reacting the carbonyl group of *N,N′*-bis(hydroxyethyl)benzophenonetetracarboxylic diimide with DOPO (Petreus et al. 2008) (Scheme 14.7). Another nucleophilic addition reaction of P–H bond of DOPO at the carbonyl group was performed in our group by reacting DOPO with terephthaldicarboxaldehyde (Wang et al. 2011;

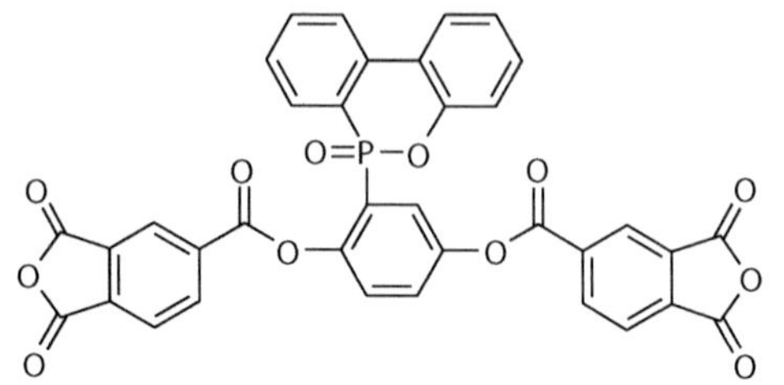

XI

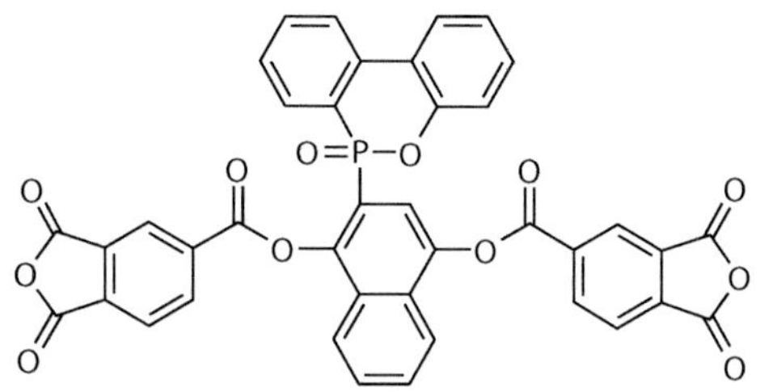

XII

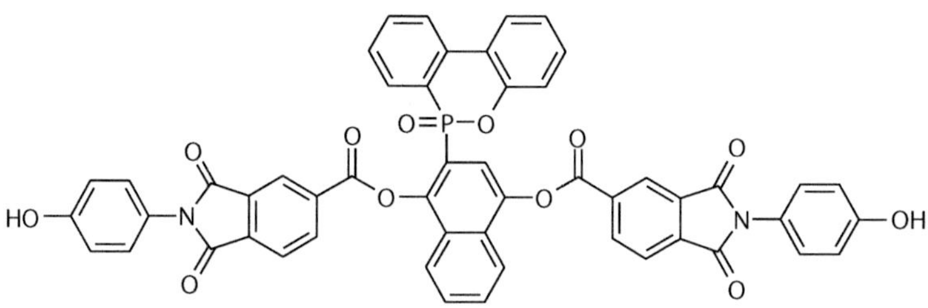

XIII

SCHEME 14.6 Chemical structures of DOPO-containing dianhydrides XI–XII and bisphenol XIII.

Carja et al. 2014b) yielding 1,4-phenylene-bis((6-oxido-6*H*-dibenz[c,e][1,2]oxaphosphorinyl)carbinol) XV (Scheme 14.7).

A new DOPO-containing compound with complex architecture, namely bis((6-oxido-6H-dibenz[c,e][1,2]oxaphosphorinyl)-(4-hydroxyaniline)-methylene)-1,4-phenylene XVI, was prepared by us in a two-step procedure, treating 4-aminophenol with terephthalaldehyde in the first step, followed by the phosphorylation reaction of its imine linkages with DOPO molecules (Scheme 14.8).

14.4 DOPO-CONTAINING ADDITIVE FLAME RETARDANTS

In contrast to reactive flame retardants, the additive (nonreactive) flame retardants have no functionality to interact with the epoxy matrix. These types of flame retardants can

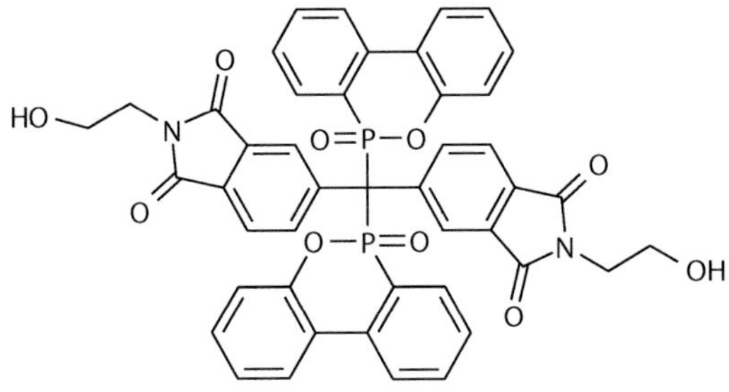

XIV

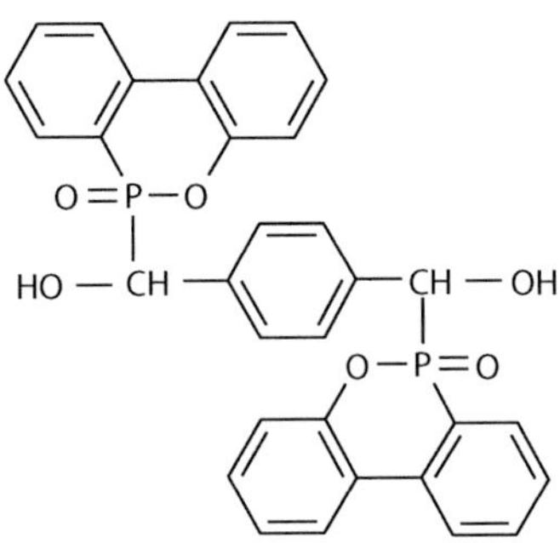

XV

SCHEME 14.7 DOPO-containing aromatic-aliphatic diols XIV and **XV**.

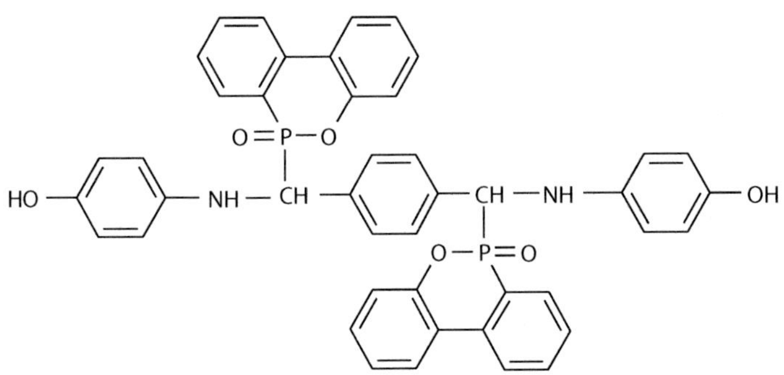

XVI

SCHEME 14.8 Chemical structure of DOPO-containing bisphenol XVI.

be introduced into an epoxy polymer system by physically mixing the resin with the flame retardant before adding an accelerator and hardener. One should note that, without the interaction of the flame retardant with the epoxy resin, the flame retardant shows the same disadvantage as any other additive. It is necessary to disperse them well in the epoxy resin to prevent phase separation and bleeding out of the later product.

Additives (particularly small molecules) could leach over time, leading to a decrease in flame retardancy. They could also plasticize the polymer matrix (Price et al. 2001, 2002). Mechanical properties are often strongly reduced by the addition of additives, generally at high loading. Transparency is also most often impaired. Nevertheless, additives can be easily incorporated through classic processes, whereas the synthesis of new phosphorus-containing polymers is cost-consuming. Moreover, the presence of phosphorus in the macromolecule could also lead to detrimental changes: undesirable modification of glass transition temperature, lower cross-linking, and so on (Tibiletti et al. 2012). Despite all these facts, Döring et al. demonstrated, in a comprehensive study, the efficiency of additive flame retardants based on DOPO either in DGEBA-based epoxy resins that requires a higher loading of the flame retardant or in novolac systems (Döring et al. 2012).

The authors have shown that the flame-retardant effect of phosphorus is enhanced by adding a nitrogen containing compound, which itself has only low flame-retardant activity. Combining an organophosphate with a nitrogen containing organic in one compound shows a synergistic effect, which allows one to reduce the loading of the usually more expensive organophosphate flame retardants. By adding additional inorganic fillers such as aluminum oxide hydroxide or melamine polyphosphate, the loading of the organophosphate could be decreased even more. This not only benefits the thermal properties of the material, such as the Tg, but also reduces the processing time of the epoxy materials.

14.4.1 Polymeric Additives

To overcome some problems with low molecular additives such as leaching or low thermal stability, DOPO-containing polymeric additives could be used. Phosphorus is incorporated into macromolecules used as an additive to improve the flame retardancy of a polymer.

In this respect, during the last decade, our group focused on the preparation and characterization of DOPO-containing polymers starting from some of the reactive DOPO-containing monomers discussed in the previous section. We have prepared DOPO-containing polyesters by the Higashi polycondensation reaction using reactive flame retardants III and IV and different aromatic diacids (Vlad-Bubulac et al. 2006; Hamciuc et al. 2008). The polymers exhibited glass transition in the range 180–226°C. The introduction of pendant DOPO units into the polymer backbone preserved the high thermal stability characteristic for aromatic polyesters with no significant weight loss below 400°C in nitrogen. Moreover, the thermal oxidative stability significantly increased with the introduction of phosphorus atoms in the polymer structure. For instance, the char yield measured at 700°C for the polyester based on III and terephthalic acid was with about 2.3 times higher than the one corresponding to a similar system but lacking the pendant DOPO unit (Hamciuc et al. 2007; Vlad-Bubulac et al. 2007). This result indicates that the presence of phosphorus atom and a high degree of aromaticity can induce good flame-retardant properties. Using dianhydride XII, new DOPO-containing poly(ester-imide)-polydimethylsiloxane copolymers have been prepared (Hamciuc et al. 2011b). By incorporation of both phosphorus and silicon atoms into the macromolecular structures, a further enhancement of flame-retardant properties can be achieved (Liu and Chou 2005; Zhang et al. 2012; Gao et al. 2013; Hamciuc et al. 2013).

Novel DOPO-containing polyesterimides XVII were prepared from a mixture of bisphenol XIII with aliphatic diols such as 1,3-propanediol, 1,4-butanediol, 1,5-pentanediol, 1,6-hexanediol, and 1,12-dodecanediol—and an aromatic diacid chloride containing two preformed ester groups, namely terephthaloylbis-(4-oxibenzoyl-chloride), via high-temperature polycondensation reaction (Serbezeanu et al. 2010c) (Scheme 14.9).

The effect of the phosphorus content on the thermal properties and flame retardancy was evaluated by thermogravimetric analysis (TGA), TGA–FTIR, and scanning electron microscopy. The polymers were stable up to 340°C, showing a 5% weight loss in the range of 340–395°C and a 10% weight loss in the range of 370–415°C. The char yields at 700°C were in the range of 13.6%–38%, increasing with the content of XIII.

Other DOPO-containing polyesterimides have been prepared from double DOPO-containing diol XIII and different aromatic diacide chlorides (Petreus et al. 2008). The polymers did not show significant weight loss below 340°C; they began to decompose in the range of 345–360°C; they showed 10% weight loss in the range of 375–385°C. The char yield at 700°C was in the range of 20%–37%. Using the empirical relationship of Davies and Horrocks (1986), the authors established that the LOI values of these polymers were in the range of 40–54.

A new oligophosphonate with phosphorus atoms both in the main and side chains, XVIII, was prepared by a polycondensation reaction between compound XV and phenylphosphonic dichloride (Carja et al. 2014a, 2014b) (Scheme 14.10).

DOPO-containing epoxy resins have been prepared by thermal cross-linking using oligophosphonate XVIII and DGEBA-type epoxy resin, in the presence of DDM as a curing agent. Due to the low molecular weight and moderate glass transition temperature, XVIII displayed a low melt viscosity such that it could be processed without adding any solubilizer. SEM investigations upon the fracture surfaces of the

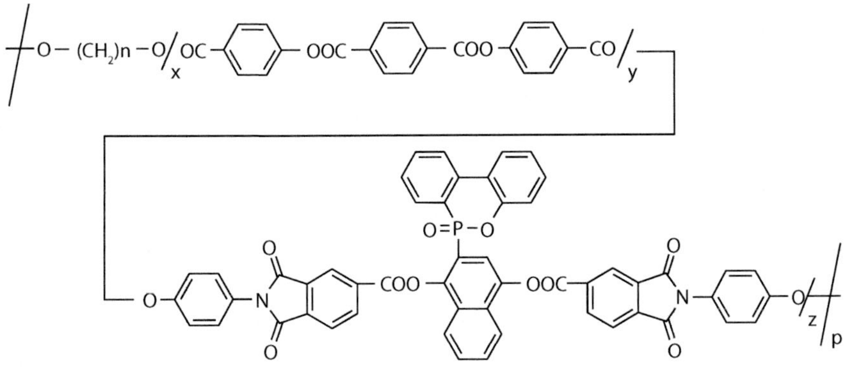

XVII

SCHEME 14.9 Chemical structure of DOPO-containing polyesterimide XVII.

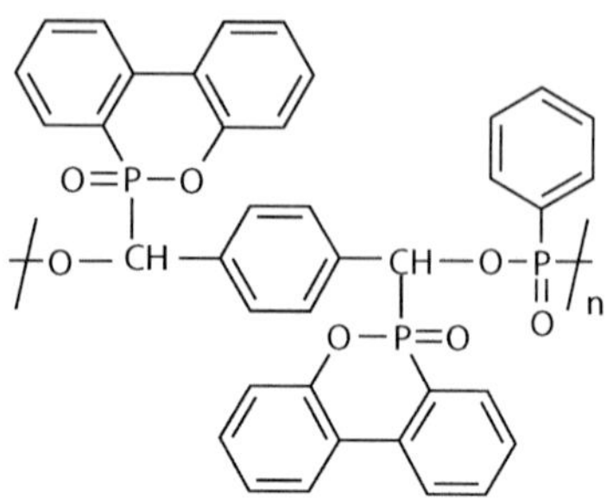

XVIII

SCHEME 14.10 DOPO-containing oligophosphonate XVIII.

neat DGEBA/DDM system and XVIII–DGEBA/DDM formulations highlighted the uniform dispersion of the flame-retardant additive into the epoxy matrix. The addition of XVIII into the epoxy matrix resulted in a homogeneous mixture, with a single glass transition. The *Tg* values of the cured epoxy resins decreased almost linearly with increases in the phosphorus content introduced into the epoxy matrix, ranging from 126°C to 146°C. The LOI values increased linearly with increases in the XVIII content and reached 42.20%, when 3% P (wt) was added into the system. These results indicate that DOPO-containing additive XVIII induced good flame retardancy to epoxy resin. Indeed, Wang et al. obtained a LOI value equal to 38.2% for the same phosphorus content added to a similar epoxy system (Wang et al. 2011) but using an aliphatic spirocyclic phosphorus containing dichloride derivative, which supports the previous affirmation. Thus, a concentration of 1% P (wt) was enough to increase the LOI value to about 30%. The amount of carbonized char produced after LOI tests was consistent with the TGA results. The UL-94 V-0 materials were achieved when 2% and 3% P (wt) were added into the epoxy matrix. These results indicate that XVIII imparts excellent flame-retardant performance to epoxy resins. The non–flame-retardant DGEBA/DDM system showed no UL-94 rating, whereas XVIII–DGEBA/DDM containing 1% P (wt) reached only UL 94 V-1 rating of flammability. The authors evaluated the combustion behavior of DOPO-containing epoxy resins by cone calorimetry. This technique provides a comprehensive insight into the fire risks through a series of parameters such as ease of ignition, flame spread, fire endurance, heat release rate, mass loss rate, ease of extinction, smoke evolution, and toxic and corrosive gas release (Carja et al. 2014b). The non-flame-retardant DGEBA/DDM system and XVIII–DGEBA/DDM containing 2% or 3% P (wt) formulations displayed similar time-to-ignition (TTI) values. In the case of XVIII–DGEBA/DDM containing 1% P (wt) formulation, a significant increase in TTI value was detected. To ignite this material, a higher amount of volatiles is required, which implies rather lower flammability.

The apparent contradiction with the LOI value obtained for this composition could be explained by the release, upon heating, of more inert gases (such as CO_2) and less flammable volatiles, which is advantageous for increasing the TTI value. According to the heat release rate (HRR) curves, the DGEBA/DDM sample burned quickly after ignition, reaching the peak of heat release rate (p-HRR) value at 275.5 kW m^{-2}. The p-HRR values decreased progressively with increases in the amount of XVIII incorporated into the epoxy matrix. In addition, the p-HRR values of XVIII–DGEBA/DDM containing 2% or 3% P (wt) formulations were reduced up to 45% that of the non–flame-retardant DGEBA/DDM system, depending on the content of the flame-retardant additive. This behavior is typical for the samples with charring tendency.

The time to reach the peak of heat release rate values also increased almost linearly with increasing the XVIII content. From a practical point of view, this delay implies that there would be more time for escaping during a real fire accident. Also, the incorporation of XVIII reduced the heat release capacity and total heat release (THR). The THR value of XVIII–DGEBA/DDM with 3% P (wt) formulation decreased by 28% when the phosphorus amount introduced into the epoxy matrix increased from 1% to 3%. The increase in char yield with increases in the phosphorus content suggests that the condensed phase mechanism is responsible for the improved flame-retardant properties. By studying the morphologies and chemistry of the charred layer using powerful methods such as SEM, EDX, and XPS, the authors confirmed the aforementioned hypothesis.

14.5 COMPARATIVE STUDIES ON SOME NEW DOPO-CONTAINING EPOXY RESINS

The present chapter is a contribution to the growing field of research in this area. The main objective is to highlight in a few pages the influence of phosphorus on the characteristic parameters that govern the response of epoxy resins upon heating and burning. Several additive mixtures prepared in our group and investigated in an epoxy relevant in the printed wired board industry will be discussed (Scheme 14.11).

The epoxy system investigated in this study is based on the diglycidyl ether of bisphenol A (DGEBA, D.E.R. 331, epoxy equivalent 182–192 g mol^{-1}, DOW Chemical Company), dicyandiamide (DICY, Aldrich) as hardener, and fenuron (FEN) (1,1-dimethyl-3-phenylurea, Aldrich). The epoxy resin used has lower epoxy functionality and a higher content of aliphatic substructures. As a consequence, it forms weaker networks, with lower T_g (127°C) and burns vigorously. Hence, a higher loading of DOPO-based flame retardant is necessary to render this material flame retardant. In this respect, for the current study, we proposed the use of some flame retardants with increased phosphorus content. The first one, PFR-1, is a reactive DOPO-containing flame retardant prepared by reacting DOPO with benzaldehyde. By reaction of PFR-1 with $POCl_3$, a new triple DOPO-containing additive flame retardant, PFR-2, was obtained. Finally, starting from bisphenol XVI, a new oligophosphonate flame retardant, PFR-3, containing phosphorus both in the side chain and the main chain was obtained by polycondensation reaction with phenylphosphonic dichloride, using *N*-methyl-2-pyrrolidone (NMP) as solvent, in the presence of triethylamine, as acceptor acid (Scheme 14.12). The oligomer PFR-3 was soluble in polar aprotic solvents such as NMP, *N,N*-dimethylformamide (DMF), or *N,N*-dimethylacetamide (DMAc). GPC was used to determine the molecular weights and

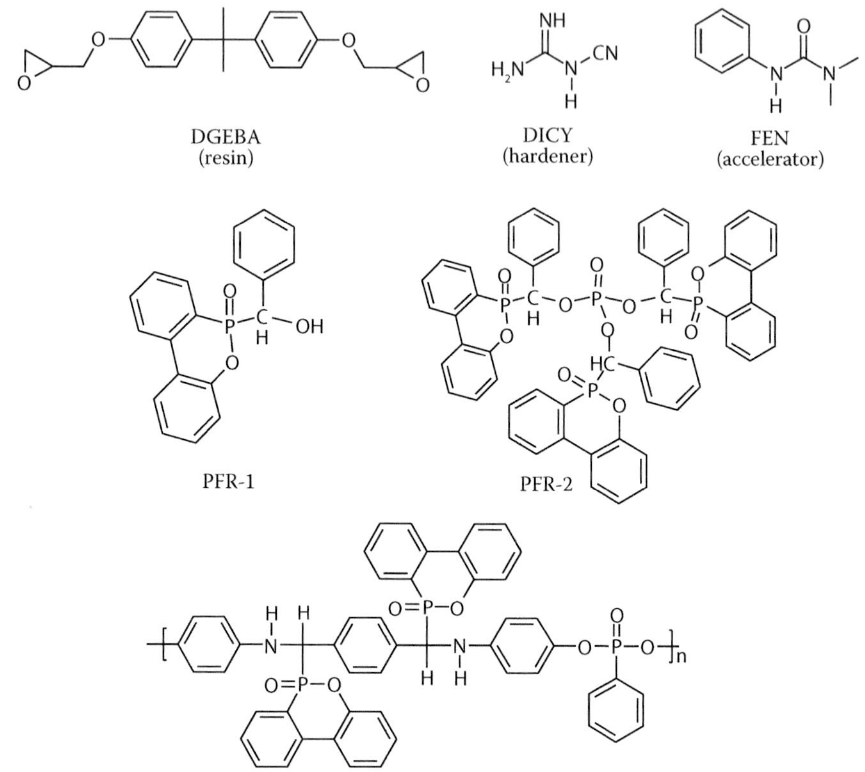

PFR-3

SCHEME 14.11 Epoxy resin formulation and the flame retardants investigated in this study.

their distribution. The oligomer has the number average molecular weight (Mn) of 5,400 g mol^{-1}, the weight average molecular weight (Mw) of 6,100 g mol^{-1}, and the polydispersity (Mw/Mn) of 1.13. The GPC curve displays narrow molecular weight distribution. The relative low molecular weight values are due to the presence of voluminous DOPO groups that led to a decrease of functional group reactivity and to a hindered propagation of polycondensation reaction. The semi-interpenetrating polymer networks are based on DGEBA epoxy resin, PFR-1 with one DOPO group, PFR-2 having three DOPO groups, and the oligophosphonate PFR-3 having a high content of phosphorus, combined with nitrogen, DICY as cross-linking agent, and FEN as an accelerator (Scheme 14.13).

The formulations of the pre-curing mixtures of DGEBA, DICY, FEN, and PFR are listed in Table 14.1. Certain amounts of epoxy resin were mixed with PFRs under continuous stirring at 130°C until complete dissolution was achieved. After cooling the mixtures down to 80°C, DICY and FEN were added in the system. All the samples were cured at 90°C for 1.5 h, at 120°C for 2 h, and then at 155°C for 2.5 h. The resulting thermosets were slowly cooled to the room temperature to prevent cracking. The mass ratio ER:DICY:FEN was 94:5:1, being established after some

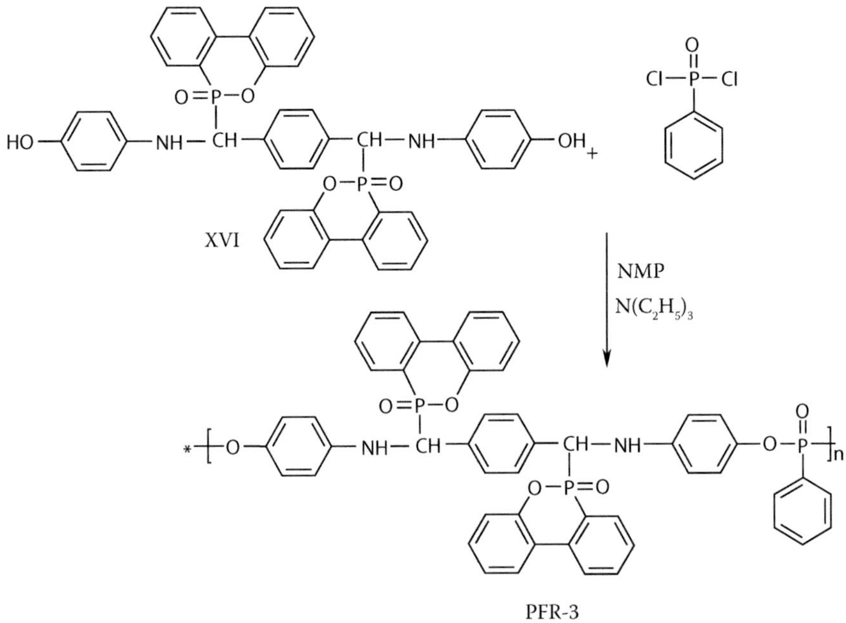

SCHEME 14.12 Synthesis of oligophosphonate PFR-3.

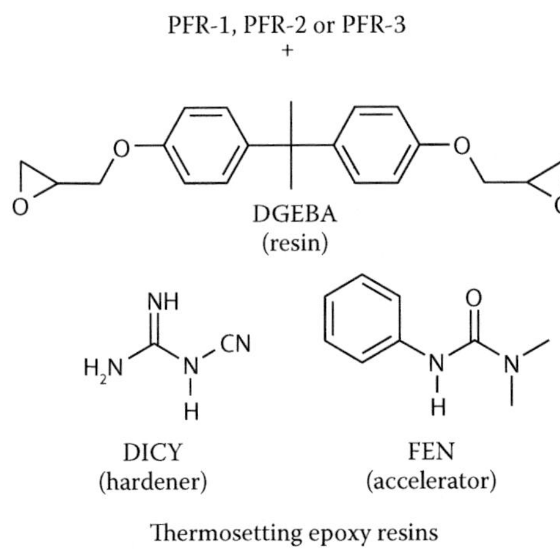

SCHEME 14.13 Preparation of DOPO-containing thermosetting epoxy resins.

preliminary experiments made in order to find the optimum conditions for the preparation of thermosets with the highest glass transition temperature values.

FTIR spectra were recorded for all cured formulations. All FTIR spectra of DOPO-containing epoxy resins exhibited characteristic bands due to the presence of PFR at 1,476 cm^{-1} (aromatic P-C stretching vibrations), 1,240 cm^{-1} (aromatic

TABLE 14.1
Preparation of Cured Epoxy Resins Based on DOPO-Containing PFRs

Sample	Epoxy Resin (g)	DICY (g)	FEN (g)	PFR (g)	P (%)
ER-0	18.80	1.000	0.200	0	0
ER/PFR-1(1%)	13.44	0.71	0.14	1.56	1
ER/PFR-1(2%)	15.84	0.84	0.16	4.15	2
ER/PFR-2(1%)	15.30	0.81	0.16	1.35	1
ER/PFR-2(2%)	16.74	0.89	0.17	3.25	2
ER/PFR-3(0.5%)	15.86	0.84	0.17	0.78	0.5
ER/PFR-3(1%)	15.10	0.8	0.15	1.56	1.0
ER/PFR-3(2%)	16.28	0.86	0.17	3.74	2.0

P=O stretching vibrations), and 751 cm^{-1} (deformation vibration caused by the 1,2-disubstituted aromatic DOPO rings), while the phosphorus-free resin EP-0 spectrum presented characteristic bands only at 2,970 cm^{-1} and 2,930 cm^{-1} (aliphatic C–H asymmetric stretching vibrations), 2,875 cm^{-1} (aliphatic symmetric stretching vibration), 1,610 cm^{-1} and 1,501 cm^{-1} (aromatic C=C stretching vibration), and at 1,240 cm^{-1} and 1,040 cm^{-1} ($–C_6H_4–O–CH_2–$asymmetric and symmetric stretching vibrations, respectively).

Studies on the compatibility and morphology of the DOPO-containing epoxy resins have been performed by SEM and revealed that either the reactive flame retardant PFR-1 or the additive flame retardants PFR-2 or PFR-3 were uniformly dispersed into the epoxy matrix. The incorporation of PFRs into epoxy resin modifies the morphology of the samples. Thus, ER/PFR-3(1%) fracture surfaces indicate extensive crazing. They are full of branches and fibrils disposed almost parallel. In the case of ER/PFR-3(2%), some aggregations of PFR-3 are observed in epoxy resins. Differential scanning calorimetry measurements performed on the ER/PFR-3 series revealed that the introduction of oligophosphonate PFR-3 in the epoxy matrix resulted in a homogenous mixture with a single glass transition, suggesting a proper dispersion of the flame retardant into the epoxy resin. The T_g values of the DICY-cured epoxy resins based on PFR-1 and PFR-2 slightly decreased, due to the moderate plasticizing effect of the flame retardants, which reduced the cross-linking density and thereby increased the mobility of molecular chains (Su et al. 2010; Carja et al. 2014b). On the other hand, the non reactive additive having oligomeric, bridging, and rather rigid molecular structure that PFR-3 does, preserved and even increased to some degree the glass transition. Similar behaviors have been observed and discussed in the work of Ciesielski et al. (2008).

The effect of phosphorus content on the thermal stability and decomposition mechanism was investigated by TGA carried out under air atmosphere, with a constant heating rate of 10°C min^{-1}. The main TG data as determined from the TG and DTG curves are given in Table 14.2. The thermal degradation of ER-0 is carried out in two stages in air atmosphere—the temperature corresponding to a 5% weight loss being around 350°C, while the temperature corresponding to a 10% weight loss was 375°C. Flame-retardant epoxy resins ER/PFR-3(0.5%), ER/PFR-3(1%),

TABLE 14.2
TG Parameters of Non-Flame-Retarded Epoxy Resin ER-0 and of Flame-Retarded Epoxies, as Determined from TG and DTG Curves Recorded in Air

Sample	T_5 (°C)	T_{10} (°C)	Char Yield at 600°C (%)	LOI
ER-0	353	375	17.7	24
ER/PFR-1(1%)	324	355	29	24
ER/PFR-1(2%)	312	344	31.6	25
ER/PFR-2(1%)	345	365	32	25
ER/PFR-2(2%)	325	345	34.4	27
ER/PFR-3(0.5%)	350	371	27.7	26
ER/PFR-3(1%)	340	365	31.0	28
ER/PFR-3(2%)	336	355	35.5	30

and ER/PFR-3(2%) have a degradation mechanism similar to the reference epoxy, ER-0, the degradation occurring in two main stages under air. A moderate decrease of the thermal stability was observed by increasing the content of PFR-3 in the epoxy matrix, similar to results obtained by other research groups (Rwei et al. 2005; Ho et al. 2006; Ménard et al. 2014). Regarding the char yield, which was measured at 600°C, it increased by raising the content of DOPO-containing PFRs, as was expected, with values between 31.6 and 35.5 for the formulations containing 2% P (wt). Char, a carbonaceous porous residue, results from the thermal degradation of the material being pyrolyzed. This char layer can act as an insulating barrier between the external heat source and the unpyrolyzed material beneath the char. In the most phosphorus-containing compounds, the decomposition leads to a phosphorus-rich char layer. The presence of phosphorus reduces the production of combustible gases while increasing the char yield in the solid phase.

SEM investigations revealed that the char of PFR-3 is dense and compact, while the char of the phosphorus-free sample ER-0 is loose and porous. With the introduction of PFR-3 to epoxy resin, the charring properties have been improved. Thus, the char of ER/PFR-3(2%) is dense and compact, suggesting a better barrier against the heat transfer between the flame zone and the burning substrate in comparison with that of ER-0 char. Also, it can inhibit the release of flammable or toxic volatiles, resulting in an enhancement of flame retardancy. The high concentration of phosphorus in the chars of PFR-3 and ER/PFR-3(2%) was evidenced by EDX analyses. The maximum phosphorus concentration on the residue surface was observed in the case of PFR-3, but high phosphorus content was also observed in the case of ER/PFR-3(2%). These results, which are in accordance with the TGA data and LOI testing, demonstrate that phosphorus-containing PFR has an important role in improving the fire resistance of epoxy resins. The LOI measurements were performed in order to evaluate the flame retardancy of the thermosets. The LOI values were 24 for ER-0, 26 for ER/PFR-3(0.5%), 28 for ER/PFR-3(1%), and 30 for ER/PFR-3(2%).

A good result was also obtained for ER/PFR-2(2%) formulation (27). Taking into account that a material can be considered as fire resistant when the LOI value is higher than 26, it can be concluded that in the series ER/PFR-3, the samples containing 1% and 2% phosphorus [ER/PFR-3(1%) and ER/PFR-3(2%)] correspond from this point of view. These results show that the novel PFR-3 is an effective flame retardant for this type of epoxy resin.

14.6 CONCLUSIONS

The main goal of the current work is to bring quick insight into the development of powerful DOPO-containing flame retardants suitable for epoxy resins relevant in composites. To this aim, a progressive literature survey was done, revealing most relevant DOPO-containing compounds, from simple molecules to more complex architectures, that have been taken into consideration for preparation of flame-retardant epoxy thermosets. The flame retardant properties of such cured samples were determined by coupled techniques (LOI, UL-94, cone calorimetry, SEM-TGA-FTIR, char versus phosphorus content correlations, etc.). Recent work in the field was then performed by preparing new flame-retardant epoxy thermosets using a reactive DOPO-containing flame retardant, a new additive triple DOPO-containing flame retardant, and new a oligophosphonate flame retardant, containing phosphorus in the side chain as well as in the main chain. Dicyandiamide was used as an aminic hardener and 1,1-dimethyl-3-phenylurea as an accelerator for the curing process. SEM, TGA, and DSC results demonstrated good dispersion of PFRs into the epoxy network. The char residue at 600°C increased by raising the content of DOPO-containing PFRs (31.6–35.5 for the formulations containing 2% P (wt). LOI values of the thermosets ranged from 24% in the case of phosphorus-free sample ER-0 to 30% for the sample containing 2% P [ER/PFR-3(2%)]. SEM measurements of the char residue showed a more compact structure for the samples containing an increasing ratio of phosphorus.

ACKNOWLEDGMENTS

The financial support offered by CNCSIS-UEFISCDI, project number 28/29.04.2013, code PNII-RU-TE-2012-3-0123 is gratefully acknowledged. Dr. Hamciuc expresses appreciation to CNCSIS-UEFISCDI, project H2Sense, number 43/2014, code PN-II-PT-PCCA-2013-4, for the financial support.

REFERENCES

Aslzadeh Mohammad M., Abdouss Majid, and Sadeghi Gity M. M. Preparation and characterization of new flame retardant polyurethane composite and nanocomposite. *J. Appl. Polym. Sci.* 127 no. 3 (2013): 1683–1690.

Becker Luz, Lenoir Dieter, Matuschek Georg, and Kettrup Antonius. Thermal degradation of halogen-free flame retardant epoxides and polycarbonate in air. *J. Anal. Appl. Pyrol.* 60 no. 1 (2001): 55–67.

Benin Vladimir, Durganala Sravanthi, and Morgan Alexander B. Synthesis and flame retardant testing of new boronated and phosphonated aromatic compounds. *J. Mater. Chem.* 22 no. 3 (2012): 1180–1190.

Braun Ulrike, Balabanovich Aliaksandr I., Schartel Bernhard, et al. Influence of the oxidation state of phosphorus on the decomposition and fire behaviour of flame-retarded epoxy resin composites. *Polymer.* 47 no. 26 (2006): 8495–8508.

Brehme Sven, Schartel Bernhard, Goebbels Jürgen, et al. Phosphorus polyester versus aluminium phosphinate in poly(butylene terephthalate) (PBT): Flame retardancy performance and mechanisms. *Polym. Degrad. Stab.* 96 no. 5 (2011): 875–884.

Cai Sheng Xiong, and Lin Ching Hsuan. Flame-retardant epoxy resins with high glass-transition temperatures from a novel trifunctional curing agent: Dopotriol. *J. Polym. Sci. A. Polym. Chem.* 43 no. 13 (2005): 2862–2873.

Carja Ionela-Daniela, Serbezeanu Diana, Lisa Gabriela, Vlad-Bubulac Tachita, and Hamciuc Corneliu. Thermal degradation and kinetic studies of new flame-retardant phosphorus-containing polymers with liquid crystalline properties. *Int. J. Polym. Anal. Charact.* 19 no. 4 (2014a): 372–382.

Carja Ionela-Daniela, Serbezeanu Diana, Vlad-Bubulac Tachita, et al. A straightforward, eco-friendly and cost-effective approach towards flame retardant epoxy resins. *J. Mater. Chem. A.* 2 no. 38 (2014b): 16230–16241.

Carja Ionela-Daniela, Serbezeanu Diana, Vlad-Bubulac Tachita, and Hamciuc Corneliu. Eco-friendly phosphorus flame retardants: From rational design towards applications In: *Advances in Materials Science Research, Wythers* Maryann C. Editor, pp. 111–134. New York: Nova Science, 2016.

Chen Dan-Qi, Wang Yu-Zhong, Hu Xiao-Ping, Wang De-Yi, Qu Ming-Hai, and Yang Bing. Flame-retardant and anti-dripping effects of a novel char-forming flame retardant for the treatment of poly(ethylene terephthalate) fabrics. *Polym. Degrad. Stab.* 88 no. 2 (2005): 349–356.

Chen Ming-Jun, Shao Zhu-Bao, Wang Xiu-Li, Chen Li, and Wang Yu-Zhong. Halogen-free flame-retardant flexible polyurethane foam with a novel nitrogen–phosphorus flame retardant. *Ind. Eng. Chem. Res.* 51 no. 29 (2012): 9769–9776.

Ciesielski Michael, Schäfer Alexander, and Döring Manfred. Novel efficient DOPO-based flame-retardants for PWB relevant epoxy resins with high glass transition temperatures. *Polym. Adv. Technol.* 19 no. 6 (2008): 507–515.

Cochez Marianne, Ferriol Michel, Weber Jean Victor, Chaudron Paul Verlaine, Oget Nicolas, and Mieloszynski Jean-Luc. Thermal degradation of methyl methacrylate polymers functionalized by phosphorus-containing molecules I. TGA/FT–IR experiments on polymers with the monomeric formula $CH_2C(CH_3)C(O)OCHRP(O)(OC_2H_5)_2$ (R=H, $(CH_2)_4CH_3$, C_6H_5Br, $C_{10}H_7$). *Polym. Degrad. Stab.* 70 no. 3 (2000): 455–462.

Davies Duncan, and Horrocks A. Richard. Nitrogen–phosphorous antagonism in flame-retardant cotton. *J. Appl. Polym. Sci.* 31 no. 6 (1986): 1655–1662.

Deng Yi, Wang Yu-Zhong, Ban Da-Ming, Liu Xiu-Hua, and Zhou Qian. Burning behavior and pyrolysis products of flame-retardant PET containing sulfur-containing aryl polyphosphonate. *J. Anal. Appl. Pyrolysis.* 76 no. 1–2 (2006): 198–202.

Döring Manfred, Ciesielski Michael, and Heinzmann Christian. Synergistic flame retardant mixtures in epoxy resins. In: *Fire and Polymers VI: New Advances in Flame Retardant Chemistry and Science*, Morgan Alexander B., Wilkie Charles A., Nelson Gordon L. (eds.), pp. 295–310. Washington: ACS, 2012.

Duquesne Sophie, Lefebvre Jérôme, Seeley Gordon, Camino Giovani, Delobel Rene, and Le Bras Michel. Vinyl acetate/butyl acrylate copolymers: Part 2: Fire retardancy using phosphorus-containing additives and monomers. *Polym. Degrad. Stab.* 85 no. 2 (2004): 883–892.

Ebdon John R., Price Dennis, Hunt Barry J., et al. Flame retardance in some polystyrenes and poly(methyl methacrylate)s with covalently bound phosphorus-containing groups: Initial screening experiments and some laser pyrolysis mechanistic studies. *Polym. Degrad. Stab.* 69 no. 3 (2000): 267–277.

Gao Ming, Wu Weihong, and Xu Zhi-Qiang. Thermal degradation behaviors and flame retardancy of epoxy resins with novel silicon-containing flame retardant. *J. Appl. Polym. Sci.* 127 no. 3 (2013): 1842–1847.

Hamciuc Corneliu, Hamciuc Elena, Serbezeanu Diana, and Vlad-Bubulac Tachita. Thermal and optical properties of some phosphorus-containing poly(1,3,4-oxadiazole-ester-imide)s. *Polym. Adv. Technol.* 22 no. 12 (2011a): 2458–2468.

Hamciuc Corneliu, Hamciuc Elena, Serbezeanu Diana, Vlad-Bubulac Tachita, and Cazacu Maria. Phosphorus-containing poly(ester-imide)-polydimethylsiloxane copolymers. *Polym. Int.* 60 no. 2 (2011b): 312–321.

Hamciuc Corneliu, Serbezeanu Diana, Carja Ionela-Daniela, et al. Effect of DOPO units and of polydimethylsiloxane segments on the properties of epoxy resins. *J. Mater. Sci.* 48 no. 24 (2013): 8520–8529.

Hamciuc Corneliu, Vlad-Bubulac Tachita, Petreus Oana, and Lisa Gabriela. Kinetics of thermal degradation in non-isothermal conditions of some phosphorus-containing polyesters and polyesterimides. *Eur. Polym. J.* 43 no. 3 (2007): 980–988.

Hamciuc Corneliu, Vlad-Bubulac Tachita, Petreus Oana, and Lisa Gabriela. Synthesis and characterization of new aromatic polyesters and poly(ester-imide)s containing phosphorous cyclic bulky groups. *Polym. Bull.* 60 no. 5 (2008): 657–664.

Hamerton Ian, Tang Winnie, Anguita Jose V., and Silva S. Ravi P. Towards the rational design of polymers using molecular simulation: Predicting the effect of cure schedule on thermo-mechanical properties for a cycloaliphatic amine-cured epoxy resin. *React. Funct. Polym.* 74 no. (2014): 1–15.

Hergenrother Paul M., Thompson Craig M., Smith Joseph G., Jr., et al. Flame retardant aircraft epoxy resins containing phosphorus. *Polymer.* 46 no. 14 (2005): 5012–5024.

Ho Tsung-Han, Leu Tsu-Shang, Sun Yih-Min, and Shieh Jeng-Yueh. Thermal degradation kinetics and flame retardancy of phosphorus-containing dicyclopentadiene epoxy resins. *Polym. Degrad. Stab.* 91 no. 10 (2006): 2347–2356.

Hoang Dongquy, and Kim Jinhwan. Synthesis and applications of biscyclic phosphorus flame retardants. *Polym. Degrad. Stab.* 93 no. 1 (2008): 36–42.

Hoang Dongquy, Kim Jinhwan, and Jang Bok Nam. Synthesis and performance of cyclic phosphorus-containing flame retardants. *Polym. Degrad. Stab.* 93 no. 11 (2008): 2042–2047.

Hörold Sebastian. Phosphorus flame retardants in thermoset resins. *Polym. Degrad. Stab.* 64 no. 3 (1999): 427–431.

Jian Rong-Kun, Chen Li, Chen Si-Yang, Long Jia-Wei, and Wang Yu-Zhong. A novel flame-retardant acrylonitrile-butadiene-styrene system based on aluminum isobutylphosphinate and red phosphorus: Flame retardance, thermal degradation and pyrolysis behavior. *Polym. Degrad. Stab.* 109 no. 0 (2014): 184–193.

Joseph Paul, and Tretsiakova-Mcnally Svetlana. Reactive modifications of some chain- and step-growth polymers with phosphorus-containing compounds: Effects on flame retardance–a review. *Polym. Adv. Technol.* 22 no. 4 (2011): 395–406.

Kandola Baljinder Kaur, and Deli Dario. Flame-retardant thermoset nanocomposites for engineering applications. In: *Polymer Green Flame Retardants: A Comprehensive Guide to Additives and Their Applications*, Papaspyrides Constantine D., Kiliaris Pantelis (eds.), pp. 503–550, Oxford: Elsevier, 2014.

Kilinc Mert, Cakal Gaye O., Bayram Goknur, Eroglu Inci, and Özkar Saim. Flame retardancy and mechanical properties of pet-based composites containing phosphorus and boron-based additives. *J. Appl. Polym. Sci.* 132 no. 22 (2015): 42016–42021.

Levchik Sergei, Camino Giovanni, Costa Luigi, and Levchik Galina F. Mechanism of action of phosphorus-based flame retardants in nylon 6. I. Ammonium polyphosphate. *Fire Mater.* 19 no. 1 (1995): 1–10.

Levchik Sergei, Piotrowski Andrew, Weil Edward, and Yao Qiang. New developments in flame retardancy of epoxy resins. *Polym. Degrad. Stab.* 88 no. 1 (2005): 57–62.
Levchik Sergei V., and Weil Edward D. Thermal decomposition, combustion and flame-retardancy of epoxy resins–a review of the recent literature. *Polym. Int.* 53 no. 12 (2004): 1901–1929.
Liang Bing, Cao Jun, Hong Xiao Dong, and Wang Chang Song. Synthesis and properties of a novel phosphorous-containing flame-retardant hardener for epoxy resin. *J. Appl. Polym. Sci.* 128 no. 5 (2013): 2759–2765.
Lin Ching Hsuan, Chang Sheng Lung, Wei Tsai Pei, Ding Shiue Huei, and Su Wen Chiung. Synthesis and properties of phosphorus-containing advanced epoxy resins. II. *J. Appl. Polym. Sci.* 78 no. 1 (2000): 228–235.
Lin Ching Hsuan, Chang Sheng Lung, Wei Tsai Pei, Ding Shiue Huei, and Su Wenchiung. Facile, one-pot synthesis of phosphinate-substituted bisphenol A and its alkaline-stable diglycidyl ether derivative. *Polym. Degrad. Stab.* 95 no. 7 (2010): 1167–1176.
Liu Weichang, Varley Russell J., and Simon George P. Understanding the decomposition and fire performance processes in phosphorus and nanomodified high performance epoxy resins and composites. *Polymer.* 48 no. 8 (2007): 2345–2354.
Liu Xiaoqing, Xin Wenbo, and Zhang Jinwen. Rosin-derived imide-diacids as epoxy curing agents for enhanced performance. *Bioresource Technol.* 101 no. 7 (2010): 2520–2524.
Liu Ying Ling. Flame-retardant epoxy resins from novel phosphorus-containing novolac. *Polymer.* 42 no. 8 (2001): 3445–3454.
Liu Ying Ling. Phosphorous-containing epoxy resins from a novel synthesis route. *J. Appl. Polym. Sci.* 83 no. 8 (2002b): 1697–1701.
Liu Ying-Ling, and Chou Ching I. The effect of silicon sources on the mechanism of phosphorus–silicon synergism of flame retardation of epoxy resins. *Polym. Degrad. Stab.* 90 no. 3 (2005): 515–522.
Lu Shui-Yu, and Hamerton Ian. Recent developments in the chemistry of halogen-free flame retardant polymers. *Prog. Polym. Sci.* 27 no. 8 (2002): 1661–1712.
Luo Chaoyun, Zuo Jiandong, and Zhao Jianqing. Synthesis and property of epoxy prepolymer and curing agent with high refractive index. *High Perform. Polym.* 25 no. 8 (2013): 986–991.
Ménard Raphaël, Negrell-Guirao Claire, Ferry Laurent, Sonnier Rodolphe, and David Ghislain. Synthesis of new flame-retardants by radical chain transfer copolymerization of glycidyl methacrylate and dimethoxy-phosphorylmethyl methacrylate. *Eur. Polym. J.* 57 no. 7 (2014): 109–120.
Morgan Alexander B, and Wilkie Charles A. The Non-halogenated Flame Retardant Handbook. New Jersey: John Wiley & Sons, 2014.
Pascault Jean-Pierre, and Williams Roberto J.J. *Epoxy Polymers: New Materials and Innovations.* Weinheim: Wiley-VCH, 2010.
Petreus Oana, Avram Ecaterina, Lisa Gabriela, and Serbezeanu Diana. Synthesis and characterization of phosphorus-containing polystyrene. *J. Appl. Polym. Sci.* 115 no. 4 (2010a): 2084–2092.
Petreus Oana, Avram Ecaterina, and Serbezeanu Diana. Synthesis and characterization of phosphorus-containing polysulfone. *Polym. Eng. Sci.* 50 no. 1 (2010b): 48–56.
Petreus Oana, Lisa Gabriela, Avram Ecaterina, and Rosu Dan. Thermal degradation and pyrolysis study of phosphorus-containing polysulfones. *J. Appl. Polym. Sci.* 120 no. 6 (2011): 3233–3241.
Petreus Oana, Vlad-Bubulac Tachita, and Hamciuc Corneliu. Synthesis and characterization of new polyesters with enhanced phosphorus content. *Eur. Polym. J.* 41 no. 11 (2005): 2663–2670.

Petreus Oana, Vlad-Bubulac Tachita, and Hamciuc Corneliu. New Polyesterimides Containing Bulky, Cyclic Phosphorus Groups. *High. Perform. Polym.* 20 no. 6 (2008): 588–600.

Petrie Edward. *Epoxy Adhesive Formulations.* New York: McGraw-Hill Chemical Engineering, 2005.

Price Dennis, Pyrah Kelly, Hull T. Richard, et al. Flame retarding poly(methyl methacrylate) with phosphorus-containing compounds: Comparison of an additive with a reactive approach. *Polym. Degrad. Stab.* 74 no. 3 (2001): 441–447.

Price Dennis, Pyrah Kelly, Hull T. Richard, et al. Flame retardance of poly(methyl methacrylate) modified with phosphorus-containing compounds. *Polym. Degrad. Stab.* 77 no. 2 (2002): 227–233.

Purser David. The performance of flame retardants in relation to toxicity, toxic hazard, and risk in fires. In: *Fire Retardancy of Polymeric Materials*, Wilkie Charles A., Morgan Alexander B. (eds.), pp. 449–500. New York: Marcel Dekker Inc., 2000.

Rwei Syang Peng, Cheng C. Y., Liou Guey-Sheng, and Cheng Kuo Chung. Curing and pyrolysis of cresol novolac epoxy resins containing BABODOPN. *Polym. Eng. Sci.* 45 no. 4 (2005): 478–486.

Salamone Joseph C. *Polymeric Materials Encyclopedia.* Boca Raton: CRC Press, 1996.

Serbezeanu Diana, Vlad-Bubulac Tachita, and Hamciuc Corneliu. Synthesis and characterization of new aromatic-aliphatic poly(ester-imide)s containing phosphorus. *Rev. Roum. Chim.* 55 no. 10 (2010a): 667–672.

Serbezeanu Diana, Vlad-Bubulac Tachita, Hamciuc Corneliu, and Aflori Magda. Structure and properties of phosphorous-containing thermotropic liquid-crystalline aliphatic-aromatic copolyesters. *Macromol. Chem. Phys.* 211 no. 13 (2010b): 1460–1471.

Serbezeanu Diana, Vlad-Bubulac Tachita, Hamciuc Corneliu, and Aflori Magdalena. Synthesis and properties of novel phosphorus-containing thermotropic liquid crystalline copoly(ester imide)s. *J. Polym. Sci. A Polym. Chem.* 48 no. 23 (2010c): 5391–5403.

Shieh Jeng-Yueh, and Wang Chun-Shan. Synthesis and properties of novel phosphorus-containing hardener for epoxy resins. *J. Appl. Polym. Sci.* 78 no. 9 (2000): 1636–1644.

Su Chung Hwei, Chiu Yi Pang, Teng Chih Chun, and Chiang Chin Lung. Preparation, characterization and thermal properties of organic–inorganic composites involving epoxy and polyhedral oligomeric silsesquioxane (POSS). *J. Polym. Res.* 17 no. 5 (2010): 673–681.

Tibiletti Lucie, Ferry Laurent, Longuet Claire, Mas André, Robin Jean-Jacques, and Lopez-Cuesta José-Marie. Thermal degradation and fire behavior of thermoset resins modified with phosphorus containing styrene. *Polym. Degrad. Stab.* 97 no. 12 (2012): 2602–2610.

Toldy Andrea, Szolnoki Beáta, Csontos István, and Marosi György. Green synthesis and characterization of phosphorus flame retardant crosslinking agents for epoxy resins. *J. Appl. Polym. Sci.* 131 no. 7 (2014): 40105–40112.

Vlad-Bubulac Tachita, Hamciuc Corneliu, and Petreus Oana. Synthesis and properties of some phosphorus-containing polyesters. *High Perform. Polym.* 18 no. 3 (2006): 255–264.

Vlad-Bubulac Tachita, Hamciuc Corneliu, Petreus Oana, and Lisa Gabriela. Thermal stability of some phosphorus-containing polyesters and polyesterimides. *Mater. Plast.* 44 no. 4 (2007): 284–288.

Wan Jintao, Li Cheng, Bu Zhi-Yang, Xu Cun-Jin, Li Bo-Geng, and Fan Hong. A comparative study of epoxy resin cured with a linear diamine and a branched polyamine. *Chem. Eng. J.* 188 (2012): 160–172.

Wang Chun Shan, and Lee Ming-Chun. Synthesis and properties of epoxy resins containing 2-(6-oxid-6H-dibenz(c,e)(1, 2) oxaphosphorin-6-yl) 1, 4-benzenediol (II). *Polymer.* 41 no. 10 (2000): 3631–3638.

Wang Chun Shan, and Lin Ching Hsuan. Properties and curing kinetic of diglycidyl ether of bisphenol A cured with a phosphorus-containing diamine. *J. Appl. Polym. Sci.* 74 no. 7 (1999): 1635–1645.

Wang Chun Shan, and Lin Ching Hsuan. Synthesis and properties of phosphorus containing advanced epoxy resins. *J. Appl. Polym. Sci.* 75 no. 3 (2000): 429–436.

Wang Chun-Shan, and Shieh Jeng-Yueh. Synthesis and properties of epoxy resins containing 2-(6-oxid-6H-dibenz⟨c,e⟩ ⟨1,2⟩ oxaphosphorin-6-yl)1, 4-benzenediol. *Polymer.* 39 no. 23 (1998): 5819–5826.

Wang Chun Shan, and Shieh Jeng Yueh. Phosphorus-containing epoxy resin for an electronic application. *J. Appl. Polym. Sci.*73 no. 3 (1999): 353–361.

Wang Xin, Hu Yuan, Song Lei, Yang Hongyu, Xing Weiyi, and Lu Hongdian. Synthesis and characterization of a DOPO-substituted organophosphorus oligomer and its application in flame retardant epoxy resins. *Prog. Org. Coat.* 71 no. 1 (2011): 72–82.

Weber Lutz W. D., and Greim Helmut. The toxicity of brominated and mixed-halogenated dibenzo-p-dioxins and dibenzofurans: An overview. *J. Toxicol. Env. Health.* 50 no. 3 (1997): 195–215.

Weil Edward D., and Levchik Sergei. Current practice and recent commercial developments in flame retardancy of polyamides. *J. Fire Sci.* 22 no. 3 (2004): 251–264.

Xiang Hong Fa, Xu Hua-Yun, Wang Zheng-Zhou, and Chen Chun-Hua. Dimethyl methylphosphonate (DMMP) as an efficient flame retardant additive for the lithium-ion battery electrolytes. *J. Power Sources.* 173 no. 1 (2007): 562–564.

Xu Wei, Wang Guojian, and Zheng Xiaorui. Research on highly flame-retardant rigid PU foams by combination of nanostructured additives and phosphorus flame retardants. *Polym. Degrad. Stab.* 111 no. 0 (2015): 142–150.

Zhang Wenchao, Li Xiangmei, Fan Haibo, and Yang Rongjie. Study on mechanism of phosphorus–silicon synergistic flame retardancy on epoxy resins. *Polym. Degrad. Stab.* 97 no. 11 (2012): 2241–2248.

Zhang Yi, Ni Yan-Peng, He Ming-Xin, Wang Xiu-Li, Chen Li, and Wang Yu-Zhong. Phosphorus-containing copolyesters: The effect of ionic group and its analogous phosphorus heterocycles on their flame-retardant and anti-dripping performances. *Polymer.* 60 no. 0 (2015): 50–61.

15 Magnetic Polymer Composites for Sensor Applications

Simona Luminita Nica and Silvia Ioan

CONTENTS

15.1 INTRODUCTION

Introduction of magnetic elements into electronic devices made of polymer composites appears as a new technique in the field of electrical engineering and biomedical research for sensor applications.

The correlation between the magnetic properties of soft magnetic materials and their elastic or electrical response causes different phenomena that can be used for the construction of magneto-impedance sensors (namely the skin-effect-induced change of impedance caused by an applied magnetic field)—especially for geomagnetic and biosensing applications, magnetoelastic resonance sensors (occurring when a mechanical resonant response is excited through the magnetostrictive effect of soft magnetic materials)—when applied as remote sensors and as magneto-electric laminated composite sensors (obtained from combining magnetostrictive amorphous ferromagnetic ribbons with piezoelectric materials), show an extremely high sensitivity for magnetic field detection (García-Arribas et al. 2014; Prinz 1999).

On the other hand, magnetic polymer composites are components of sensor devices, due to the processability of polymers and to the enhanced magnetic attributes of the embedded nanostructures (Ergeneman et al. 2014).

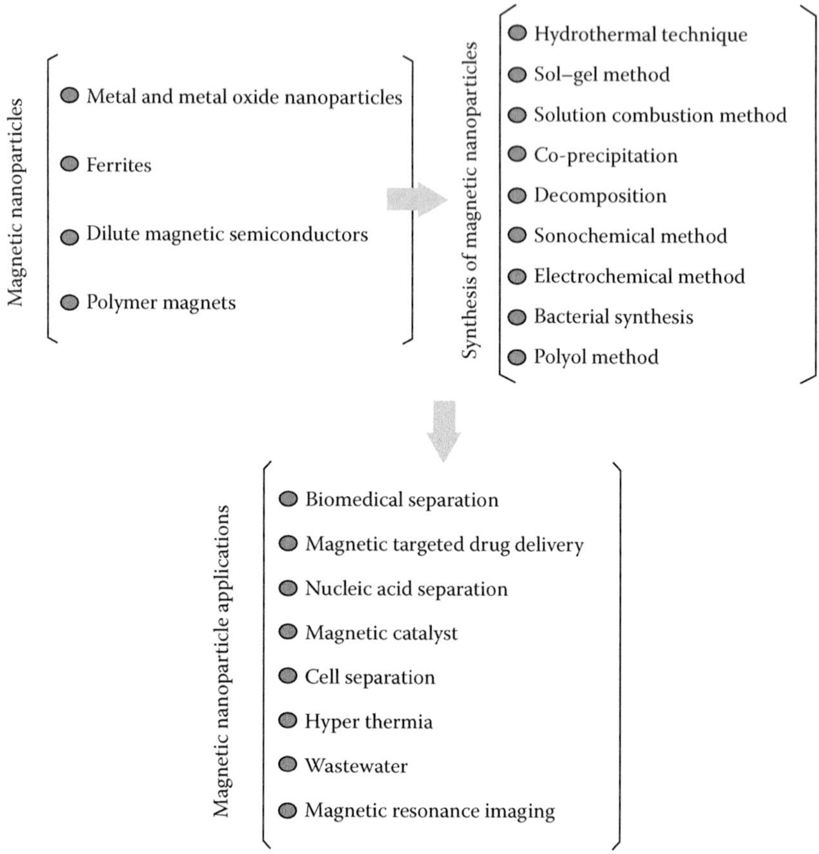

SCHEME 15.1 Types, synthesis, and applications of nanoparticles. (Adapted from Kalia Susheel et al. *Colloid Polym. Sci.* 292 no. 9 (2014): 2025–2052.)

Magnetic nanoparticles (Scheme 15.1) and magnetic nanocomposites (Scheme 15.2) have aroused significant scientific and technological interest due to their potential applications in biomedicine, information technology, magnetic resonance imaging, catalysis, telecommunication, and environmental remediation (Kalia et al. 2014).

Different assembly strategies for preparation of magnetic polymer composites were established by multiphase polymer studies during their evolution from micro- to nanostructures.

15.2 SYNTHESIS OF MAGNETIC POLYMERS

Kalia et al. (2014) have reviewed the synthesis and developments in the field of organic magnets (which possess both magnetic and structural, optical, electrical, and gas-sensing properties) by the hydrothermal technique, sol–gel method, solution

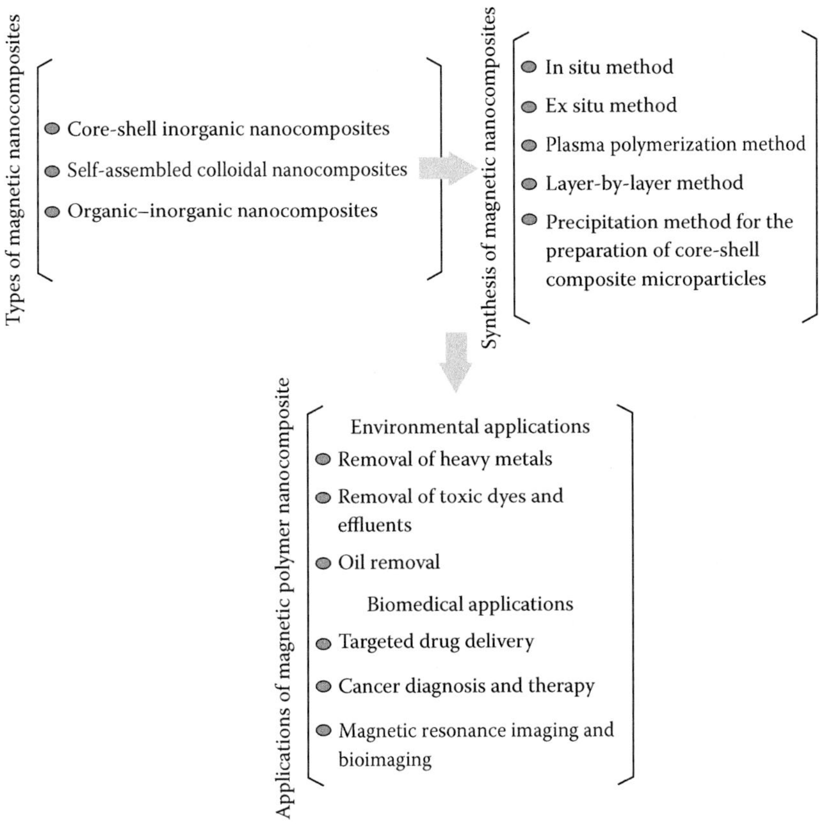

SCHEME 15.2 Types, synthesis, and applications of magnetic nanocomposites. (Data from Kalia Susheel et al. *Colloid Polym. Sci.* 292 no. 9 (2014): 2025–2052.)

combustion method, thermal decomposition, sonochemical, electrochemical, bacterial synthesis, or the polyol method. Magnetic polymers are organized into different types of magnetic nanocomposites, such as core–shell inorganic nanocomposites, self-assembled colloidal nanocomposites, and organic–inorganic nanocomposites, so that the synthesis of magnetic polymer nanocomposites can be classified into two major categories—ex situ and in situ processes.

15.3 INKJET-PRINTED SUPERPARAMAGNETIC POLYMER COMPOSITES

Superparamagnetic polymer composites (SPMPC) are recommended for microsystems and nanorobotics due to the functionality of the polymers used and the magnetic properties of the included nanostructures. These composites can perform functions wirelessly in different environments, using external magnetic fields, for sensing the indicated physical or chemical changes, for drug or biomass delivery, or for manipulation of small objects. Various studies have been devoted to preparation of composites capable of performing a complex magnetic action, such as:

- Superparamagnetic polymeric microcomponents with programmable magnetic anisotropy with specific actuation mechanisms, where the embedded nanoparticles are aligned by applying magnetic fields to set a preferred magnetic direction (Kim et al. 2011).
- Directed nanoparticle self-organization and two-photon polymerization combined to enable three-dimensional soft-magnetic microactuators with complex shapes and shape-independent magnetic properties. By applying a magnetic field during the fabrication of magnetic polymer composite-based helical swimmers by two-photon polymerization, the corkscrew propulsion of the microhelices could be significantly enhanced (Peters et al. 2014).
- Superparamagnetic magnetite nanoparticles dispersed in an epoxy monomer (SU-8) to form magnetic anisotropic hemispheres—by applying a magnetic field during curing and a standard SPMPC. The distribution of nanoparticles inside the polymer matrix and the magnetic characteristics of the superparamagnetic polymer composite permit different assembly strategies, in the form of lines or hemispheres, as well as a control of these formations dynamic. A two-interface assembly strategy is shown to combine hemispheres into complete spheres for advanced self-assembly tasks (Ergeneman et al. 2014).

Generally, complex composite microstructures consisting of a matrix containing functional nanoparticles can be manufactured with ink formulations. Literature makes mention of high resolution on print deposition, patterned multicolored thin films of luminescent colloidal quantum dot (QD)-polymer composites, and use of printed patterns in fabricating robust, bright, full-color, alternating current (AC)-driven displays (Wood et al. 2009). The use of quantum dot luminescent centers in thin-film light-emitting diode (LED) structures is indicated because it offers a narrow-band luminescence that can be tuned across the visible spectrum by varying size and chemical composition. To deposit thin films of colloidally synthesized quantum dots in an AC electroluminescent device, authors introduced an inkjet-print quantum dot into a polyisobutylene matrix that shows an efficient photoluminescence over thin films.

15.4 MAGNETIC POLYMER NANOCOMPOSITES FOR ENVIRONMENTAL AND BIOMEDICAL APPLICATIONS

Magnetic polymers represent a new class of magnetic materials prepared from an organic polymer. These magnetic materials, possessing nanoscale dimensions, have been applied for targeted drug delivery due to their high potential for improving diagnosis and treatment. Magnetic nanoparticles are especially important as drug carriers because, under the influence of the magnetic field, they can be directed and localized (Kalia et al. 2014). Nanomaterials including metallic or bimetallic nanoparticles, metal oxides, ferrites, and superparamagnetic iron oxide nanoparticles are used the most as diagnostic and therapeutic agents. Addition of bioactive and organic molecules to the anorganic surface provides tailored treatments, producing contrast agents that differentiate between affected and healthy tissues. For treating complex diseases, the polymer nanocomposites (consisting of the carrier, the bound or encapsulated bioactive payload, and surface modifiers) with superparamagnetic behavior are important for targeted drug delivery using an external magnetic field.

In this context, the literature mentions:

- Biocompatible magnetic nanocomposites of silica, iron oxide, and polymethacrylic acid used as anticancer drug doxorubicin model drugs (Chen et al. 2010). This nanocomposite was fabricated by grafting fluorescent polymethacrylic acid and folic acid, thus imparting to nanoparticles fluorescence and magnetic resonance. These nanoparticles can be used in cancer targeting and drug delivery. In another study, the composite was also present as an in vitro antibacterial agent (Bardajee et al. 2013).
- Superparamagnetism, which is very useful in drug delivery because the superparamagnetic iron oxide nanoparticles are transported to the desired site by electrical field effects remain at target site even after removal of the magnetic field (Kalia et al. 2014).
- Polymer nanocomposites of superparamagnetic iron oxide nanoparticles and polyvinyl alcohol modified with polymethyl methacrylate used for the delivery of the drug ciprofloxacin. The release was accelerated in the presence of a magnetic field, yet with lower amounts of poly(vinyl alcohol) and higher amounts of polymethyl methacrylate (Bajpai and Gupta 2011). Accordingly, this composite is recognized as a promising tool for targeted drug delivery because it was an external magnetic field for treating complex diseases such as cancers and tumors. As known, magnetic targeted drug delivery involves the attachment of a cytotoxic drug to a biocompatible carrier (magnetic nanoparticles functionalized by biocompatible molecules), followed by intravenous injection of these carriers in the form of a colloidal suspension, application of a magnetic field to direct the carrier to the target site, and release of the drug (Kalia et al. 2014).

15.5 SENSORS BASED ON CONDUCTING POLYMER NANOMATERIALS

Some studies have demonstrated that conducting polymer nanomaterials can be used in the construction of sensors, the performance of which includes response/recovery time, sensitivity, selectivity, and stability. Noble metals such as gold, silver, and platinum have been utilized as chemical sensors in biotechnology and microelectronics or as electrode materials in emerging all printed, flexible electronic, optoelectronic, and energy-related devices—even though still vulnerable to chemical and electrochemical attacks during operation as well as upon storage in an ambient environment (Chang et al. 2012).

The most representative conducting polymers used as sensors and in fabrication of conducting polymer nanomaterials and sensor applications are listed in Scheme 15.3 (Yoon 2013). In addition, organic/inorganic nanoparticles, anodic alumina membranes, track-etched polymer membranes, and mesoporous silica, along with surfactants, liquid crystals, and polyelectrolytes, are examples given in literature for solid-phase templates and molecular templates. The solid phase-templates used for fabricating nanostructures are generally nanospheres, porous matrices, and nanorods. In addition, the molecular templates used for obtaining nanostructures are spherical micelles, rod-like micelles, and liquid crystals. These solid-phase and molecular templates might be suitable for obtaining nanospheres, nanocapsules, nanorods/fibers, nanofibers, and nanotubes.

Generally, studies focus on the synthesis and design of building blocks that can assemble into nanostructures under special conditions. Thus, mention is made of:

- A site-specific electrochemical method for the fabrication of polypyrrole, polyaniline, and a derivate of polythiophene nanowire frameworks on microelectrode junctions (Alam et al. 2005).

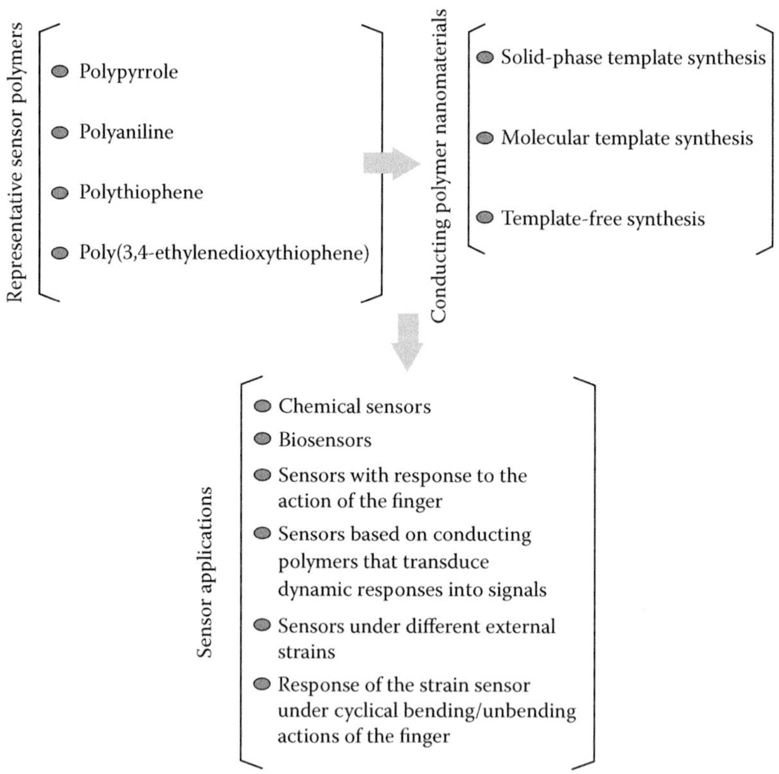

SCHEME 15.3 Representative conducting polymers for sensor applications, fabrication of conducting polymer nanomaterials, and sensor applications. (Adapted from Yoon Hyeonseok. *Nanomaterials.* 3 no. 3 (2013): 524–549.)

- The one-step interfacial polymerization of 3,4-ethylenedioxythiophene (in organic phase) and $HAuCl_4$ (in aqueous phase) leads to the synthesis of Au-poly(3,4-ethylenedioxythiophene) core-shell nanocables (Lu et al. 2007).
- Synthesis of Ag-polypyrrole core-shell nanoparticles offers an efficient and simple route for the fabrication of a nanostructured metal/conducting polymer complex (Chang et al. 2012).

Various conducting polymer nanostructures have been used to detect the presence of toxic gases, volatile organic compounds, alcohol, or humidity. To produce highly efficient magnetic polymer composites, incorporation of magnetic nanoparticles (iron oxide Fe_3O_4, $\gamma - Fe_2O_3$ or different spinel ferrites of transition metals—e.g., Mn, Mg, Zn, Ni, Co, and Cd) into different polymer matrices seems to be the most attractive solution. Among these types of polymers, polyaniline (PANI)—a material with a highly π-conjugated polymeric chain (Heeger 2001)—attracted much attention when compared with metal oxides because of its characteristics: environmental stability, high flexibility, relatively simple synthesis, high conductivity, and high absorption capability, due to its porous nature and low power consumption (Bandgar et al. 2015). In addition, PANI can be easily oxidized and reduced (Bavane et al. 2013). From this perspective, the choice of conductive polymers as sensing materials motivates scientists to focus on the fabrication of different kinds of sensors. Along these lines, the literature makes special mention of the development of a nanocomposite system—consisting of crystalline tin oxide (SnO_2) nanoparticles and a polyaniline matrix prepared by the in situ polymerization method—as a resistive-type humidity sensor (Shukla et al. 2015). Researchers analyzed the humidity-sensing behavior for a PANI/SnO_2 nanocomposite and have reported a continuous decrease in electrical resistance, from 127.5 to 11.4 kΩ, as relative humidity (RH) increased from 5% to 95%, the obtained results confirming improvement in sensing (a ten times higher sensitivity, i.e., 0.22% RH) as well as an increase in both the response (26 s) and recovery time (30 s).

Generally, the sensitivity of humidity sensors based on conducting polymers is determined by the differences recorded for a specific property before and after exposure to a test target molecule (capacitance—for capacitive humidity sensors, or electrical resistance—for resistive humidity sensors). In this respect, the specific surface area of the conducting polymers influences sensitivity. According to the literature, a "good sensor" has to rapidly modify (in structure, shape, conductivity) under different environmental stimuli (water molecules for humidity sensors or different toxic gases for gas sensors) by returning to a pristine state over a short time interval and by completely recovering the initial energy level (Coiai et al. 2015).

Besides the utilization of PANI/SnO_2 as a humidity sensor device, many scientific works emphasized that nanocomposites developed from polyaniline (PANI) and metal oxides—especially SnO_2—are good candidates as gas sensor devices. Taking into account all these observations, the literature supports fabrication of PANI/SnO_2 hybrid materials with different amounts of SnO_2 content prepared by in situ chemical oxidative polymerization method, with the sensitivity of materials toward toluene gas being measured at room temperature (Murugan et al. 2014). Researchers claim that the sensitivity of the investigated hybrid materials is enhanced and that sensor efficiency becomes maximal at a 40% loading of SnO_2.

Using the spin-coating preparation technique, the literature recommends preparation of PANI/SnO_2 nanocomposites as gas sensor elements (Khuspe et al. 2013). Test results have shown that, at room temperature, the highest selectivity along with a maximum response (72%), fast response time (167 s), and increased stability (86%) was recorded for NH_3.

Nambiar et al. have undertaken a comprehensive review of the conducting polymers used for fabricating electrochemical biosensors, tactile sensing "skins," and thermal sensors for applications in biomedical domains (Nambiar and Yeow 2011).

Significant efforts have been made to design and develop magnetic responsive materials (or smart materials) based on their biomimetic behavior and potential applications in biomedicine, coatings, and microfluidic and microelectronic fields. According to findings in the literature, smart materials are defined as those that can respond to pH, light, ionic strength, and an electric or magnetic field by modifying their specific properties—changes in size, solubility, permeability, and shape (Thévenot et al. 2013). As researchers have pointed out, magnetic responsive polymer materials (composites made of polymer and magnetic nanoparticles) can be divided into three classes according to the mechanisms involved in their response to the magnetic field. These classes are as follows:

- The first group (ferrogels, microsized hydrogels, and elastomeric materials) can be deformed upon exposure to a magnetic field.
- The second group is composed of polymers that can interact with biomolecules or biostructures and are used in the construction of biosensors, the magnetic guidance of drug delivery systems, and in cell separation.
- The third group includes thermoresponsive polymers. The strategy to apply magnetic induction to these materials allows the development of both implantable and shape-memory devices (Thévenot et al. 2013).

The technological demands for obtaining nanoscale materials with uniformly shaped particles and narrow particle distribution require various preparation strategies and a careful selection of materials. In this respect, both the nature of functional groups in the polymeric phase and the nature of the nanoparticles dispersed in different polymer matrices have a strong influence on the properties of the final magnetic polymer composites. The literature presents two composite systems, polyaniline (Fe_2O_3) and polyaniline ($MgFe_2O_4$), prepared by an in situ polymerization technique, as elements for gas-sensing devices. Researchers have demonstrated that introduction of polyaniline improves the conductivity of iron oxide particles and the dispersion of ferrites and that composites prepared from PANI and Fe_3O_4 are the most sensitive to acetone vapors. The particle size of the PANI–Fe_3O_4 composite was 10 μm (Tudorache and Grigoras 2010).

Generally, the electrophysical properties of materials are strongly dependent on their microstructure. In this respect, sensing properties are controlled by their grain size. In the case of inorganic particles embedded into a polymer matrix, the applicability of a "grain" model depends decisively on the preparation route of metal oxides or on the sintering conditions (time and temperature) (Yamazoe 1991).

In the case of humidity sensors, a decrease in grain size generates a significant increase in sensitivity to humidity, which is a very interesting effect (Williams and Coles 1998). The literature notes that the humidity influence of sensor parameters is structure-dependent and may be controlled through structural engineering of the metal oxide used (Korotcenkov 2008). In this respect, with a loading level of excess filler content in the polymer matrix, a percolation transition can be applied. According to findings in the literature, the percolation process is a consequence of geometrical phase transition in a random medium as well as of the modifications in the transport and physical properties of the systems—which follow a universal power law and scaling laws (Mukherjee 2014).

On the other hand, the influence of crystallographic structure and morphology on the gas-sensing characteristics of metal oxide conductometric (resistive)-type sensors has been widely investigated (Korotcenkov 2008). Korotcenkov concluded that optimization of the structural parameters of metal oxides (grain size, porosity, texture), in order to meet the requirements of sensing technology, is essential for improving the gas-sensing characteristics of resistive-type sensors.

Formation of one-dimensional structured nanomaterials (nanowires, nanotubes) attracted the interest of scientists (and researchers) due to the unique combination of electronic properties of conducting polymers and the large surface area of nanomaterials (Lu et al. 2011), which makes them suitable as sensing devices. For example, a group of scientists reported fabrication of a magnetic polymer flow sensor made of eight cylindrical polydimethylsiloxane (PDMS) pillars (500 μm long and 100 μm in diameter) that contains iron nanowires (NWs) (6 μm in length and 35 nm in diameter) used on a giant magneto-impedance (GMI) thin film sensor (Alfadhel et al. 2014). The researchers stated that the fabricated flow sensor can be used for both water and air flow measurement applications because of the following features:

- Detection of air flow rates up to 190 mm/s with a sensitivity of 24 mΩ/(mm/s) and a resolution of 0.56 mm/s at a power consumption of 31.6 μW.
- Detection of water flow rates up to 7.8 mm/s with a sensitivity of 0.9 Ω/(mm/s) and a resolution of 15 μm/s. With regard to power consumption, these characteristics can be reduced to 80 nW, with a small decrease in resolution (i.e., 32 μm/s, in the case of water flows).

A detailed comprehensive review described the spectral, optical, magnetic, and electrical properties of one-dimensional conducting polymer composites, as well as their applicability in electronic devices as chemical and biological sensors, gas sensors, electronic nanodevices, and energy devices (Lu et al. 2011).

Tremendous efforts have been made in the development of nanotechnology-based sensors for the detection of molecular contaminants in complex food matrices. In this respect, nanosensors can be defined as an array of thousands of nanoparticles that fluoresce on contact with food pathogens. On the other hand, tests for the detection of spoilage reveal modification of the color among the metal nanoparticles solution and analytes (Ai et al. 2009).

The use of nanomaterials as conducting polymers in a variety of sensing technologies will open new horizons for their applications. This is due primarily to their advantages, such as large surface area, easy processing, and adjustable transport properties (Sree et al. 2002). The advantages and disadvantages of different synthetic routes for the preparation of nanostructured conducting polymers and polymer composites are described in the literature (Rajesh et al. 2009).

15.6 SENSORS BASED ON NONCONDUCTING POLYMER NANOMATERIALS

The development of magnetic polymer composites attracted a lot of attention from the part of specialists. One comprehensive review describes the synthesis, properties, and applications of magnetic polymer composites, with the most studied materials being magnetic metal oxides (especially magnetite and maghemite) (Philippova et al. 2011). Generally, magnetic nanoparticles tend to agglomerate into host polymer matrices. It is possible for the dispersion of metal oxides inside the host material to occur through different mechanisms (Sarkar et al. 2012). On the other hand, as already highlighted, the morphology, size, surface area, porosity, and shape of the micropores of the host material determine the size and nature of the magnetic metal oxide nanoparticles dispersed inside the polymer matrices (Rozenberg and Tenne 2008) as well as the synthesis route for the development of hybrid organic–inorganic systems (Yu et al. 2014).

For this reason, creation of hybrid organic–inorganic materials is a difficult objective because it depends on a variety of factors, as is shown in the literature (Coey J. Michael David et al. 1998).

On the other hand, most sensing polymers are either hydrophilic or hydrophobic. In the case of humidity sensors, the absorption of water will result in the modification of conductive or dielectric properties with humidity.

Generally, humidity sensors are classified into two types: resistive and capacitive (Sakai et al. 1996). The construction of capacitive sensors is preferable to resistive sensors due to the advantages of the former, as illustrated in Scheme 15.4.

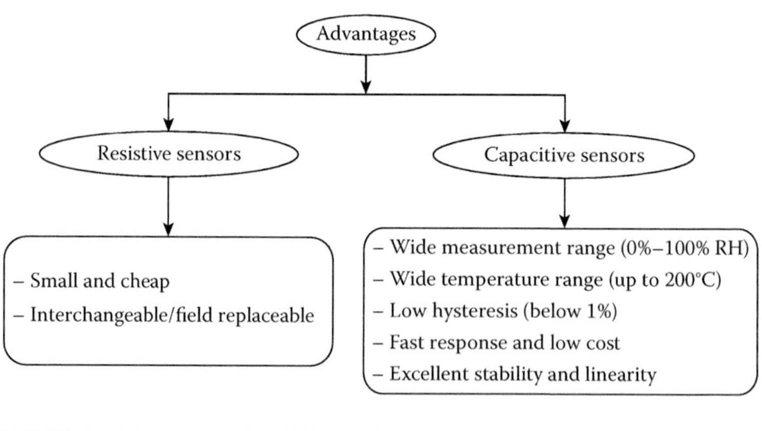

SCHEME 15.4 Advantages of resistive and capacitive humidity sensors. (Adapted from Sakai Yohko et al. *Sensor Actuat. B*. 35 no. 1–3 (1996): 85–90.)

Another classification, as well as a listing of the drawbacks of humidity sensors, is outlined by Sikarwar and Yadav (2015). In addition, a comprehensive review of humidity sensor principle mechanism and fabrication technologies is provided (Farahani et al. 2014).

An important issue in the construction of humidity sensors is the selection of materials. Among the polymeric humidity-sensitive materials, polyimides represent a good choice because they exhibit excellent thermal stability, low dielectric constant, and low equilibrium moisture content. Conversely, polyimides are highly resistant to irradiation and are mechanically strong and chemically stable in the presence of most common contaminants.

Besides their excellent characteristics, polyimides can be considered fully compatible with standard electronic processing procedures, an important cost control consideration (Strijkova and Georgieva 2013). On the other hand, iron oxides have been used for the manufacture of humidity sensors devices. Therefore, maghemite appears as a humidity sensor material chosen by many researchers due to its high sensitivity, simple design, and low cost.

The preparation of polyimides with magnetic filler content as humidity sensor devices represents a new research area. Because aromatic polyimides are more frequently utilized in sensing technology due to their excellent mechanical and thermal properties or radiation resistance (Sroog 1991) as well as their favorable thermal stability and dielectric properties (Hu et al. 2009), the development of new, more flexible polyimides requires different strategies. In order to increase the solubility of polyimides, two structural modifications of the polymer chain were proposed:

- The addition of flexible linkages (-O-) that reduce the rigidity of the polymer backbone and inhibit close packing of the chain, thereby reducing interchain interactions (Yang et al. 2003)
- Incorporation of an alicyclic dianhydride: 5-(2,5-dioxotetrahydrofurfuryl)-3-methyl-3-cyclohexene-1,2-dicarboxylic acid anhydride (DOCDA, 98% purity), produced by Merck, Darmstadt, Germany (Matsumoto 1999)

As a result, two composite systems derived from partially aliphatic polyimides—5-(2,5-dioxotetra-hydrofurfuryl)-3-methyl-3-cyclohexene-1,2-dicarboxylic acid anhydride reacting with 4,4′-oxydianiline (DOCDA–ODA) and with iron oxide as Fe_3O_4—were prepared at different temperatures, 250 and 300°C (Nica et al. 2015). Correlations among the structural, morphological, thermal, and magnetic properties were established for possible application as humidity sensor devices. In addition, the influence of humidity on the electrical properties of poly(DOCDA–ODA)/iron oxide composite were discussed, which led to an analysis of the humidity sensitivity of pure polyimide and of a poly(DOCDA–ODA)/Fe_3O_4 with different filler contents of iron oxides at various values of relative humidity (RH).

Generally, the literature shows that, for magnetic composite systems, relative permittivity decreases as frequency increases, this behavior being attributed to the interfacial polarization predicted by Maxwell–Wagner model, derived for the permittivity of a homogeneous two-phase system. In addition, a decrease in electrical resistivity for these magnetic composites was recorded with an increase in frequency in the range of 0–100% RH, a behavior attributed to the typical Maxwell–Wagner–Debye relaxation. Reduction of electrical resistivity under the influence of humidity depends on composition changes as well as microstructure. A relevant sensitivity of samples is recorded at low-frequency values (200 Hz). Test results have shown that the poly(DOCDA–ODA)/iron oxide composite synthesized at 300°C with 20 wt% iron oxide loadings results in optimum humidity sensitivity, as indicated in Figure 15.1 (Nica et al. 2015).

From these findings, another group of researchers reported fabrication of water-resistant humidity sensors based on sulfonated polyimides (SPI) derived from 1,4,5,8-naphthalenetetracarboxylic dianhydride (NTDA) and various sulfonated diamines (Ueda et al. 2007). They reported that the humidity sensors prepared by a dip-coating sensing membrane of NTDA-4,4′-bis(4-aminophenoxy)-biphenyl-3,3′-disulfonic acid (BAPBDS) in triethylammonium salt formed on alumina substrates pretreated with an amine-type silane-coupling reagent, show excellent resistance to water and heat, high temperature and high humidity, as well as to heat shock and dewy atmosphere. SPI sensors showed a fast (90%) response time of about 1 min for humidification from 30% to 90% RH but a fairly slow response (about 4 min) for dehumidification from 90% to 30% RH. In addition, a large hysteresis of 5%–6% RH in the R versus RH relationship between humidification and dehumidification was observed. The 90% response time was significantly faster at 75°C; however, hysteresis was only slightly improved.

Carbon nanotubes (CNTs) represent a unique class of materials that generate interest for the development of CNT-based devices, including various types of sensors in many technological industries. A serious concern regarding humidity sensors based on carbon nanotubes is the limitation of selectivity, possibly generated by the non selective character of the absorption process. In this respect, the possibility of using plasma-functionalized (pf) multiwall carbon nanotube (MWCNT) networks

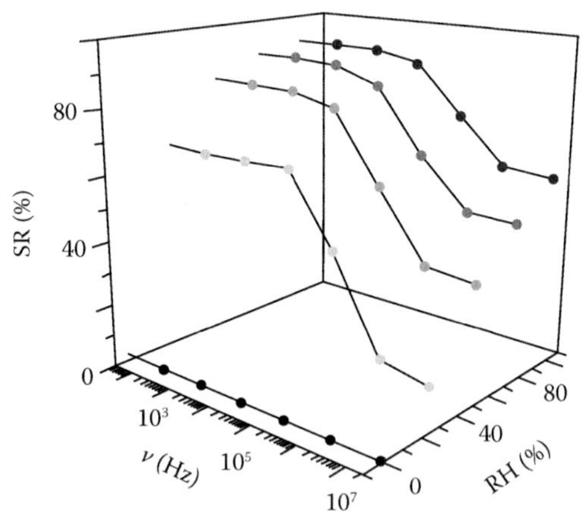

FIGURE 15.1 Dependence of sensitivity on frequency and humidity of poly(DOCDA ODA)/Fe_3O_4 composite films synthesized at 300°C, at 20 wt% Fe_3O_4 contents. (Adapted from Nica Simona Luminita et al. *Polym. Int.*, 64, no. 9, 1172–1181, 2015.)

as the upper electrode for poly(pyromellitimide-1,4-diphenyl ether) (PDMA-ODA) capacitive humidity sensors has been studied (Lee et al. 2008). Researchers have found that polyimide film sensors with a plasma functionalized-MWCNT electrode have an absorption time of 1.5 s and a recovery time of 1.75 s—about six times faster than those of chromium electrode sensors. In addition, the pf-MWCNT electrode sensor exhibited a sensitivity of 0.65 pf/% RH. As researchers pointed out, these results may be attributed partially to the percolated pore structure of the pf-MWCNT electrode polyimide sensor, which is more accessible to water molecules and partially to the oxygenated surface of the pf-MWCNT films.

In addition, the development of a resistive-type humidity sensor based on multiwall carbon nanotube (MWCNT)/polyimide composite films attracted more attention for development of sensing technologies (Lee et al. 2013). The relative humidity sensors showed a good linearity (1.10) over a wide humidity range, with a sensitivity of 60 Ω% RH and a response time of 10 min. In addition, an increase of resistance to water absorption was observed. This effect was explained by the charge transfer between the adsorbed water molecules and the MWCNTs.

15.7 GENERAL REMARKS

Research in the field of magnetic polymer composites and their applications in sensing technology has been focused mainly on the challenge of creating organic–inorganic composite materials containing magnetic nanoparticles, spinel ferrite, or different inorganic materials embedded into conductive or non conductive polymer matrices as well as on the development of various strategies that can be used to integrate the investigated materials in functional sensing devices. From this perspective, the literature in this field provides information on preparation methods, structure, and properties of magnetic polymer composite systems.

The possibility of modifying the different shape/dimension relation (nano- or microsized magnetic nanoparticles, nanotubes, nanowires) with the commitment to produce and create a new generation of materials exhibiting enhanced physical, mechanical, thermal, electrical, and magnetic properties has been intensively investigated. Despite the multiple purposes and effects given to the final multiphase polymer systems by the filler content, there are still many unsolved problems in this research area. From this point of view, the influence of the shape, size, and dispersion of particles from the polymer matrix (either conductive or nonconductive) on the macroscopic behavior of the resulting product represents the main issue.

From this perspective, extensive investigations in this research field cover different topics—from polymer blends and composite systems to nanocomposite systems. In this respect, smart materials occupy a special place in multiphase polymer systems, where the success of their preparation depends on the ability to exploit the potential of nanostructuring in the resulting product. The possibility of analyzing and optimizing the material performance and also designing new materials at different (nano- and microsized) length scales is strongly related to the structure–properties relationships of final multiphase polymer systems. Most of the applications of smart materials indicate their powerful impact on sensor technologies, which are used frequently in our daily life.

ACKNOWLEDGMENTS

This work was supported by a grant from the Romanian National Authority for Scientific Research, CNCS-UEFISCDI, project no. PN-II-ID-PCE-2011-3-0937, 2016 stage.

REFERENCES

Ai Kelong, Liu Yanlan, and Lu Lehui. Hydrogen-bonding recognition-induced color change of gold nanoparticles for visual detection of melamine in raw milk and infant formula. *J. Am. Chem. Soc.* 131 no. 27 (2009): 9496–9497.

Alam M. Maksudul, Wang Jun, Guo Yaoyao, Lee P. Stephanie, and Tsen Hsian-Rong. Electrolyte-gated transistors based on conducting polymer nanowire junction arrays. *J. Phys. Chem. B.* 109 no. 26 (2005): 12777–12784.

Alfadhel Ahmed, Li Bodong, and Kosel Jurgen. Magnetic polymer nanocomposites for sensing applications. *IEEE Sensors.* no. 6985442 (2014): 2066–2069.

Bajpai Anil Kumar, and Gupta Rashmi. Magnetically mediated release of ciprofloxacin from polyvinyl alcohol based superparamagnetic nanocomposites. *J. Mater. Sci. Mater. Med.* 22 no. 2 (2011): 357–369.

Bandgar D. K., Navale T. Shankararao, Naushad U. Mu, Mane Rajaram Sakharam, Stadler J. Florian, and Patil B. Vikas. Ultra-sensitive polyaniline–iron oxide nanocomposite room temperature flexible ammonia sensor. *RSC Adv.* 5 no. 84 (2015): 68964–68971.

Bardajee Ghasem Rezanejade, Hooshyar Zari, and Rastgo Freshteh. Kappa carrageenan-g-poly (acrylic acid)/SPION nanocomposite as a novel stimuli-sensitive drug delivery system. *Colloid Polym. Sci.* 291 no. 12 (2013): 2791–2803.

Bavane G. Ravindra Kumar, Mahajan M. Ashok, Shirsat D. Mahendra, and Gore B. Ratnakar. Ammonia gas sensing characteristics of spin coated polyaniline films. *J. Adv. Phys.* 3 no. 3 (2013): 241–248.

Chang Mincheol, Kim Taejoon, Park Hyun-Woo, Kang Minjeong, Reichmanis Elsa, and Yoon Hyeonseok. Imparting chemical stability in nanoparticulate silver via a conjugated polymer casing approach. *ACS Appl. Mater. Interfaces.* 4 no. 8 (2012): 4357–4365.

Chen Dongyun, Jiang Mengjun, Li Najun, et al. Modification of magnetic silica/iron oxide nanocomposites with fluorescent polymethacrylic acid for cancer targeting and drug delivery. *J. Mater. Chem.* 20 no 31 (2010): 6422–6429.

Coey J. Michael David, Berkowitz E. Ami, Balcells I. Lluís, Putris F. Firas, and Parker T. Frederick. Magnetoresistance of magnetite. *Appl. Phys. Lett.* 72 no. 6 (1998): 734–736.

Coiai Serena, Passaglia Elisa, Pucci Andrea, and Ruggeri Giacomo. Nanocomposites based on thermoplastic polymers and functional nanofiller for sensor applications. *Materials.* 8 no. 6 (2015): 3377–3427.

Ergeneman Olgaç, Peters Christian, Gullo R. Maurizio, et al. Inkjet printed superparamagnetic polymer composite hemispheres with programmed magnetic anisotropy. *Nanoscale.* 6 no. 18 (2014): 10495–10499.

Farahani Hamid, Wagiran Rahman, and Hamidon N. Mohd. Humidity sensors principle, mechanism, and fabrication technologies: A comprehensive review. *Sensors.* 14 no. 5 (2014): 7881–7939.

García-Arribas Alfredo, Gutiérrez Jon, Kurlyandskaya V. Galina, et al. Sensor applications of soft magnetic materials based on magneto-impedance, magneto-elastic resonance and magneto-electricity. *Sensors.* 14 no. 5 (2014): 7602–7624.

Heeger J. Alan. Semiconducting and metallic polymers: The fourth generation of polymeric materials. *Angew. Chem. Int. Ed. (English).* 40 no. 14 (2001): 2591–2611.

Hu Longfei, Li Meishuan, Xu Caihong, Luo Yongming, and Zhou Yanchun. A polysilazane coating protecting polyimide from atomic oxygen and vacuum ultraviolet radiation erosion. *Surf. Coat. Technol.* 203 no. 22 (2009): 3338–3343.

Kalia Susheel, Kango Sarita, Kumar Amit, Haldorai Yuvaraj, Kumari Bandna, and Kumar Rajesh. Magnetic polymer nanocomposites for environmental and biomedical applications. *Colloid Polym. Sci.* 292 no. 9 (2014): 2025–2052.

Khuspe G. D., Navale T. Shankararao, Chougule A. Manik, et al. Facile method of synthesis of polyaniline-SnO_2 hybrid nanocomposites: Microstructural, optical and electrical transport properties. *Synth. Met.* 178 (2013): 1–9.

Kim Jiyun, Chung Eun Su, Choi Sung-Eun, Lee Howon, Kim Junhoi, and Kwon Sunghoon. Programming magnetic anisotropy in polymeric microactuators. *Nat. Mater.* 10 no. 10 (2011): 747–752.

Korotcenkov S. Ghenadii. The role of morphology and crystallographic structure of metal oxides in response of conductometric-type gas sensors. *Mater. Sci. Eng. R.* 61 no. 1–6 (2008): 1–39.

Lee Myung Jin, Hong Hyun Pyo, Kwon Kwang Ho, Park Chan Won, and Min Nam Ki. Fast-speed, high-sensitivity polyimide humidity sensors with superhydrophilic carbon nanotube network electrodes. *Sens. Actuator B.* 185 (2013): 97–104.

Lee Myung Jin, Lee Cheoljin, Singh R. Ved, Kum-Pyo Yoo, and Nam-Ki Min. Humidity sensing characteristics of plasma functionalized multiwall carbon nanotube-polyimide composite films. *IEEE Sensors.* 4716470 (2008): 430–433.

Lu Gewu, Li Chun, Shen Jiaoyan, Chen Zhaojia, and Shi Gaoquan. Preparation of highly conductive gold-poly (3, 4-ethylenedioxythiophene) nanocables and their conversion to poly (3, 4-ethylenedioxythiophene) nanotubes. *J. Phys. Chem. C.* 111 no. 16 (2007): 5926–5931.

Lu Xiaofeng, Zhang Wanjin, Wang Ce, Wen Ten-Chin, and Wei Yen. One-dimensional conducting polymer nanocomposites: Synthesis, properties and applications. *Prog. Polym. Sci.* 36 no. 5 (2011): 671–712.

Matsumoto Toshihiko. Nonaromatic polyimides derived from cycloaliphatic monomers. *Macromolecules.* 32 no. 15 (1999): 4933–4939.

Mukherjee Rupam. Correlation effects in nanoparticle composites: Percolation, packing and tunneling, Dissertation at Wayne State University, 2014, http://digitalcommons.wayne.edu/oa_dissertations

Murugan Chinnaraj, Subramanian Easwara, and PathinettamPandiyan Duraipandi. Sensor functionality of in situ synthesised polyaniline-SnO_2 hybrids towards toluene vapour. *J. Environ. Nanotechnol.* 3 no.1 (2014): 37–42.

Nambiar Shruti, and Yeow T. W. John. Conductive polymer-based sensors for biomedical applications. *Biosens. Bioelectron.* 26 no. 5 (2011): 1825–1832.

Nica Simona Luminita, Nica Valentin, Grigoras Vasile Cristian, et al. Influence of two structural phases of Fe_3O_4 and γ-Fe_2O_3 on the properties of polyimide/iron oxide composites. *Polym. Int.* 64 no. 9 (2015): 1172–1181.

Peters Christian, Ergeneman Olgaç, García D. Wendel Pedro, et al. Superparamagnetic twist-type actuators with shape-independent magnetic properties and surface functionalization for advanced biomedical applications. *Adv. Funct. Mater.* 24 no. 33 (2014): 5269–5276.

Philippova Olga, Barabanova Anna, Molchanov Vyacheslav, and Khokhlov Alexei. Magnetic polymer beads: Recent trends and developments in synthetic design and applications. *Eur. Polym. J.* 47 no. 4 (2011): 542–559.

Prinz A. Gary. Magnetoelectronics applications. *J. Magn. Magn. Mater.* 200 no. 1–3 (1999): 57–68.

Rajesh, Ahuja Tarushee, and Kumar Devendra. Recent progress in the development of nano-structured conducting polymers/nanocomposites for sensor applications. *Sensor. Actuat. B.* 136 no. 1 (2009): 275–286.

Rozenberg A. Boris, and Tenne Reshef. Polymer-assisted fabrication of nanoparticles and nanocomposites. *Prog. Polym. Sci.* 33 no. 1 (2008): 40–112.

Sakai Yohko, Sadaoka Yoshihiko, and Matsuguchi Motoyuki. Humidity sensors based on polymer thin films. *Sensor Actuat. B.* 35 no. 1–3 (1996): 85–90.

Sarkar Sudipta, Guibal Eric, Quignard Françoise, and SenGupta K. Arup. Polymer-supported metals and metal oxide nanoparticles: Synthesis, characterization, and applications. *J. Nanopart. Res.* 14 no. 2 (2012): 715–727.

Shukla K. Saroj, Shukla K. Sudheesh, Govender P. Penny, and Agorku S. Eric. A resistive type humidity sensor based on crystalline tin oxide nanoparticles encapsulated in polyaniline matrix. *Microchim. Acta.* 183 no. 2 (2016): 573–580.

Sikarwar Samiksha, and Yadav C. Balchandra. Opto-electronic humidity sensor: A review. *Sensor. Actuat. A.* 233 no. 9178 (2015): 54–70.

Sree Usha, Yamamoto Yojiro, Deore Bhavana, Shiigi Hiroshi, and Nagaoka Tsutomu. Characterization of polypyrrole nano-films for membrane-based sensors. *Synt. Met.* 131 no. 1–3 (2002): 161–165.

Sroog E. Cyrus. Polyimides. *Prog. Polym. Sci.* 16 no. 4 (1991): 561–694.

Strijkova Velichka, and Georgieva Biliana. Vacuum deposited polyimide layers as humidity sensors. *Bulg. J. Phys.* 40 no. 3 (2013): 224–228.

Thévenot Julie, Oliveira Hugo, Sandre Olivier, and Lecommandoux Sébastien. Magnetic responsive polymer composite materials. *Chem. Soc. Rev.* 42 no. 17 (2013): 7099–7116.

Tudorache Florin, and Grigoras Mircea. Study of polyaniline—Iron oxides composites using for gas detection. *Optoelectron. Adv. Mater.* 4 no. 1 (2010): 43–47.

Ueda Masahiro, Nakamura Kazumasa, Tanaka Kazuhiro, Kita Hidetoshi, and Okamoto Kenichi. Water-resistant humidity sensors based on sulfonated polyimides. *Sens. Actuat. B.* 127 (2007): 463–470.

Williams Geraint, and Coles S. V. Gary. Gas sensing properties of nanocrystalline metal oxide powders produced by a laser evaporation technique. *J. Mater Chem.* 8 no. 7 (1998): 1657–1664.

Wood Vanessa, Panzer J. Matthew, Chen Jianglong, et al. Inkjet-printed quantum dot–polymer composites for full-color AC-driven displays. *Adv. Mater.* 21 no. 21 (2009): 2151–2155.

Yamazoe Noboru. New approaches for improving semiconductor gas sensors. *Sens. Actuat. B.* 5 no. 1–4 (1991): 7–19.

Yang Chin-Ping, Hsiao Sheng-Huei, and Wu Kuen-Lin. Organosoluble and light-colored fluorinated polyimides derived from 2, 3-bis (4-amino-2-trifluoromethylphenoxy) naphthalene and aromatic dianhydrides. *Polymer.* 44 no. 23 (2003): 7067–7078.

Yoon Hyeonseok. Current trends in sensors based on conducting polymer nanomaterials. *Nanomaterials.* 3 no. 3 (2013): 524–549.

Yu Lili, Wu Hui, Wu Beina, et al. Magnetic Fe_3O_4—Reduced graphene oxide nano—Composites—Based electrochemical biosensing. *Nano-Micro Lett.* 6 no. 3 (2014): 258–267.

16 Development of Imide-Type Polymer Fibers Containing Metal Nanoparticles

Diana Serbezeanu, Ionela-Daniela Carja, Ion Sava, René Michel Rossi, and Giuseppino Fortunato

CONTENTS

16.1 INTRODUCTION

Polyimides (PIs) have attracted much attention due to their good thermal stability, chemical resistivity, and excellent mechanical and dielectric properties in fabricating multichip module packaging, ultra-large scale integrated circuit technology, and printed circuit board fabrication (Gosh and Mittal 1996; Weiser et al. 2000; Ghosh et al. 2015). However, aromatic polyimides in their imidized form present low solubility in common organic solvents and infusibility, which make them difficult to process. Different approaches are used to improve the solubility and processability of aromatic polyimides without sacrificing the chemical, physical, and mechanical properties. These methods include the incorporation of flexible linkages, alicyclic or nonlinear moieties into the main chain, or bulky lateral substituents along the polymer backbone (Shang et al. 2006; Wang et al. 2006; Hamciuc et al. 2011, 2013; Carja et al. 2012; Sava et al. 2015).

Electrospinning technology has been shown to be a facile and versatile method to manufacture nanoscale fibrous membranes. Electrospun fiber membranes have gained much attention due to their properties such as smaller pore size, higher porosity, and air permeability, compared to the conventional nonwovens. Recently, organic (Chen et al. 2011), inorganic (Li and Xia 2003; Ding et al. 2004; Park and Lee 2010), and organic–inorganic structured nanofibers (Guo et al. 2008; Ouyang et al. 2013; Tang et al. 2013), with uniform diameters and large surface area-to-volume ratio, have been produced because of their potential applications in membrane technology, tissue engineering, optical sensors, biosensors, and drug delivery (Min et al. 2004; Ma et al. 2006; Wang et al. 2009; Schiffman and Elimelech 2011). A variety of materials such as polymers, polymer blends, composites, and so on can be processed into nanofibers by using the electrospinning process (Ramakrishna et al. 2005). The aspect ratios and morphology of the nanofibers can be controlled by varying the solution properties (solution concentration, polymer molecular weight, solution viscosity, surface tension of solution, and conductivity), processing conditions (applied electric potential, feed rate, and working distance), and the environmental conditions (temperature, humidity) (Bhardwaj and Kundu 2010; Kim et al. 2013).

Gold nanoparticles (AuNPs) have attracted much attention because of the wide range of their potential applications in different fields such as electronics (Kumar Jena and Retna Raj 2008), catalysis (Herves et al. 2012), biotechnology (Yokota et al. 2008), and medicine (Daniel and Astruc 2004; O'Neal et al. 2004). However, the potential applications of AuNPs are strongly connected with their size, distribution, and morphology, and all these characteristics can be controlled by the proper choice of preparation conditions. The most used methods for the preparation of AuNPs are derived from chemical, physical, or even biosynthesis approaches. The common reducing agents for a gold precursor are sodium citrate (Nath et al. 2010; Polte et al. 2010), sodium boron hydride (Sakai and Alexandridis 2004), and block copolymers (Hou et al. 2007; Mendoza et al. 2009). However, these reducing agents show potential environmental risks which could restrict the practical applications of AuNPs. Therefore, the demand for nontoxic and eco-friendly AuNP synthesis methods are very crucial for a sustained environment. An alternative method for chemical synthesis of the AuNPs could be green chemistry-based, eco-friendly methods using biological systems such as microbes, fungi, and several plant extracts. Zhu et al. (2012) have reported a facile and green synthesis of well-dispersed small AuNPs embedded in the electrospun polyacrylonitrile nanofibers by combining an electrospinning technique and in situ reduction approach using tea polyphenols as the reductant. Deniz et al. described for the first time the use of a laser ablation and electrospinning technique for the fabrication of poly(vinylpyrrolidone)/AuNP nanofibrous composites (Deniz et al. 2011). The advantages of this method are related with the process itself which is a clean, chemically safe, less time-consuming, low cost, and highly applicable strategy for the production of AuNPs. Liu et al. prepared AuNP–gelatin fibers with bright photoluminescence emission under excitation with potential values in the research area of functional fibers (Liu et al. 2014b).

The design and development of multiphase polymer systems are strongly dependent on the control of the interface and the control of the morphology. The fabrication of hybrid materials is a benefit of both scientific and industrial interest due to added functionalities as well as for their enhanced mechanical and related physical properties. Among all the polymers used as support, limited attention has been directed to polyimides. Meyer et al. (1988) reported a related study regarding gold–polyimide interactions at room temperature. Travaly et al. have studied the growth and aggregation of gold atoms onto polyimide surfaces. The authors selected two main parameters (the cluster surface density and the surface coverage), with the aim to follow the nucleation and growth process of the AuNPs/PI system. They observed that the adsorption of gold atoms on PI substrates led to particle nucleation that covers only a fraction of the surface, even after a nominal coverage of 20 monolayers (Travaly et al. 1999). Gold-nanoparticle-dispersed fluorinated polyimide films were prepared by Koizumi et al. (2002), starting from a solution of poly(amic acid) (PAA) (pyromellitic dianhydride and 2,2′-bis(trifluoromethyl)-4,4′-diaminobiphenyl) in *N,N*-dimethylacetamide (DMAc) to which $HAuCl_4 \cdot 3H_2O$ was added as a dopant. The conversion of gold salt into zero-valent gold particles was conducted after thermal imidization. The authors observed that the films containing AuNPs showed much higher absorption in the visible region than the host polyimide as a result of surface plasmon absorption of gold nanoparticles.

Zhang et al. described a method to prepare multilayered AuNPs/PI composite films using layer-by-layer assembly of a freshly prepared solution of naked AuNPs and an in-house synthesized amino poly(amic acid). Thermal imidization of poly(amic acid) was performed after layer-by-layer assembly (Zhang and Srinivasan 2007). Vanherck et al. have demonstrated the incorporation of different amounts of AuNPs into a PI membrane matrix. After that, the membranes were locally heated during filtrations by irradiation with continuous green laser light. They observed that the photothermal heating effect on the temperature and the permeance behavior of the membrane depended on the concentration of gold nanoparticles in the membrane, AuNPs aggregation, and the laser intensity (Vanherck et al. 2011). Bauld et al. demonstrated the incorporation of AuNPs in thermosetting PI thin films using Au_{25}^- molecules as precursors (Bauld et al. 2012).

This chapter discusses some recent advances in the preparation of new composite nanofibers based on polyimide-bearing phenol units and gold nanoparticles by using an in situ electrospinning approach. The influence of AuNPs on the morphology of the PI nanofibers was investigated in detail.

16.2 PREPARATION OF POLYIMIDE NANOFIBERS BY ELECTROSPINNING TECHNIQUE

Nanofibrous PI membranes obtained by using the electrospinning technique have been a subject of increasing interest. Hou et al. used the electrospinning technique to make high-strength fibers (Huang et al. 2006a, 2006b). Also, the dielectric constant of the PI fibers based on pyromellitic dianhydride and diaminodiphenyl ether was investigated and found to be 1.53–1.56 (Liu et al. 2007). Fukushima et al. investigated the effects of electrospinning parameters such as flow rate, applied voltage, molecular weight, viscosity of the polymer solution, salt addition, and humidity on the average fiber diameters and uniformity of fluorinated polyimide electrospun nanofibers. Investigation of the parameters mentioned led to successfully prepared bead-free, ultrafine, uniform diameters with a narrow range of nanoscaled diameters having the potential for application in catalysis, electronic and photonic devices, and tissue scaffolds (Fukushima et al. 2010). In order to obtain PI membranes with good processability, high thermal stability, and water absorption, Chen et al. used fluorinated polyimides and the electrospinning technique. Thus, they obtained PI mats with low dielectric constant, high glass transition temperature, and high hydrophobicity with possible application as insulating materials in interlayer dielectrics (Chen et al. 2012). Until recently, high-strength electrospun PI fibers were obtained using two steps—electrospinning the PI precursor followed by imidization of PAA fibers at high temperature. Karube et al. fabricated novel nanofibrous fluorinated polyimide membranes by the electrospinning technique using a PI solution without thermal imidization (Karube and Kawakami 2010). Chisca et al. studied the properties of a PI solution to obtain the optimal conditions for processing the PI into thin fibers by the electrospinning technique. The polyimide chosen for that study was synthesized by a two-step polycondensation reaction in solution using 3,3′,4,4′-benzophenonetetracarboxylic dianhydride and 3,3′-dimethyl-4,4′-diaminodiphenylmethane. The authors found a value of 18.3 wt% as the critical entanglement concentration (defined as the minimum concentration required for electrospinning of beaded nanofibers) and a value of 30 wt% as the minimum concentration required for electrospinning of uniform, bead-free fibers (Chisca et al. 2012). In our previous studies, we reported the preparation of PI nanofibers by using an in situ electrospinning approach. The polyimides were synthesized by a two-step polycondensation reaction, using 4,4′-(4,4′-isopropylidenediphenyl-1,1′-diyldioxy) dianiline and benzophenone-3,3′,4,4′-tetracarboxylic dianhydride to produce PI-1 or 4,4′-diamino-4″-hydroxy triphenylmethane and two different dianhydrides and benzophenone-3,3′,4,4′-tetracarboxylic dianhydride for PI-2 or 4,4′-oxydiphthalic anhydride for PI-3 (Aflori et al. 2015; Serbezeanu et al. 2015a, 2015b). The structures of the investigated polyimides are shown in Figure 16.1.

The synthesized PIs were soluble in common organic solvent such as *N,N*-dimethylformamide (DMF), *N,N*-dimethylacetamide (DMAc), dimethyl sulfoxide (DMSO), or *N*-methyl-2-pyrrolidone (NMP). The good solubility in the case of PI-1 is attributed to the presence of a higher number of flexible ether and isopropylidene groups, which leads to an increased flexibility of the macromolecular chains. For PI-2 and PI-3, the presence of the pendant phenol groups from the diamine units reduces the interchain interactions and thus increases the solubility.

To investigate the influence of PI solution concentrations on respective fiber morphologies, PI solutions in DMF were prepared and subjected to rheological measurements by cone–plate assembly. The solution viscosity represents one of the most important parameters in electrostatic spinning in terms of homogeneous fiber formation due to the need for a minimum concentration of polymer chain entanglements (McKee et al. 2005). Shear rate viscosity was measured for all PI solutions and, in all cases, a Newtonian flow character of the individual solutions was observed. The viscosity of the 10 wt% PI-1 solution measured at 20°C and a shear rate of 100 s^{-1} using DMF as the solvent was 0.35 Pa · s (Figure 16.2). For the PI-2 solution, the viscosity measured at 20°C and a shear rate of 100 s^{-1} was in the range of 0.002–3.60 Pa · s for solution concentrations between 2 and 20 wt%. In this case, the critical lower concentration of polyimide PI-2 for the formation of fibers without beads was found to be 15 wt%, corresponding to a viscosity of 0.52 Pa · s (Figure 16.2).

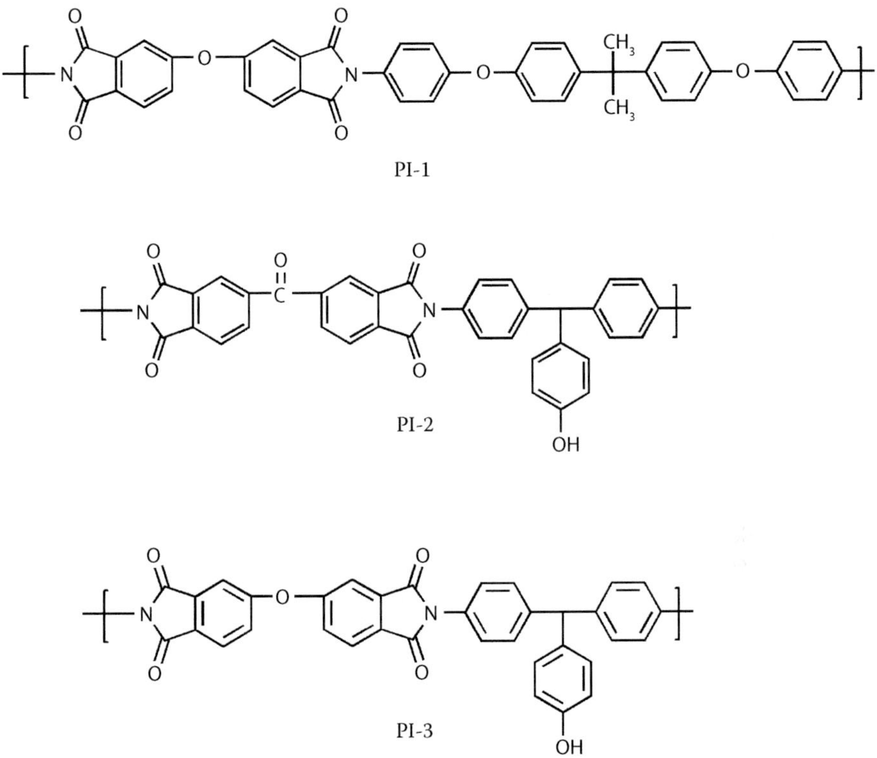

FIGURE 16.1 Chemical structures of PIs. (Adapted from Aflori M., et al. *Chem. Lett.* 44 no. 10: 1440–1442, 2015; Serbezeanu D., et al. *Text. Research J.* 85 no. 17 1763–1775, 2015a; Serbezeanu D., et al. *Eur. Polym. J.*, 64, 10–20, 2015b.)

From Figure 16.3, it can be observed that the viscosity of the solution increased linearly with the increase of shear rate, showing a Newtonian fluid behavior for PI-3 in the analyzed regime. The viscosity measured at a shear rate of 100 s^{-1} increased from 0.074 to 4.17 Pa · s as the concentration of PI-3 increased from 12 up to 25 wt%.

Table 16.1 shows the average fiber diameter distribution of the electrospun PI-1, PI-2, and PI-3 meshes spun at different concentrations and at a distance of 10 and 15 cm, respectively. The average fiber diameter at the optimal condition (10 wt% PI-1 solution with a 0.8 mm gauge needle at a feed rate of 20 μL/min^{-1}, the collecting distance of 10 cm, and the voltage of +15 kV/−2 kV) for PI-1 was observed to be 444 ± 57 nm. At concentrations below 10 wt%, mostly beads were noted. In the case of PI-2 at a concentration below 7 wt%, the beads-on-strings structure was obtained. At concentrations higher than 10 wt%, the average fiber diameter increased drastically to 401 nm. From Table 16.1, it can be observed that the average fiber diameter of the PI-2 increased from 174 ± 11 to 500 ± 12 nm as the concentration of PI-2 in DMF solution increased from 10 to 15 wt%, respectively. At 20 wt%, continuous electrospinning was not possible because the polymer solution flow was not constant due to high viscosity. The experiment was set at different needle-to-collector

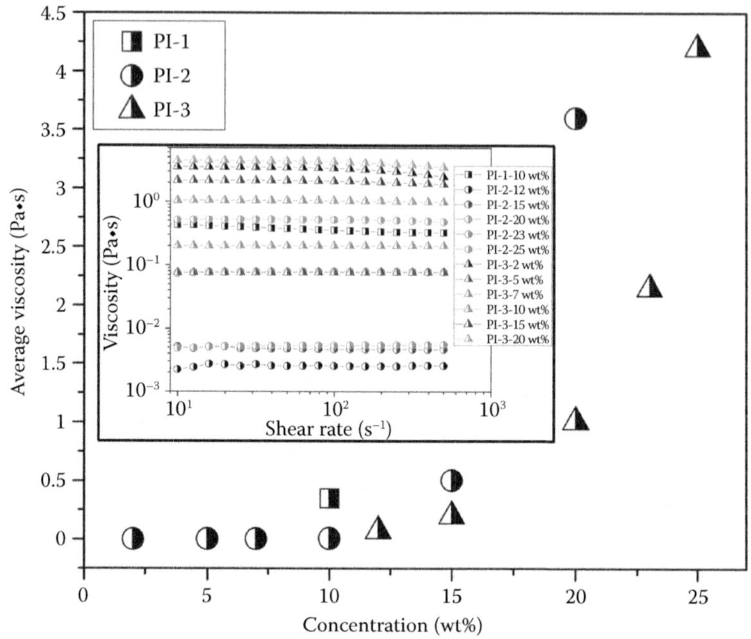

FIGURE 16.2 The average viscosity versus concentration for PI-1, PI-2, and PI-3 measured at a shear rate of 100 s^{-1}. (Adapted from Serbezeanu D., et al. *Eur. Polym. J.*, 64, 10–20, 2015b.)

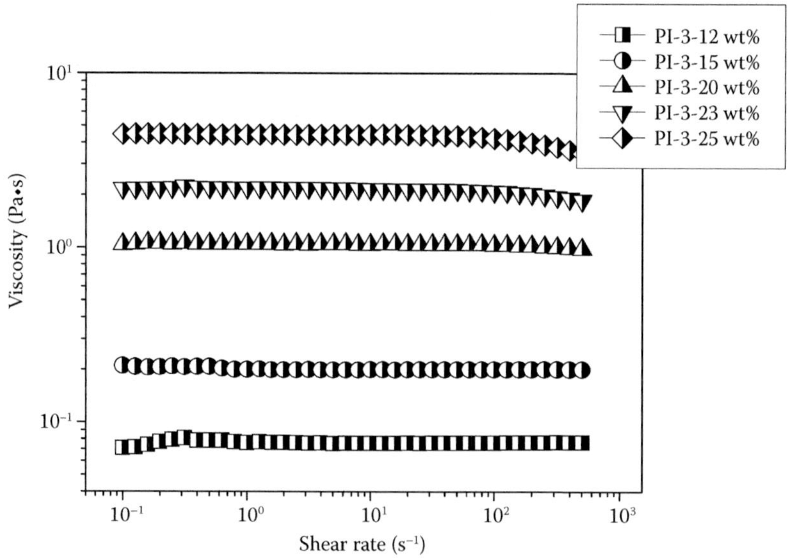

FIGURE 16.3 Logarithmic plot of viscosity as a function of shear rate and concentration for polyimide PI-3 solutions.

TABLE 16.1
Average Fiber Diameters as a Function of the Needle-to-Collector Distance for PI-1, PI-2, and PI-3 e-Spun Solutions

		Average Fiber Diameter (nm)	
Polymer	**Solution Concentration (wt%)**	**10 cm**	**15 cm**
PI-1	10	444 ± 57	–
PI-2	10	174 ± 11	370 ± 7
	15	500 ± 12	640 ± 11
	20	–	1464 ± 38
PI-3	12	157 ± 52	293 ± 72
	15	351 ± 114	297 ± 103
	20	550 ± 115	493 ± 138
	23	–	730 ± 89
	25	1049 ± 280	875 ± 266

Source: Adapted from Aflori M., et al. *Chem. Lett.*, 44, no. 10, 1440–1442, 2015; Serbezeanu D., et al. *Text. Research J.*, 85, no. 17, 1763–1775, 2015a; Serbezeanu D., et al. *Eur. Polym. J.*, 64, 10–20, 2015b.

distances (10, 15 cm) (Table 16.1) while maintaining a constant wt% PI-2/DMF solution. The distance in the electrospinning process was investigated because it is closely correlated with the evaporation time of the solvent. To obtain a dry product with homogeneous fiber morphology, the needle-tip-to-collector distance needs to be chosen appropriately to allow for full solvent evaporation (Cramariuc et al. 2013). Furthermore, a direct dependence of the fiber diameter on the needle-to-collector distance was revealed. By an increase from 10 to 15 cm, the average diameter of the fibers increased from 174 ± 11 to 370 ± 7 nm and 500 ± 12 to 640 ± 11 nm, respectively.

In the case of PI-3 at 12 wt%, a mixture of beads and fibers was obtained with an average fiber diameter of 157 ± 52 nm at 10 cm (Figure 16.4a) and 293 ± 72 nm when the distance was increased to 15 cm. When the concentration was increased to 15 wt%, more fibers appeared; in this case, the average fiber diameter increased to 351 ± 114 nm (Figure 16.4b). For PI-3 at 20 wt%, the shape of the beads changed from spindle-like to more uniform fibers with an average diameter of 550 ± 115 nm (Figure 16.4c). As the concentration was further increased to 25 wt%, uniform fibers also were obtained with an increased average diameter of 1049 ± 280 nm (Figure 16.4d). It can be concluded that the solution concentration is a key factor in determining the resulting nanofiber diameters. At concentrations higher than 25 wt%, the formation of continuous fibers was impossible to produce because of the inability to maintain a constant flow of the solution at the tip of the needle.

With increasing concentration of the polyimide solution, the viscosity increased significantly (Figure 16.3), leading to the transformation of sample morphology from spherical beads to homogeneous fibers (Figure 16.4). The SEM micrographs of PI-3 at 23 wt% spun at a constant voltage of +10/−2 kV, a feed rate

FIGURE 16.4 SEM morphology of electrospun PI-3 fibers at different concentrations: (a) 12 wt%, (b) 15 wt%, (c) 20 wt%, and (d) 23 wt%.

of 5 μL/min, and a distance of 10 cm (a) and 15 cm (b) are shown in Figure 16.5. Thereby, a decrease in the average fiber diameter for the samples electrospun at 15 cm is observed—the longer distance probably led to a higher drawing ratio of the wet jet.

The best conditions for obtaining PI fibers with small diameters and narrow size distribution without beads were PI-1: 10 wt%, 0.8 mm gauge needle, +15 kV/−2 kV, 10 cm spinning distance at 20 μL/min feed rate (Butnaru et al. 2015; Serbezeanu et al. 2015a, 2015b); PI-2: 15 wt%, 0.8 mm gauge needle, +10 kV/−2 kV, 15 cm spinning distance at 5 μL/min feed rate (Serbezeanu et al. 2015b); and PI-3: 23 wt%, +10 kV/−2 kV voltage, 10 cm spinning distance at 5 μL/min feed rate. Using these parameters, the average fiber diameters for PI-1, PI-2, and PI-3 were equal with 444 ± 57, 640 ± 11, and 730 ± 89 nm, respectively.

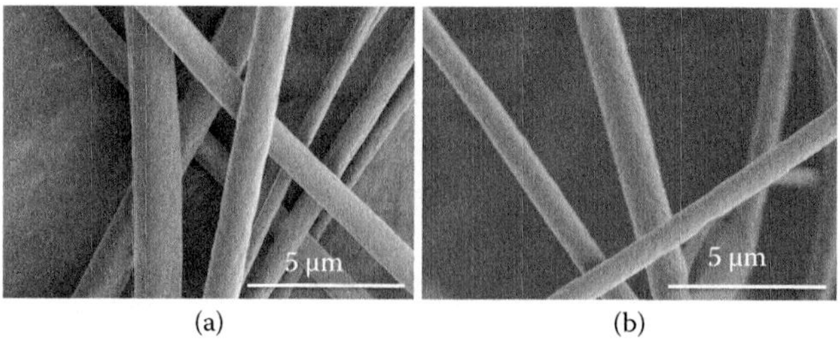

FIGURE 16.5 SEM morphology of electrospun PI-3-23 wt% fibers at (a) 10 cm and (b) 15 cm spinning distance.

16.3 PREPARATION OF GOLD NANOPARTICLES/POLYIMIDE NANOFIBERS BY ELECTROSPINNING TECHNIQUE

The fabrication of AuNPs/PI is rarely reported in the literature and even less often for PI hybrids filled with magnetic nanoparticles. However, until recently, no work has explored the application of PI by the electrospinning method for producing AuNPs/PI nanofibers. An interesting alternative application for the obtained PI-nanofibers that were explored in the first part of this chapter is their use as templates for the synthesis of gold nanoparticles. Because the PI substrate displays a very high thermal stability, it allows the thermally assisted reduction of gold salts without any induced fiber deformation. Such metal–polymer composites are interesting for sensing and catalytic applications (Alex and Tiwari 2015; Kumar et al. 2015).

Polyimide fibers incorporating the gold salt ($HAuCl_4 \cdot 3H_2O$) were produced by solution electrospinning technique. $HAuCl_4$ blended with PI was electrospun from its DMF solution to form fibers with 0.5–3 wt% $HAuCl_4$ within the fibers. Two different techniques were used in order to reduce the Au^{3+} to Au^0. One of them was thermal treatment (200°C for 1, 3, and 6 h) and the other was chemical reduction by the use of ascorbic acid.

Upon the addition of $HAuCl_4$ to the PI-2 solutions, the optical properties of the latter were drastically modified—the color changed from yellowish-orange to bright red. The ultraviolet–visible (UV–Vis) spectra of the gold precursor-containing solutions displayed an absorption band centered at around 520 nm, which is characteristic of metallic gold nanoparticles. Sawada et al. (2014) have previously shown that $HAuCl_4$ undergoes a partial spontaneous disproportionation in DMF solution at room temperature and in the presence of stabilizing agents, so that colloidal gold particles are formed. It can be assumed that the stabilization and subsequent growth of the nascent gold particles occurs due to the interaction of the pendant phenol groups of the polyimide PI-2 with the precursors yielded from $HAuCl_4$ disproportionation. However, in the case of PI-1 solution containing gold salt, no changes in the optical properties of the solutions were observed (Serbezeanu et al. 2015b).

SEM measurements were carried out to study the morphologies of the PI-2-15 wt%/AuNP hybrid fibers. The samples were observed by SEM without a conductive coating to avoid the overlapping of the metal layer on the nanoparticles. The viscosities, average fiber diameter, and the average diameters of the gold nanoparticles are listed in Table 16.2. The addition of a small amount of $HAuCl_4$ to the polymer solution was found to decrease the average fiber diameter of the PI nanofibers. Also, the viscosity of the polymer solution decreased with the addition of $HAuCl_4$. Figure 16.6 shows the shear rate dependency of viscosity for the PI-2 solution containing different amounts of gold salt. The shear stress increases almost linearly with the increase of shear rate, indicating a Newtonian fluid behavior. The low viscous solution leads to the formation of beaded fibers. The viscosity decreases gradually from 0.08 to 0.03 Pa · s as the gold salt loading increases from 0.5 to 3 wt%. It can be assumed that the nanoparticles induced a reduction of the PI-2 polymer chain interactions leading to less shear resistance and thus lower viscosities.

TABLE 16.2
$HAuCl_4$ Content, Apparent Viscosity, and Average Fiber Diameters for PI-2-15 wt% and PI-2-15 wt%/AuNP

Sample	$HAuCl_4$ (wt%)	Zero Shear Viscosity (Pa · s)	Average Fiber Diameter (nm)	Gold Nanoparticles (nm)	
				1 h	6 h
PI-2-15 wt%	0	0.52	401 ± 51	–	–
PI-2-15 wt%/Au^{3+} 0.5 wt%	0.5	0.08	58 ± 8	9.8 ± 7	10.3 ± 9
PI-2-15 wt%/Au^{3+} 1 wt%	1	0.06	50 ± 6	13.5 ± 12	22.1 ± 16
PI-2-15 wt%/Au^{3+} 3 wt%	3	0.03	56 ± 6	19.2 ± 8	19.5 ± 16

Source: Adapted from Serbezeanu D., et al. *Eur. Polym. J.*, 64, 10–20, 2015b.

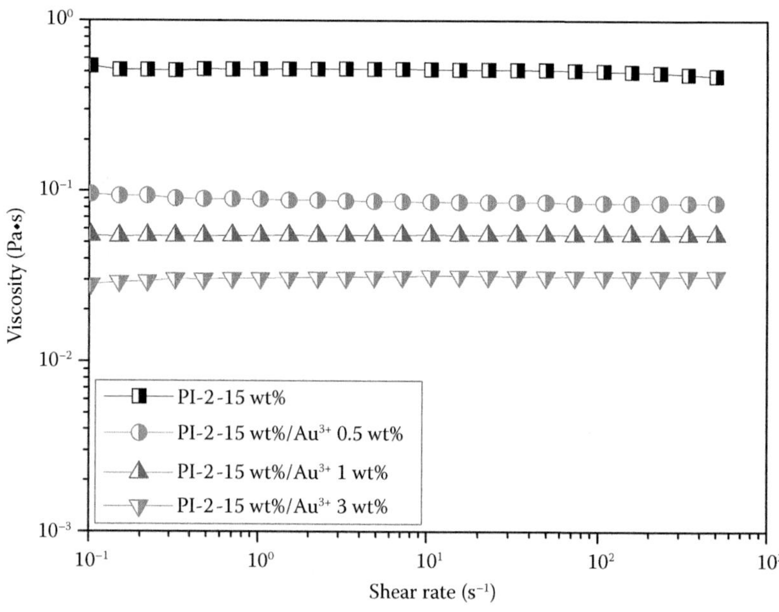

FIGURE 16.6 Shear rate versus viscosity for the PI-2-15 wt% solution with different contents of Au^{3+}. (Reprinted from *Eur. Polym. J.*, 64, Serbezeanu D., et al., Design and synthesis of polyimide—Gold nanofibers with tunable optical properties, 10–20, Copyright (2015), with permission from Elsevier.)

Some typical SEM micrographs of the PI-2-15 wt%/Au^{3+} 1 wt% and PI-2-15 wt%/Au^0 1 wt% fibers are shown in Figure 16.7. The fiber surface of PI-2-15 wt%/Au^{3+} appears smooth and uniform, indicating a good dispersion of the gold salt within the PI fibers. After thermal reduction at 200°C for 6 h, the AuNPs show spherical shapes. From SEM micrograph, the average diameter of the PI-2-15 wt%/AuNPs fibers was 60 nm. The fiber diameter or fiber arrangements were not affected

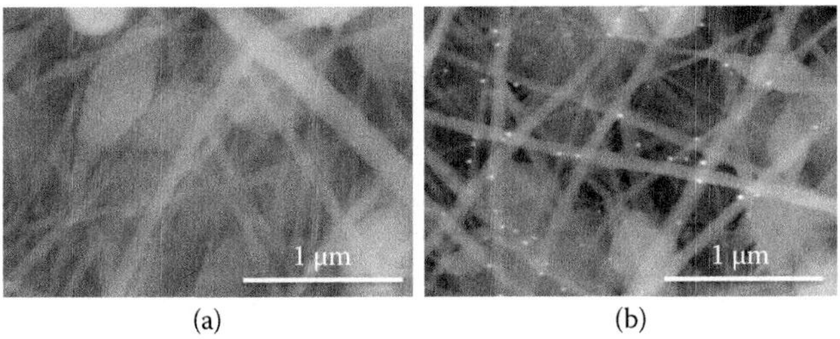

FIGURE 16.7 SEM micrographs of electrospun (a) PI-2-15 wt%/Au^{3+} 1 wt% at room temperature and (b) PI-2-15 wt%/Au^0 1 wt% after being thermally treated at 200°C for 6 h. (Adapted from Serbezeanu D., et al. *Eur. Polym. J.*, 64, 10–20, 2015b.)

by the exposure to temperature. The average diameters of the gold nanoparticles on the PI-2-15 wt% fibers electrospun with 0.5, 1, and 3 wt% of $HAuCl_4$ were estimated to be in the range of 9–19 nm. The AuNPs were mostly homogeneously dispersed into the PI-2-15 wt% matrix.

The average fiber diameters of the PI-2-15 wt%/AuNPs remained constant with increasing Au concentration, whereas the viscosity decreased with the increase of Au^{3+} content added to the PI-1-15 wt% (Table 16.2). The gold salt added to the PI solution probably hindered the electrostatic repulsive forces along the polymer backbone and enabled the PI to adopt a flexible coil-like conformation, ultimately leading to a systematically decreased viscosity with the $HAuCl_4$ addition.

The PI-3 fibers containing gold salt (PI-3-23 wt%/Au^{3+} 3 wt%) were electrospun at best conditions obtained from electrospinning tested for the pure solutions: +15 kV/–2 kV voltage, 10 cm spinning distance at 1 μL/min feed rate. The average fiber diameter of pure PI-3-23 wt% was 730 nm; the addition of gold nanoparticles into the 23 wt% PI-3 solution resulted in improved fiber morphology with a decreased average diameter 380 nm for 3 wt% $HAuCl_4$ (Figure 16.8). The addition of the $HAuCl_4$ to the spinning solution might lead to a higher charge density in the ejected jet, increasing the columbic repulsive forces which leads to a higher drawing ratio and thus thinner fibers (Zong et al. 2002; Tombe et al. 2013). The reduction of gold salt in the metallic gold was performed using two different modes, thermally (200°C for 6 h) and chemically (ascorbic acid). Figure 16.8 shows the SEM images of PI-3-23 wt%/Au^0 3 wt% nanofibers annealed at 200°C for 6 h in comparison with the fibers treated with ascorbic acid. During annealing, gold ions are reduced to metallic gold. If, in the first case, the gold nanoparticles were almost homogeneously dispersed in the PI-2-15 wt% matrix, in the case of reduction with ascorbic acid, some aggregations of the AuNPs could be observed, as shown in Figure 16.8b.

It can be concluded that uniform gold nanoparticles can be easily generated in PI electrospun fibers only by combining an electrospinning process and a thermal reduction of PI/$HAuCl_4$ fibers.

PI and AuNPs/PI were characterized by X-ray photoelectron spectroscopy (XPS) measurements in order to determine the elements present on the fiber surfaces and the nature of the gold particles. Figure 16.9 shows that the Au/O peak ratio decreased

FIGURE 16.8 SEM images of electrospun (a) PI-3-23 wt%/AuNP 3 wt% thermal treatment at 200°C for 6 h and (b) chemical reduction with ascorbic acid.

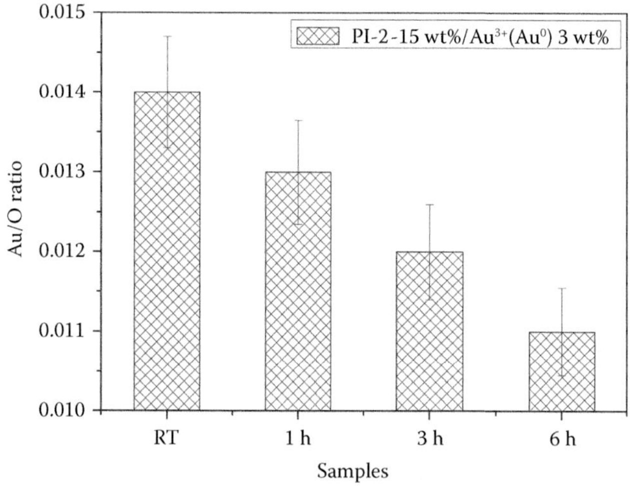

FIGURE 16.9 Au/O peak ratio obtained from XPS measured for PI-2-15 wt%/ Au^{3+} 3 wt% and the PI-2-15 wt%/Au^0 3 wt% treated at 200°C for different periods of time.

from 0.014 to 0.011. It is clear that the oxygen was implicated in the reduction of Au^{3+}, even at a relatively low temperature. The chemical state of gold species in the PI fibers was also examined by XPS. Two emission components were observed at 84 eV ($4f_{5/2}$) and 87.7 eV ($4f_{7/2}$). The peak-to-peak distance measured between the binding energies for the Au4f doublet was ~3.7 eV. No peak or even a shoulder was observed around 84 eV (Au^0 $4f_{5/2}$), which means that Au^+ is not present in the sample that was thermally treated. These results are in agreement with those reported in the literature (Leff et al. 1995; Kumar et al. 2003; Zhang and Srinivasan 2007; Kumar Jena and Retna Raj 2008).

The XPS survey spectra for the PI-3-23 wt% fibers, PI-3-23 wt% fibers containing Au^{3+}, and the PI-3-23 wt% fibers containing Au^0 (thermally and chemically treated) are shown in Figure 16.10. For the sample noted, PI-3-23 wt%/Au^{3+}, the surface gold atomic content was 0.19%. After curing at 200°C for 6 h, the surface gold atomic

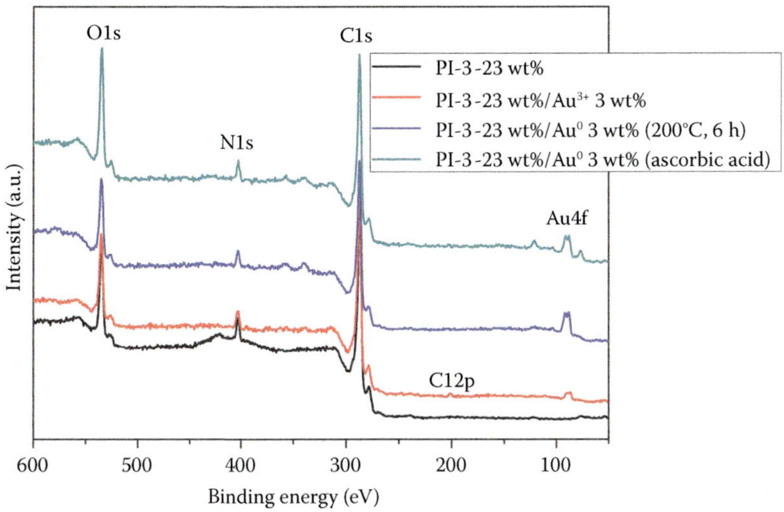

FIGURE 16.10 XPS survey spectra of electrospun PI fibers (PI-3-23 wt%), untreated polyimide fibers containing $HAuCl_4$ (PI-3-23 wt%/Au^{3+} 3 wt%), and polyimide fibers containing gold nanoparticles (PI-3-23 wt%/AuNP 3 wt%) at 200°C for 6 h/ascorbic acid.

content was 0.83%, while the carbon atomic content was still more than 78%. Further chemical treatment induces the decomposition of the polyimide fibers, as demonstrated by the XPS, with a decrease in the atomic concentration of carbon from 79.98 to 74.72%. The appearance in the XPS survey spectra of the PI fibers of Au4f and Cl2p in the case of PI-3 23 wt%/Au^{3+} 3 wt% revealed the presence of salt precursors in the PI-3 matrix. When the samples were heated at 200°C for 6 h, the binding energy characteristic to the Cl2p disappeared, suggesting a successful thermal reduction of gold salt to AuNPs.

The (photo)-catalytic activity of the PI-3 fibers containing gold nanoparticles was investigated using the formaldehyde degradation process at room temperature (Ritter et al. 2010; Michalow et al. 2013). The principle of this analysis was designed to determine the concentration of formaldehyde after 24 h of treatment and to compare the differences in concentrations with the initially added amounts of formaldehyde, both for pure PI and PI/Au meshes. Figure 16.11 shows the catalytic degradation profiles of formaldehyde for the PI-3-23 wt% fibers and the PI-3-23 wt% fibers containing gold nanoparticles. The highest decomposition rates of formaldehyde were about 76% (dark)/75% (light) for PI-3-23 wt% fibers containing gold nanoparticles, while the degradation rate for the PI-3-23 wt% fibers reached 46% (dark)/41% (light), respectively. No differences were found for the Au-containing fibers with respect to the degradation rate from the illuminated and the dark exposure. Therefore, it can be assumed that the degradation was driven by a pure heterogeneous catalytic reaction mechanism. It is well known that the AuNP exhibits high catalytic activities for certain chemical reactions (Alvaro et al. 2010; Ma and Dai 2011; Sarina et al. 2013; Zhang et al. 2013; Liu et al. 2014a, 2014b).

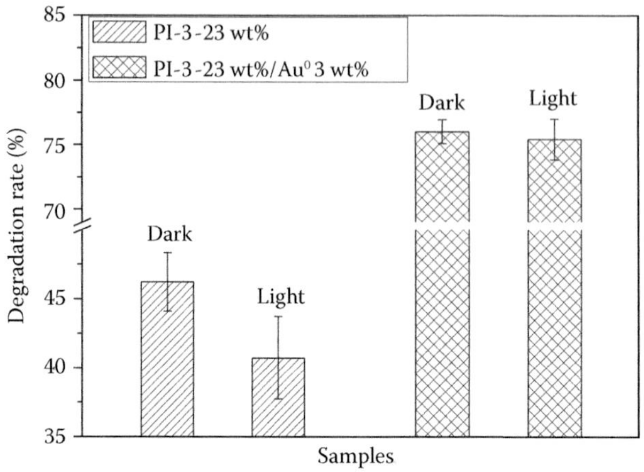

FIGURE 16.11 Degradation rate of formaldehyde after 24 h for PI-3-23 wt% wt% and PI-3-23 wt%/Au^0 3 wt% in the dark and exposed to light.

16.4 CONCLUSION

This study focused on the generation of gold nanoparticles on a benzophenone-3,3′,4,4′-tetracarboxylic dianhydride or 4,4′-oxydiphthalic anhydride/4,4′-diamino-4″-hydroxy triphenylmethane-based polyimide matrix via electrospinning technique and using a gold salt, gold(III) chloride hydrate, as a precursor. The synthesis of gold nanoparticles in polyimide solutions and the successful processing of these solutions into composite nanofibers with interesting optical properties were investigated. The nanofiber diameter for pure PIs ranged from 174 ± 11 to 1049 ± 280 nm and increased with polymer solution concentration and the needle-tip-to-collector distance. The resulting PI fibers containing gold nanoparticles were found to exhibit remarkable catalytic performance with respect to the decomposition of formaldehyde. Also, the PI nanofibers stabilized the nanoparticles and prevented aggregation. Therefore, the incorporation of AuNPs into PI by electrostatic spinning procedures offers a straightforward way to obtain new nanoscaled hybrids incorporating promising properties—those that could be used, for example, in heterogeneous catalysis.

ACKNOWLEDGMENTS

The authors acknowledge the financial support of CNCSIS–UEFISCSU, Project no. PN-II-RU-TE-0123 nr. 28/29.04.2013, and SCIEX Program, Grant no. 5211.00737.

REFERENCES

Aflori M., Serbezeanu D., Carja I.-D., and Fortunato G. Gold nanoparticles incorporated into electrospun polyimide fibers. *Chem. Lett.* 44 no. 10 (2015): 1440–1442.

Alex S., and Tiwari A. Functionalized gold nanoparticles: Synthesis, properties and applications; A review. *J. Nanosci. Nanotechnol.* 15 no. 3 (2015): 1869–1894.

Alvaro M., Cojocaru B., Ismail A. A., et al. Visible-light photocatalytic activity of gold nanoparticles supported on template-synthesized mesoporous titania for the decontamination of the chemical warfare agent Soman. *Appl. Catal. B.* 99 no. 1–2 (2010): 191–197.

Bauld R., Hesari M., Workentin M. S., and Fanchini G. Thermal stability of Au25—Molecular precursors and nucleation of gold nanoparticles in thermosetting polyimide thin films. *Appl. Phys. Lett.* 101 no. 24 (2012): 243114.

Bhardwaj N., and Kundu S. C. Electrospinning: A fascinating fiber fabrication technique. *Biotechnol. Adv.* 28 no. 3 (2010): 325–347.

Butnaru I., Serbezeanu D., Bruma M., Sava I., Gaan S., and Fortunato G. Physical and thermal properties of poly(ethylene terephthalate) fabric coated with electrospun polyimide fibers. *High Perform. Polym.* 27 no. 5 (2015): 616–624.

Carja I.-D., Serbezeanu D., Vlad-Bubulac T., Hamciuc C., and Brumă M. Synthesis and liquid crystalline behavior of phosphorus-containing aliphatic–aromatic copoly(ester-imide) s. *Polym. Bull.* 68 no. 7 (2012): 1921–1934.

Chen D., Wang R., Tjiu W. W., and Liu T. High performance polyimide composite films prepared by homogeneity reinforcement of electrospun nanofibers. *Compos. Sci. Technol.* 71 no. 13 (2011): 1556–1562.

Chen F., Bera D., Banerjee S., and Agarwal S. Low dielectric constant polyimide nanomats by electrospinning. *Polym. Adv. Technol.* 23 no. 6 (2012): 951–957.

Chisca S., Barzic A. I., Sava I., Olaru N., and Bruma M. Morphological and rheological insights on polyimide chain entanglements for electrospinning produced fibers. *J. Phys. Chem. B.* 116 no. 30 (2012): 9082–9088.

Cramariuc B., Cramariuc R., Scarlet R., Manea L. R., Lupu I. G., and Cramariuc O. Fiber diameter in electrospinning process. *J. ELECTROSTAT.* 71 no. 3 (2013): 189–198.

Daniel M.-C., and Astruc D. Gold nanoparticles: Assembly, supramolecular chemistry, quantum-size-related properties, and applications toward biology, catalysis, and nanotechnology. *Chem. Rev.* 104 no. 1 (2004): 293–346.

Deniz A. E., Vural H. A., Ortaç B., and Uyar T. Gold nanoparticle/polymer nanofibrous composites by laser ablation and electrospinning. *Mater. Lett.* 65 no. 19–20 (2011): 2941–2943.

Ding B., Kim C., Kim H., Seo M., and Park S. Titanium dioxide nanofibers prepared by using electrospinning method. *Fiber Polym.* 5 no. 2 (2004): 105–109.

Fukushima S., Karube Y., and Kawakami H. Preparation of ultrafine uniform electrospun polyimide nanofiber. *Polym J.* 42 no. 6 (2010): 514–518.

Ghosh A., Mistri E. A., and Banerjee S. 3—Fluorinated polyimides: Synthesis, properties, and applications. In: Banerjee S (ed). *Handbook of Specialty Fluorinated Polymers*, pp. 97–185. Chadds Ford, Pennsylvania, USA: William Andrew Publishing, 2015.

Gosh M. K., and Mittal K. L. *Polyimides; Fundamentals and Applications.* New York: Marcel Dekker, 1996.

Guo B., Zhao S., Han G., and Zhang L. Continuous thin gold films electroless deposited on fibrous mats of polyacrylonitrile and their electrocatalytic activity towards the oxidation of methanol. *Electrochim. Acta.* 53 no. 16 (2008): 5174–5179.

Hamciuc C., Carja I.-D., Hamciuc E., Vlad-Bubulac T., and Ignat M. Phthalonitrile-containing aromatic polyimide thin films with nano-actuation properties. *Polym. Adv. Technol.* 24 no. 2 (2013): 258–265.

Hamciuc C., Hamciuc E., Serbezeanu D., and Vlad-Bubulac T. Thermal and optical properties of some phosphorus-containing poly(1,3,4-oxadiazole-ester-imide)s. *Polym. Adv. Technol.* 22 no. 12 (2011): 2458–2468.

Herves P., Perez-Lorenzo M., Liz-Marzan L. M., Dzubiella J., Lu Y., and Ballauff M. Catalysis by metallic nanoparticles in aqueous solution: Model reactions. *Chem. Soc. Rev.* 41 no. 17 (2012): 5577–5587.

Hou G., Zhu L., Chen D., and Jiang M. Core–shell reversion of hybrid polymeric micelles containing gold nanoparticles in the core. *Macromolecules.* 40 no. 6 (2007): 2134–2140.

Huang C., Chen S., Reneker D. H., Lai C., and Hou H. High-strength mats from electrospun poly(p-phenylene biphenyltetracarboximide) nanofibers. *Adv. Mater.* 18 no. 5 (2006a): 668–671.

Huang C., Wang S., Zhang H., et al. High strength electrospun polymer nanofibers made from BPDA–PDA polyimide. *Eur. Polym. J.* 42 no. 5 (2006b): 1099–1104.

Karube Y., and Kawakami H. Fabrication of well-aligned electrospun nanofibrous membrane based on fluorinated polyimide. *Polym. Adv. Technol.* 21 no. 12 (2010): 861–866.

Kim S. D., Lee S., Heo J., Kim S. Y., and Chung I. S. Soluble polyimides with trifluoromethyl pendent groups. *Polymer.* 54 no. 21 (2013): 5648–5654.

Koizumi S., Matsuda S.-I., and Ando S. Synthesis, characterization, and optical properties of uniaxially drawn and gold nanoparticle dispersed fluorinated polyimide films. *J. Photopolym. Sci. Technol.* 15 no. 2 (2002): 231–236.

Kumar A., Mandal S., Pasricha R., Mandale A. B., and Sastry M. Investigation into the Interaction between surface-bound alkylamines and gold nanoparticles. *Langmuir.* 19 no. 15 (2003): 6277–6282.

Kumar S., Ahlawat W., Kumar R., and Dilbaghi N. Graphene, carbon nanotubes, zinc oxide and gold as elite nanomaterials for fabrication of biosensors for healthcare. *Biosens. Bioelectron.* 70 (2015): 498–503.

Kumar Jena B., and Retna Raj C. Morphology dependent electrocatalytic activity of Au nanoparticles. *Electrochem. Commun.* 10 no. 6 (2008): 951–954.

Leff D. V., Ohara P. C., Heath J. R., and Gelbart W. M. Thermodynamic control of gold nanocrystal size: Experiment and theory. *J. Phys. Org. Chem.* 99 no. 18 (1995): 7036–7041.

Li D., and Xia Y. Fabrication of titania nanofibers by electrospinning. *Nano Lett.* 3 no. 4 (2003): 555–560.

Liu J., Hu C., Xu J., Jiang F., and Chen F. Enhanced photocatalytic performance of partially reduced graphene oxide under simulated solar light through loading gold nanoparticles. *Mater. Lett.* 134 (2014a): 134–137.

Liu J., Min Y., Chen J., Zhou H., and Wang C. Preparation of the ultra-low dielectric constant polyimide fiber membranes enabled by electrospinning. *Macromol. Rapid Commun.* 28 no. 2 (2007): 215–219.

Liu S., Tan L., Li X., Fu J., Chronakis I. S., and Ge M. Gold nanoparticles–gelatin hybrid fibers with bright photoluminescence. *Mater. Lett.* 135 no. 0 (2014b): 1–4.

Ma Z., and Dai S. Development of novel supported gold catalysts: A materials perspective. *Nano Research.* 4 no. 1 (2011): 3–32.

Ma Z., Masaya K., and Ramakrishna S. Immobilization of cibacron blue F3GA on electrospun polysulphone ultra-fine fiber surfaces towards developing an affinity membrane for albumin adsorption. *J. Membrane Science.* 282 no. 1–2 (2006): 237–244.

McKee M. G., Hunley M. T., Layman J. M., and Long T. E. Solution rheological behavior and electrospinning of cationic polyelectrolytes. *Macromolecules.* 39 no. 2 (2005): 575–583.

Mendoza C., Pietsch T., Gutmann J. S., Jehnichen D., Gindy N., and Fahmi A. Block copolymers with gold nanoparticles: Correlation between structural characteristics and mechanical properties. *Macromolecules.* 42 no. 4 (2009): 1203–1211.

Meyer H. M., Anderson S. G., Atanasoska L., and Weaver J. H. Summary abstract: X-ray photoemission investigation of electron injection and trapping at the gold/polyimide interface. *J. Vac. Sci. Technol., A.* 6 no. 3 (1988): 1002–1004.

Michalow K. A., Otal E. H., Burnat D., et al. Flame-made visible light active TiO2:Cr photocatalysts: Correlation between structural, optical and photocatalytic properties. *Catal. Today.* 209 (2013): 47–53.

Min B.-M., Lee G., Kim S. H., Nam Y. S., Lee T. S., and Park W. H. Electrospinning of silk fibroin nanofibers and its effect on the adhesion and spreading of normal human keratinocytes and fibroblasts in vitro. *Biomaterials*. 25 no. 7–8 (2004): 1289–1297.

Nath S., Jana S., Pradhan M., and Pal T. Ligand-stabilized metal nanoparticles in organic solvent. *J. Colloid Interface Sci.* 341 no. 2 (2010): 333–352.

O'Neal D. P., Hirsch L. R., Halas N. J., Payne J. D., and West J. L. Photo-thermal tumor ablation in mice using near infrared-absorbing nanoparticles. *Cancer Letters*. 209 no. 2 (2004): 171–176.

Ouyang Z., Li J., Wang J., et al. Fabrication, characterization and sensor application of electrospun polyurethane nanofibers filled with carbon nanotubes and silver nanoparticles. *J. Mater. Chem. B*. 1 no. 18 (2013): 2415–2424.

Park J.-Y., and Lee I.-H. Characterization and morphology of prepared titanium dioxide nanofibers by electrospinning. *J. Nanosci. Nanotechnology*. 10 no. 5 (2010): 3402–3405.

Polte J., Ahner T. T., Delissen F., et al. Mechanism of gold nanoparticle formation in the classical citrate synthesis method derived from coupled In situ XANES and SAXS evaluation. *J. Am. Chem. Soc.* 132 no. 4 (2010): 1296–1301.

Ramakrishna S., Fujihara K., Teo W., Lim T., and Ma Z. *An Introduction to Electrospinning and Nanofibers*. Singapore: World Scientific Publishing Company, 2005.

Ritter A., Reifler F. A., and Michel E. Quick screening method for the photocatalytic activity of textile fibers and fabrics. *Text. Res. J.* 80 no. 7 (2010): 604–610.

Sakai T., and Alexandridis P. Single-step synthesis and stabilization of metal nanoparticles in aqueous pluronic block copolymer solutions at ambient temperature. *Langmuir*. 20 no. 20 (2004): 8426–8430.

Sarina S., Waclawik E. R., and Zhu H. Photocatalysis on supported gold and silver nanoparticles under ultraviolet and visible light irradiation. *Green Chemistry*. 15 no. 7 (2013): 1814–1833.

Sava I., Chisca S., Wolinska-Grabczyk A., et al. Synthesis and thermal, mechanical and gas permeation properties of aromatic polyimides containing different linkage groups. *Polym. Int.* 64 no. 1 (2015): 154–164.

Sawada K., Sakai S., and Taya M. Polyacrylonitrile-based electrospun nanofibers carrying gold nanoparticles in situ formed by photochemical assembly. *J. Mater. Sci.* 49 no. 13 (2014): 4595–4600.

Schiffman J. D., and Elimelech M. Antibacterial activity of electrospun polymer mats with incorporated narrow diameter single-walled carbon nanotubes. *ACS Appl. Mater. Interfaces*. 3 no. 2 (2011): 462–468.

Serbezeanu D., Popa A. M., Sava I., et al. Design and synthesis of polyimide—Gold nanofibers with tunable optical properties. *Eur. Polym. J.* 64 (2015b): 10–20.

Serbezeanu D., Popa A.-M., Stelzig T., Sava I., Rossi M. R., and Fortunato G. Preparation and characterization of thermally stable polyimide membranes by electrospinning for protective clothing applications. *Text. Research J.* 85 no. 17 (2015a): 1763–1775.

Shang Y., Fan L., Yang S., and Xie X. Synthesis and characterization of novel fluorinated polyimides derived from 4-phenyl-2,6-bis[4-(4′-amino-2′-trifluoromethyl-phenoxy) phenyl]pyridine and dianhydrides. *Eur. Polym. J.* 42 no. 5 (2006): 981–989.

Tang K., Yu Y., Mu X., van Aken P. A., and Maier J. Multichannel hollow TiO_2 nanofibers fabricated by single-nozzle electrospinning and their application for fast lithium storage. *Electrochem. Commun.* 28 (2013): 54–57.

Tombe S., Antunes E., and Nyokong T. Electrospun fibers functionalized with phthalocyanine-gold nanoparticle conjugates for photocatalytic applications. *J. Mol. Catal. A: Chem.* 371 (2013): 125–134.

Travaly Y., Zhang L., Zhao Y., et al. Nucleation, growth, and aggregation of gold on polyimide surfaces. *J. Mater. Res.* 14 no. 9 (1999): 3673–3683.

Vanherck K., Vankelecom I., and Verbiest T. Improving fluxes of polyimide membranes containing gold nanoparticles by photothermal heating. *J. Membrane Science*. 373 no. 1–2 (2011): 5–13.

Wang H.-S., Fu G.-D., and Li X.-S. Functional polymeric nanofibers from electrospinning. *Recent Patents on Nanotechnology*. 3 no. 1 (2009): 21–31.

Wang L., Zhao Z., Li J., and Chen C. Synthesis and characterization of fluorinated polyimides for pervaporation of n-heptane/thiophene mixtures. *Eur. Polym. J.* 42 no. 6 (2006): 1266–1272.

Weiser E. S., Johnson T. F., St Clair T. L., Echigo Y., Kaneshiro H., and Grimsley B. W. Polyimide foams for aerospace vehicles. *High Perform. Polym.* 12 no. 1 (2000): 1–12.

Yokota S., Kitaoka T., Opietnik M., Rosenau T., and Wariishi H. Synthesis of gold nanoparticles for in situ conjugation with structural carbohydrates. *Angew. Chem. Int. Ed.* 47 no. 51 (2008): 9866–9869.

Zhang F., and Srinivasan M. P. Multilayered gold-nanoparticle/polyimide composite thin film through layer-by-layer assembly. *Langmuir.* 23 no. 20 (2007): 10102–10108.

Zhang J., Chen G., Chaker M., Rosei F., and Ma D. Gold nanoparticle decorated ceria nanotubes with significantly high catalytic activity for the reduction of nitrophenol and mechanism study. *Appl. Catal., B.* 132–133 (2013): 107–115.

Zhu H., Du M., Zou M., Xu C., Li N., and Fu Y. Facile and green synthesis of well-dispersed Au nanoparticles in PAN nanofibers by tea polyphenols. *J. Mater. Chem.* 22 no. 18 (2012): 9301–9307.

Zong X., Kim K., Fang D., Ran S., Hsiao B. S., and Chu B. Structure and process relationship of electrospun bioabsorbable nanofiber membranes. *Polymer.* 43 no. 16 (2002): 4403–4412.

17 Electrochemical Inclusion of Metallic Clusters in Organic Polymers

Raluca Marinica Albu

CONTENTS

17.1 INTRODUCTION

The theoretical and experimental research developed in the last century, which was focused on the obtaining and characterization of nanostructure systems, witnessed significant progress. The discovery of new materials, processes, and phenomena produced at a nanoscale size as well as the unprecedented development of novel experimental techniques offer new opportunities for the utilization of nanosystems and nanostructured materials in numerous important applications. In addition, the investigations devoted to nanostructured and nanocomposite materials have aroused a great interest for different current and emerging technologies, due to their special electrochemical properties (Karttunen et al. 2008). The complementary properties of the nanometer-sized particles, combined with those of the host material, permit the manufacture of multifunctional materials used for various applications in optics, food packaging, or medicine (Emamifar et al. 2011). In addition, nanocomposites are also applied in biological systems.

For example, nacre ("mother of pearl") is a composite material produced naturally by some mollusks in their inner shell layer. In spite of its fragile structure, nacre exhibits a 1000 times higher resistance and greater hardness than that of its constituents (Barthelat et al. 2006). The presence of such unusual mechanical properties in nacre is used in the manufacture of decorative products and as an inspiration for the development of composite materials that imitate microscopic structure.

As an example, obtaining of wide range of supramolecular structures, multifunctional super molecules, carbon nanotubes, and metal semiconductor nanoparticles is essential for the development of electro-optical devices (Ambrogio et al. 2011; Bhosale et al. 2006). Moreover, metals and metal oxides are attractive for nanoscale systems due to the easiness with which metal particles can be realized and adjusted chemically according to the demands of the applications. Metal colloids, known as metal nanoparticles (MNPs), are of special technological importance, being used in a variety of applications (biological assays, sensors, paints, ceramics, and photonic crystals) due to their unique electrical, optical, magnetic, and chemical properties, which act as bridges between the bulk materials and the single molecular or atomic structures. The physicochemical properties of metal colloids differ significantly from those of free metal atoms and of bulk metals, recommending them for numerous investigations. Therefore, metal colloids may be viewed as a transition regime between two physical states with particular properties (Giannazzo et al. 2011; Macanás et al. 2011). MNPs can be prepared by synthetic methods, such as physical, chemical, or electrochemical, or they can be obtained only in the presence of stabilizers, for avoiding the agglomeration of nanoparticles (Yang et al. 2005).

The main role of stabilizers (surfactants or polymers) is to provide a steric or an electrostatic barrier between particles, thereby preventing inhibition of aggregation. Furthermore, stabilizers play an essential role in the control of both size and shape of nanoparticles. Generally, polymers are recommended as stabilizers for metal colloids due to their transparent, permeable, and nonconductive properties and also because they do not influence the optical, electrical, and catalytic properties of the nanoparticles. In addition, investigation of polymer-stabilized MNPs appears as a suitable way for solving the stability of MNPs. For this reason, great attention has been focused on the incorporation of MNPs into a polymer matrix, a procedure based on the synthesis of nanometer-sized metallic filler particles (Giannazzo et al. 2011).

These new systems represent a new type of catalyst that also may be used in various organic syntheses and in the development of polymer materials with special properties (Pomogailo 1994; Ma et al. 2010). The new functional materials obtained by inclusion of metallic clusters into a polymer matrix possess properties such as electronic conductivity (Karttunen et al. 2008), particle surface plasmon resonance (Kreibig and Vollmer 1995), magnetism (Murray et al. 2001), and granular giant magnetoresistance (Berkowitz et al. 1992). Improvement of catalytic activity is the result of the large effective surface of nanoparticles and the notable contribution of the surface energy to the chemical potential (Okumura et al. 2007). Therefore, the quantum size effect of semiconductor and metal clusters in the nanometer-sized domain imparts to them a number of interesting optical, electronic, and chemical properties (Gomez-Romero 2001; Wolf 2001).

The noble metal nanoparticles, especially gold nanoparticles, are extensively investigated for their utilization in science, technology, medicine, engineering, or decorating arts. This classical type of metal nanoparticle contains *nanocrystals* (particles with dimensions greater than 100 nm, considered as "larger nanoparticles") and *nanoclusters* (much smaller particles with a diameter below 2 nm, the so-called "molecular level") (Daniel and Astruc 2004; Jin 2010). Consequently, with the emergence of nanotechnology, gold nanoparticles have attracted a special research interest due to their extraordinary stability, tunable surface functionality

(Yang et al. 2005), and size-dependent properties. Thus, the combination of the complementary properties of metal nanoparticles with those of the polymer host materials provides new alternatives for obtaining multiphase nanostructured systems.

17.2 NANOMATERIALS AND THE GENERAL PROPERTIES OF METAL NANOPARTICLES

The most commonly used definition for a nanomaterial is "a material that has a structure in which at least one of its phases has a nanometer size in at least one dimension" (Astruc 2008). From this point of view, nanomaterials may be classified into three main categories:

1. One-dimensional nanometer-sized materials (e.g., thin films)
2. Two-dimensional nanometer-sized materials (e.g., nanowires, nanorods, and nanotubes)
3. Three-dimensional nanometer-sized materials (e.g., nanoparticles, nanoclusters, and nanocrystals) (Domènech et al. 2012)

The main categories of nanomaterials are as follows:

- Porous materials—with porous sizes in the nanometer range
- Polycrystalline materials—with nanometer-sized crystallites
- Materials with surface protrusions separated by nanometric distances or nanometer-sized metallic clusters

Among them, the most attractive—scientifically and technologically—are the metal nanoparticles (MNPs), due to their specific electrical, optical, magnetic, and chemical properties. A *nanocluster* may be considered a small measurement unit of a few nanometers, which contains a finite number of atoms. The metal particles, with sizes in the range of 1–50 nm, are realized as sols through dispersion of a solid into a liquid. Metal sols can be used as colorants and catalysts, because they possess amazing colors. In 1857, Faraday (1857) determined that such stains consist of tiny metal parts. The latest methods of synthesis allow the obtaining of sols of metals that can be dried and redissolved exactly as water colors. Because nanosized clusters exhibit a significant tendency to remain single crystalline, they are also called *nanocrystals*. Moreover, nanocrystals have a high surface area induced by the large fraction of atoms present on their surface, and they possess the ability to form one-, two-, and three-dimensional mesoscopic organizations (Reboul et al. 2012).

Reduction in the scale of bulk materials, especially metals, to a nanometric size has resulted in some new and important properties due to the size-dependent effects. These include

- An increase in the surface-to-volume ratio, leading to an increase in the total surface area and in the fraction of entities (e.g., atoms) from the surface of the material

- Modifications in the electronic structure of the entities forming nanoparticles and in the nanoparticles as a whole
- Changes in the associations (e.g., interatomic distances) of the entities forming the nanoparticle and in the occurrence of defects
- Confinement and quantic effects (due to the confinement of charge carriers in a particle with sizes comparable to the wavelength of the electron)

Moreover, composites made of nanometals with polymers are of special interest for functional applications.

The *confinement effects of charge carriers* (Koole et al. 2014) appear in nanosize metal (mesoscopic metals) domains, where the conduction of electrons moves within a very small space, comparable to their de Broglie wavelength ($\lambda = h/p$). Accordingly, their states are quantized in the same way as atoms, and these systems are called "artificial atoms." Therefore, the electrons are positioned among the atomic nuclei, behaving as a typical molecule, and the electronic energy levels are not continuous, such as in bulk. Instead, they are discrete (Klabunde et al. 2001). Thus, nanocrystals are also known as quantum dots that exhibit a size quantization effect in at least one dimension. Similar phenomena occur not only in metals but also in any material when its size decreases.

The *surface effects* occur when the size of bulk materials is reduced to a nanometer scale, so that they come to possess more and more surface atoms instead of inner atoms. In this case, the properties of the new material are influenced by the characteristics of surface atoms (Issa et al. 2013). When the size of the newly obtained material decreases, the atoms on the surface of a bulky crystalline solid, which are located on the basal planes, are all transformed into edge and corner atoms. In this context, the surface properties of nanosized metals differ from those of the bulky metals. Compared with the atoms from the basal planes, the edge and corner atoms have a very low coordination number, which makes them highly chemically reactive, supercatalytically active, and highly polarizable. In addition, the quantic effects cause particular new properties, induced by

- The size-dependent ferromagnetism and superparamagnetism that characterizes all metals (including diamagnetic metals, such as silver)
- The chromatism observed with silver, gold, and copper metals that is due to plasmon absorption
- Photo- and thermoluminescence
- The supercatalytic effect (hyperfine catalysts)

Therefore, reducing the size of metals causes disappearance of the band structure, transforming them into

- Thermal and electrical insulators.
- Highly chemically reactive. (The heterogeneous reactions become stoichiometric and new reaction schemes are possible; for example, the nanosized noble metals are very reactive.)

- Super absorbent.
- Completely different from a thermodynamic point of view. (They melt at much lower temperatures.)

For example, some physical properties of nanomaterials are significantly different from the physical properties of bulk materials, such as color, melting point, mechanical strength, electrical, or optical properties (Lubick and Betts 2008).

The *color* of nanoparticles differs considerably from that of materials from which they originated. In addition, some materials (such as gold materials) become red when their sizes attain the nanometer range. These effects appear due to the high surface electron densities of the small particles that oscillate collectively at a certain frequency. This resonant oscillation of the small monodisperse gold nanoparticles (2–150 nm), known as surface plasmon resonance (SPR) and induced by the electromagnetic field, result in absorption of the blue–green area of the spectrum (450 nm) and reflection of a red light (700 nm) (Lubick and Betts 2008).

The *melting point* decreases considerably when the particle size of materials attains the nanoscale range. Increasing the surface-to-volume ratio affects their thermodynamic and thermal properties. The melting point depression is much lower in the case of nanowires, nanotubes, and nanoparticles than the bulk form of the same materials (Lubick and Betts 2008).

The *mechanical strength* of nanomaterials is one or two times more intense in magnitude than that of conventional counterparts. Moreover, the defects in the form of atomic vacancies can decrease the tensile strength of materials up to 85%. Increases in crystal perfection of nanomaterials enhance mechanical strength (Lubick and Betts 2008).

The *electrical properties* of nanomaterials are also influenced by the reduction of material sizes. Thus, electrical conductivity can be highly altered because nanoparticles improve crystal perfection by reducing defects. Therefore, the electron scattering phenomenon is diminished with the increase of total resistivity. The continuous electronic structure of the band gap becomes wide and discrete when the size of the nanoparticles falls below a critical dimension, that is, the electron de Broglie wavelength (Vollath 2008). Modification of the electronic structure in wide and discrete band gaps appears when the size of nanoparticles becomes smaller than a critical dimension, that is, the electron de Broglie wavelength (Vollath 2008).

The *optical properties* of nanomaterials are also different from those of their bulk counterpart due to the effect of *surface plasmon resonance*. This process appears when the entry free electrons in the conduction band produce an in-phase oscillation, called surface plasmon resonance. In this context, the increase of band gap in semiconductor nanoparticles generates a shift toward shorter wavelengths of the absorption edge. At resonance, light is tightly confined to the surface of the nanostructure until it eventually is absorbed inside the metal or scattered back into photons (Vollath 2008).

The *chemical properties* of materials undergo changes through reduction at the nanosize level. Thus, the exposed surface area of nanoparticles is much larger, while their reactivity increases. In the case of nanoparticles, 50% of the atoms are surface atoms; therefore, the electric transport properties of these particles are no longer dependent on the solid-state bulk phenomenon. Moreover, the surface properties of the nanoparticles and their interactions can be modified or altered by using molecular monolayers. In this way, the unique physicochemical properties of nanosize metals remain unchanged when they are embedded into polymers, which appear as a highly interesting aspect for functional applications (Vollath 2008).

17.3 SYNTHESES AND STABILITY OF THE METAL NANOPARTICLES

From the point of view of their assembly, nanomaterials can be produced through two different routes: *top-down* and *bottom-up* (Scheme 17.1). In the *top-down* method, the macroscopic particles are reduced to the nanoscale by physical techniques and procedures. Typical examples are etching through the mask, ball milling, and application of severe plastic deformation. This method is not preferred because the small particles are not uniform. The *bottom-up* method starts from atoms and then adds in solution or in gas phase (self-assembly process), forming larger uniform nanostructures. During self-assembly, the physical forces operating at nanoscale are used to combine the basic units into larger stable structures. Typical examples are quantum dot formation during epitaxial growth and formation of nanoparticles from colloidal dispersion (Macanás et al. 2011).

Generally, the main synthetic methods used for preparation of MNPs are physical, chemical, and physicochemical (Table 17.1). The chemical methods are generally less costly and do not require equipment or instruments as specific as in the case of physical methods.

Usually, the main techniques for the preparation of metal–polymer nanocomposites can be classified as in situ and ex situ methods. The in situ methods occur in two phases: first, the monomer is polymerized in solution, with metal ions introduced before or after polymerization. Then, metal ions are thermally or chemically reduced in the polymer matrix. In the ex situ procedure, the MNPs are chemically synthesized, and their surface is organically passivated. The resulting nanoparticles are distributed into a polymer solution or liquid monomer that is further polymerized (Guo et al. 2014).

Nevertheless, most of nanosized metals are very instable, a principal drawback that limits the wide range of applications. They can self-aggregate or fuse when they touch each other due to the high surface free energy and can be oxidized or

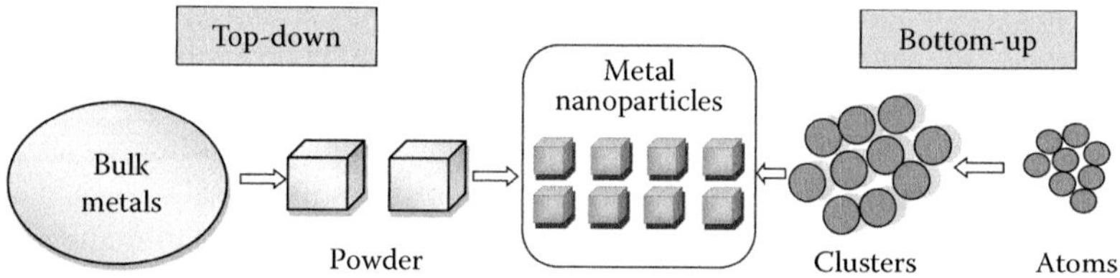

SCHEME 17.1 General representation of top-down and bottom-up approaches in the synthesis of metal nanoparticles. (Adapted from Macanás Jorge et al. Ion exchange-assisted synthesis of polymer stabilized metal nanoparticles. In *Ion exchange and solvent extraction*, pp. 1–44. Boca Raton: CRC Press, 2011.)

TABLE 17.1
The Main Classification of the Conventional Methods for the Preparation of MNPs

Chemical	Physical	Physicochemical
Traditional: impregnation, co-precipitation, precipitation–deposition	Sonochemistry	Flame spray
Microemulsions	Microwave irradiation	Sonoelectrochemistry
Photochemistry	Pulsed laser ablation	Thermal decomposition
Chemical vapor deposition	Supercritical fluids	Cyclic voltammetry
Other: ion exchange/reduction	Plasma	Laser ablation

Source: Adapted from Macanás Jorge et al. Ion exchange-assisted synthesis of polymer stabilized metal nanoparticles. In *Ion exchange and solvent extraction*, pp. 1–44. Boca Raton: CRC Press, 2011.

contaminated by air, moisture, or sulfur dioxide (SO_2) (Spagnoli et al. 2008). In this context, a loss of the nanometric size and of their special properties occurs, creating great difficulties in their manufacturing. Therefore, a good solution for preventing aggregation and stabilizing the MNPs consists of embedding these nanoscopic metals into organic dielectric matrices (Luckham 2004). Polymers are preferred in the embedding phase because they exhibit long-time stability against aggregation and oxidation and possess a wide variety of properties:

- They can be an electrical and thermal insulator or conductor.
- They may have a hydrophobic or hydrophilic nature.
- They can be mechanically hard, plastic, or rubbery.

The methods selected for stabilization of MNPs involve the following steps:

- Prevent particle aggregation
- Avoid their uncontrollable growth
- Control their final shape and size
- Allow particle solubility in various solvents
- Terminate the particle growth reaction

Some studies selected from the literature state that the mechanisms of the nanoparticles' synthesis comprise three steps: nucleation, growth, and termination with a capping agent or ligand (stabilizing agent) that acts through colloidal forces (Van der Waals, electrical double-layer, and steric interactions) (Carotenuto et al. 2005). The stabilizing agents are surfactants such as long-chain thiols, amines, or polymeric ligands—such as polyvinylpyrrolidone (PVP). Therefore, the stabilization of MNPs can be achieved in different ways. For example, the ex situ syntheses of MNPs are performed in a solid or liquid medium through various mechanochemical approaches. Thereafter, premade nanoparticles are directly dispersed into a polymer phase to form composites. During this process, some re-aggregation of the MNPs occurs, restricting the entire stabilization. In an attempt to solve these issues, sonication methods are preferred for dispersion of the nanoparticles in the matrix.

Conversely, the in situ synthesis is a simple and effective route to prepare MNPs that allows fabrication in a one-step procedure from corresponding precursors. In addition, the nanoparticles are grown inside the polymer matrix, leading to the formation of composite materials that can directly serve for a specific purpose. For this reason, in situ approaches are preferred because they prevent particle agglomeration, while maintaining a good spatial distribution in the polymer matrix.

In the stabilization mechanism of nanoparticles inside the polymer, important processes are determined by interaction of nanoparticles with the matrix. These can be explained by considering two processes that work simultaneously in the system and influence one another:

1. The considerable increase in viscosity of the polymer matrix
2. The decrease in the energy of particle–particle interaction in composite systems versus nonstabilized MNP dispersions (Visser 1981)

In the *first* process, the Brownian motion velocity and presence of electrolytes influences the increase in viscosity of the immobilizing media, the coagulation velocity, the domain of attraction forces, and the concentration of colloidal solution. Consequently, from the Smoluchowski equation (Pomogailo and Kestelman 2005), the rate constant of particle coagulation, k_c, is inversely proportional to the viscosity of the media, η:

$$k_c = \frac{8kT}{3\eta} \tag{17.1}$$

where k stands for the Boltzmann constant, and T is the temperature.

In the *second process*, the potential energy of attraction, U_r, between the surfaces of two spherical particles of radius, r, and minimum distance, l_0, can be determined by the equation (17.2):

$$U_r \approx \frac{Ar}{12l_0} \text{ when } r >> l_0 \tag{17.2}$$

where A is the effective Hamaker's constant with dimensions of energy.

The value of A is known to be close to kT for polymer particles (e.g., 6.3×10^{-20} J in the case of polystyrene), although for the metal dispersions, it is far higher (40×10^{-20} J for silver) (Visser 1981). Therefore, a large diversity of protecting agents (donor ligands, polymers, and surfactants) such as electrides, alcohols, glycols, metal borohydrides, and specialized reagents (tetrakis[hydroxymethyl] phosphonium chloride) have been selected to control the growth of the primarily formed nanoclusters and to prevent them from agglomeration.

In this context, the most common synthetic method presented in the literature, published for the first time by Faraday (1857), is based on the chemical reduction of metal salts in the presence of stabilizing agents to generate zerovalent metal colloids in aqueous or organic media. After that, Turkevich (1985) performed the first reproducible standard recipes to obtain *metal colloids* (e.g., for 20 nm gold by reduction of chloroauric acid, $[HAuCl_4]^-$, with sodium citrate). In addition, he suggested a mechanism for the gradual formation of *nanoclusters* that is still recognized. The new results and data from analytical techniques have been helpful for perfecting this model. Thus, starting from the reduction of metal salt to obtain zerovalent metal atoms in the initial stage of nucleation, one can obtain the "seed" nuclei with a diameter below 1 nm.

From this point of view, another research report explains the production of small metal particles (of varying size distributions) by reduction of metal salts dissolved in appropriate solvents (Suo et al. 2013). The *nanocrystal syntheses* suppose the use of soft templates, such as the reverse micelle in a water pool. Reverse micellar methods are successfully utilized in the preparation of Ag, Au, Co, Pt, and Co nanocrystals (Eastoe et al. 2006). Generally, more recent procedures for obtaining nanocrystals at the air–water interface, such as in Langmuir–Blodgett films or at a liquid–liquid interface (Chen et al. 1995), are preferred. For example, the literature shows that at a water–toluene interface film of metal, semiconductor and oxide nanocrystals can be obtained. A nanocrystalline film of gold can be obtained from a toluene solution of $Au(PPh_3)Cl$ partially hydrolyzed with tetrakis(hydromethyl) phosphonium chloride in water. When adding dodecanethiol to the toluene layer, the film breaks up, appearing as an organosol. Further introduction of mercaptoundecanoic acid in water results in the formation of the hydrosol (Radha and Kulkarni 2010). Usually, the control of the cluster sizes is achieved by ablation of a metal target in vacuum followed by mass selection of the plume to obtain cluster beams. For achieving nanocrystals in solution, one research method focused on controlling a selected Ag cluster beam through a toluene solution of thiol and capped the vacuum-prepared particles (Love et al. 2015). Colloids of alloys have been prepared by the chemical reduction of the appropriate salt mixture in the solution phase.

The controlled precipitation of the salt mixture is very important in the case of semiconductor nanocrystals. Therefore, Ag–Pd and Cu–Pd colloids of different composition may be obtained by alcohol reduction of mixtures of silver nitrate or copper oxide with palladium oxide. Another study reports synthesis of Fe–Pt alloy nanocrystals by thermal decomposition of the Fe and Pt acetylacetonates in high-boiling organic solvents. The co-reduction of silver nitrate and chloroauric acid with sodium borohydride led to Au–Ag alloy nanocrystals. The utilization of lasers on Au–Ag layered particles determines the Au–Ag alloying and segregation (Jia et al. 2005).

17.4 THE CONTROL OF THE METAL NANOPARTICLES' SIZE AND SHAPE

In final synthesis of nanocrystals, it is difficult to obtain a certain size distribution because the reaction temperature and reagent concentrations offer poor control of the three steps—nucleation, growth, and termination—by the capping agent or ligand. Usually, the distribution is log-normal with a standard deviation of 10% (Thøgersen et al. 2011). To achieve nanocrystals of precise dimensions with minimal size distributions, one can use size-selective precipitation either by centrifugation or by using a miscible solvent–nonsolvent liquid mixture to precipitate nanocrystals. Sattler reported a procedure for obtaining monodisperse nanocrystals that are known as "cluster compounds" (Sattler 2002). These cluster compounds are also macromolecules with a core containing metal–metal bonds, yet they are obtainable in definite stoichiometries, with typical examples being $(Pt_{38}[CO]_{44}H_2)^{2-}$ and $Au_{55}[PPh_3]_{12}C_{16}$. Therefore, it has been demonstrated that the increased stability in oxidation of Au_{55} nanocrystals is caused by the fact that Au_{55} is not oxidized under conditions that oxidize bulk Au (Sattler 2002). Generally, MNPs are represented as spheres but, in reality, they can be better described as polyhedrons. On the other hand, a cluster consists of a strictly limited number of atoms arranged in shells. In addition, nanocrystals have a nanometric size and contain a few to several thousand atoms. The nanocrystals have a special stability because they possess a "magic number" of metal atoms that allow complete enclosure of successive shells of atoms in a cubic, close-packed arrangement. When compared with clusters containing one atom more or one atom less, the "magic clusters" have interesting properties, such as high ionization potentials, low electron affinity, high symmetry, reduced reactivity, high melting points, and large highest-occupied molecular orbital–lowest-unoccupied molecular orbital (HOMO-LUMO) gaps (Sattler 2002). Therefore, the "magic number" clusters can be regarded as superatoms, capable of forming specific compounds. Limited only by size, not shape, the MNPs can present in many forms, such as spherical (most common), nanorods, nanowires, nanoplates, nanocubes, nanotetrapods, and others. The magic nuclearities offer precise information concerning the morphology and the shape of the cluster. The number of atoms necessary to form complete shells has been intensely studied from a mathematical point of view. Also, for some geometric shapes (including icosahedron and cuboctahedron), the number of atoms (N) required to form a cluster with L geometric shells is described by the equation (17.3) (Rao et al. 2007):

$$N = \frac{10L^3 + 15L^2 + 11L + 3}{3} \tag{17.3}$$

The graph from Figure 17.1 presents a full-shell cluster configuration with a magic number of atoms. The schematic representation illustrates the relationship between total number of atoms necessary in full-shell clusters with the corresponding shell number, as well as percentage of surface atoms. Thus, the magic numbers 13, 55, 147, 309, and 561 correspond to the closure of 1, 2, 3, 4, 5, and 7 shells, respectively (Nicolais 2004). Also, the surface atom percentages of these corresponding shells are 92, 76, 63, 52, 45, and 35, respectively. For other shapes, the number of atoms needed to complete the particular shell, n_L, is given by the equation (17.4):

$$N_L = 10n_{L-1}^2 + 2 \tag{17.4}$$

where $N_0 = 1$.

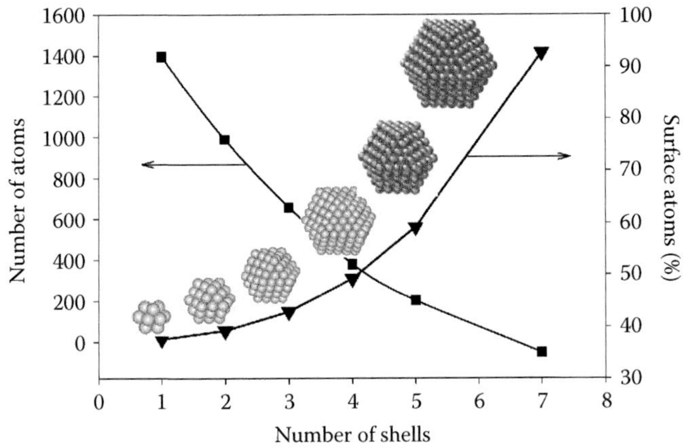

FIGURE 17.1 A schematic representation of the relationship between the total number of atoms (■) necessary in full-shell clusters with the corresponding shell number, as well as the percentage of surface atoms (▼). (Adapted from Wolf Michael O. *Adv. Mater.*, 13. no. 8, 545–553, 2001.)

An example from the literature relates obtaining the magic nuclearity nanocrystals of polyvinylpyrrolidone (PVP) capped Pd nanoparticles including Pd 561. The magic nuclearity nanocrystals obtained in this manner are monodisperse (diameter distribution < 4%) (Thomas and Kulkarni 2003).

17.5 COMPOSITION OF THE METAL NANOPARTICLES

From the point of view of their composition, nanoparticles mainly can be classified into the following categories:

- *Metal NPs* composed of one type of material (e.g., Au, Ag, Pt, Fe, Co, Cd).
- *Semiconductor NPs* composed of more than one type of atom uniformly distributed inside the material (such as zinc and oxygen atoms in ZnO NPs).
- *Magnetic NPs* composed of more than one type of atoms found only in strictly defined areas of the material (e.g., gold and silver atoms in Au [core]/Ag [shell] NPs). The core/shell NP system contains one type of atom located within the inner sphere (core), while others form the outer layer (shell) of the particle (Zhang et al. 2010).

The physical dimension of MNPs depends on the confinement of the electrons, thereby determining a variation in the MNPs' shape. Thus, it can be estimated that the light emitted by a nanorod is linearly polarized along the growth axis. This idea raised researchers' interest in developing novel synthetic strategies for preparation of nonspherical nanocrystals. The common methods allow a mixture of shapes (in addition to spherical particles)—triangular, teardrop, and many more. In a recent study, smarter synthetic schemes were proposed leading to new nanoparticle shapes, such as rods, elongated spheres, cubes, and hexagons (Sajanlal et al. 2011).

17.6 THE IMPORTANCE OF THE LIGAND SHELL ADJUSTMENT

The properties of nanocrystals in their native form are influenced by the surface species and capping agents (Wang et al. 2007). Consequently, in addition to monitoring the size and shape of the nanocrystals, it is also very important to adjust its surface with the suitable capping agent for extending the applications range. Many researchers reveal that, apart from traditional capping substances (ions, surfactants, and polymers), new compounds have been used for this purpose, including ligands-dendrimers, hydrogen-bonding fragments of protein, DNA, and dyes-with pendent thiol groups as well as silica layers (Li and Liu 2014; Sarathy et al. 1997). Sometimes, a layer of a noble metal is used as a buffer between the core nanocrystal and the ligand shell. This procedure offers a remarkable stability to the metal nanocrystals—a protection against oxidation—maintaining concomitantly the magnetic properties of metals. The adjustment of the ligand shell can be made through specific reactions that allow exchanging one set of ligands with another by transfer of nanocrystals from one phase to another. Thus, Sarathy et al. presented a new method of thiol-derivatizing hydrosols of metal sols (Sarathy et al. 1997). This procedure demands a high mixing energy of the hydrosol (including metal particles with specific size distribution) with the toluene solution of an alkane thiol in the presence of a strong acid or reducing agent. The derivatization is attained when the intense interchange of the colors from the aqueous layers to the hydrocarbon layer takes place. Therefore, the main advantage of this procedure is that well-characterized metal particles can be easily thiol-derivatized in a nonaqueous medium and in a diversity of hydrosols of Au, Ag, and Pt.

17.7 INCLUSION OF METAL NANOPARTICLES IN POLYMER MATRIX

Inclusion of the MNPs in the conducting polymers represents a good alternative for obtaining an attractive class of materials that combines redox properties of the conducting polymer and those of the metal ions (Wolf 2001). The systems formed by transition metal-containing polymers present some possible arrangements of the MNPs near and onto the backbone. From this point of view, these compounds may be classified into three major types, which are illustrated in Scheme 17.2.

1. *Type 1 polymers* containing the metal group ***tethered*** to the conjugated polymer backbone through a saturated linker, such as alkyl groups. In this case, the polymer backbone behaves as a conductive support and the metal centers are similar to untethered groups in the polymer matrix.
2. *Type II polymers* have the metal group electronically ***coupled*** to the polymer backbone through a conjugated linker group or directly by coordination to sites; the metal groups can directly influence each other's properties.

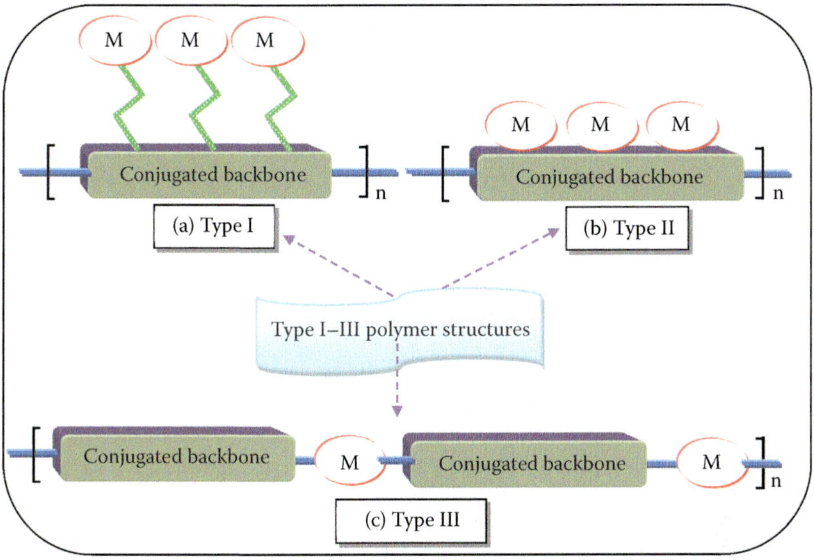

SCHEME 17.2 Schematic representation of the three types of metallopolymers: conjugated metal group/polymer backbone. (Adapted from Wolf Michael O. *Adv. Mater.* 13 no. 8 (2001): 545–553.)

The conducting polymer backbone/metal system ions can be electrochemically tuned because they are redox-active.

3. *Type III polymers* have the metal group directly *incorporated* into the conjugated backbone. The metal groups from this configuration have the greatest influence on the properties of the conducting polymer.

Various aspects concerning preparation and properties of metal-containing conjugated polymers have been reported in the literature (Roncali 1999). One of the methods used in synthesizing Type I and II is electropolymerization. The convenience of this approach is that the electrogenerated material is insoluble and can be easily deposited directly onto polymer substrate as a thin film. Moreover, the thickness and the morphology of the thin film can be controlled via variation of the electrochemical parameters, such as potential, current density, and deposition time. The in situ investigations of the electrochemical properties offer insightful hints to transform the polymer from its insulator state to its conductor one. Utilization of the heterocyclic aromatic compounds (pyrrole and thiophene) is essential for catalysts and sensors (Pineda et al. 2015). The advantage of the electropolymerization method is in the ability to grow conductive thin film on the various substrates. The chemical procedures most commonly used for preparing the Type III polymers are ring-opening and condensation. They cannot be electrochemically synthesized. Only thiophene-based Type III polymers have been successfully electropolymerized (Wolf 2001). The platinum-based rod polymers, ferrocene-based polymers, and polysilanes are a few examples of metallopolymers prepared by condensation (Pomogailo and Dzhardimalieva 2014). For this reason, the preparation and characterization of Type III polymers is not of great interest because in these systems the metal group can prevent the intrachain charge transport. Therefore, one of the most important attributes of using conducting backbones to "wire" or link redox centers is in increasing the charge transport rate—the rates in conjugated systems being up to two orders of magnitude larger than nonconjugated analogues. This improvement is due to the occurrence of a super-exchange pathway. An electron self-exchange mechanism in a conjugated redox polymer exhibits catalytic and photocatalytic activities used in sensor applications.

Many investigations concerning physicochemical properties of metallic clusters included in a polymer matrix are of great interest not only in the field of catalysis but also in the production of different high-performance polymeric materials (Black et al. 2007; El-Noura et al. 2010). These interesting systems are suitable for electrochromic devices that operate effectively in the presence of air and water. Various strategies have been proposed and applied to achieve electrochromic systems. For example, to obtain high coloration displays, novel materials were obtained through electropolymerization of poly(imide-benzidine) film onto indium-thin oxide (Black et al. 2007). These remarkable materials have a reversible electrochemical coloration and keep this property above 10^5 cycles, even in the presence of air and water.

Reversible formation of Ag nanoparticles onto transparent polymer films using an electrochemical technique has been reported (Black et al. 2007). It is known that Ag nanoparticles possess special optical properties (El-Noura et al. 2010). During the nanoparticle formation process, some colored Ag clusters appear, which can be photogenerated and stabilized inside a polymer matrix. An optically transparent film is mandatory for controlling the process of particle formation inside polymer film by means of spectroelectrochemistry. Based on this aspect, Ag nanoparticles were prepared by photoreduction of Ag^+ ions in transparent cross-linked films made up of poly(vinyl alcohol) and poly(acrylic acid) (Chalal et al. 2012). For achieving nanoparticles with greater reversibility, a redox mediator (TMB^{2+}) was used. The principal role of these species is that they supply electrons that help oxidize silver clusters (Black et al. 2007).

Another electrochemical synthesis of silver nanoparticles was performed in organic solvents by potentiostatic and galvanostatic methods (Starowicz et al. 2006). The development of these approaches is based on the chemical reduction of silver salt in aqueous solution in the presence of different reduction agents. For example, the reduction of silver acetylacetonate and palladium acetate with dimethylamine-borane in the presence of a fluorinated surfactant was reported (Kameo et al. 2003). Also, the reduction of silver iodide with alkali metal in ammonia was explained in the work of Sun et al. (2003). The size of the resulted nanoparticles is 3–12 nm, 3–6 nm, and 4–5 nm for silver, palladium, and silver–palladium, respectively. A simple method that excludes surfactants or additional reducing agents is based on the silver polarization in ethanol-containing nitrate anions (Starowicz et al. 2006). Thus, the nanocrystalline silver is obtained through the oxidation and dissolution of the anodic silver in the presence of ethanol. The final step of this mechanism consists of the reduction of the Ag_2O with aldehyde, which is obtained by oxidation of ethanol. A new effective strategy for electrochemical preparation of composite electrodes for sensing applications has been developed (Pineda et al. 2015).

One of the most important aspects of MNPs is their ability to be well-distributed on the surface of polymer support. In addition, the good electrocatalytic properties of gold nanoparticles make them attractive materials for the designing of electrochemical sensors and biosensors (An et al. 2012). Obtaining gold nanoparticles simultaneously with conducting polymer film provides good sensitivity, catalytic properties, and conductivity of the composite material. Based on these aspects, gold nanoparticles can be incorporated in polymers through several methods, such as electrodeposition on choline films formed on glassy carbon electrodes, electrodeposition on ultra-thin over-oxidized PPy films on glassy carbon, and codeposition under galvanostatic conditions. Thus, a successful optimization of the material sensing features has been implemented in the area of MNPs/conducting polymer nanocomposites after decorating polypyrrole/stainless steel electrodes with gold nanoparticles (GNPs) (Pineda et al. 2015). Composite electrodes of GNPs on PPy films have been prepared by electrodeposition. Stainless steel (SS) was used as a substrate for PPy films, preventing corrosion. The embedding of GNPs in the PPy matrix was achieved by means of an electrodeposition technique at constant potential. The electrocatalytic activity of these new composite electrodes was tested for the electro-oxidation of two molecules of interest in electrochemical sensing. Utilization of composite electrodes leads to enhanced electrocatalytic behavior compared with flat gold electrodes, owing to the larger area of the electrode and an increased sensitivity toward analytes that are electrochemically active on gold.

17.8 GENERAL REMARKS

This chapter provides a general description of the information regarding theoretical and experimental investigations of nanostructured and nanocomposite materials. The obtaining and characterization of these multiphase systems have attracted special interest due to their unusual electrochemical properties. Combination of the complementary properties of nanometer-sized particles with those of the host matrix leads to multifunctional materials with special properties. Additionally, in order to emphasize the importance of different nanoscale materials in multiple applications, an overview concerning the general aspects of nanoparticles, such as classification, properties, and their combination with some polymers, is presented. Metals and metal oxides are especially necessary for designing nanoscale systems. Due to their unique electrical, optical, magnetic, and chemical characteristics, metal nanoparticles are widely used in nanotechnology. Moreover, stabilizers play a very important role in controlling both the size and shape of metal nanoparticles. Polymers are good stabilizers for metal colloids because their transparent, permeable, and non-conductive character does not affect the optical, electrical, and catalytic properties of the clusters.

Preparation of metal–polymer nanocomposites represents a simple and effective way to use mesoscopic features of nanosized metals in different fields of technology. Moreover, organic polymers doped by atomic clusters of ferromagnetic metals show magneto-optical properties that are useful for increasing the performance of many devices, including magneto-optic modulators, optical isolators, and optical shutters. The discovery of new materials, processes, and phenomena that occur at the nanoscale level, concomitantly with the evolution of experimental techniques, provides new perspectives on the development of nanosystems and nanostructured materials for advanced technologies.

REFERENCES

Ambrogio Michael W., Thomas Courtney R., Zhao Yan Li, Zink Jeffrey I., and Stoddart J. Fraser. Mechanized silica nanoparticles: A new frontier in theranostic nanomedicine. *Acc. Chem. Res.* 44 no. 10 (2011): 903–913.

An Taechang, Choi WooSeok, Lee Eunjoo, Cho J. Seong, and Lim Geunbae. Fabrication of conducting polymer micro/nanostructures coated with au nanoparticles for electrochemical sensors. *J. Nanosci. Nanotechnol.* 12 no. 6 (2012): 4975–4978.

Astruc Didier. *Transition-metal nanoparticles in catalysis: From historical background to the state-of-the art, in nanoparticles and catalysis.* Weinheim: Wiley-VCH, 2008.

Barthelat François, Li Chun-Ming, Comi Claudia, and Espinosa Horacio D. Mechanical properties of nacre constituents and their impact on mechanical performance. *J. Mater. Res.* 21 no. 8 (2006): 1977–1986.

Berkowitz Ami E., Mitchell Jay R., Carey Matthew J., et al. Giant magnetoresistance in heterogeneous Cu-Co alloys. *Phys. Rev. Lett.* 68 no. 25 (1992): 3745–3748.

Bhosale Sheshanath, Sisson Adam L., Sakai Naomi, and Matile Stefan. Synthetic functional π-stack architecture in lipid bilayers. *Org. Biomol. Chem.* 4 (2006): 3031–3039.

Black J., Cammarata V., Gaddy G., Slaten B., and G. Mills. Reversible formation of silver nanoparticles in poly(vinyl alcohol)-based thin films via electrochemical switching. *NSTI-Nanotech* 4 (2007): 355–358.

Carotenuto Gianfranco, Nicolais Luigi, Martorana Brunetto, and Perlo Piero. Metal-polymer nanocomposite synthesis: Novel ex situ and in situ approaches. In *Metal–polymer nanocomposites*, Luigi Nicolais and Gianfranco Carotenuto (eds.), pp. 155–181. New-York: John Wiley, 2005.

Chalal Samia, Haddadine Nabila, Bouslah Naima, and Benaboura Ahmed. Preparation of poly (acrylic acid)/silver nanocomposite by simultaneous polymerization-reduction approach for antimicrobial application. *J. Polym. Res.* 19 no. 12 (2012): 1–8.

Chen Xiao, Xue Qing-Bin, Yang Kong-Zhang, and Zhang Qi-Zhen. Monolayers and Langmuir-Blodgett films of a liquid-crystalline polysiloxane with a Schiff base mesogenic unit in the side chain. *Langmuir.* 11 no. 10 (1995): 4082–4088.

Daniel Marie-Christine, and Astruc Didier. Gold nanoparticles: Assembly, supramolecular chemistry, quantum-size-related properties, and applications toward biology, catalysis, and nanotechnology. *Chem. Rev.* 104 (2004): 293–346.

Domènech Berta, Bastos-Arrieta Julio, Alonso Amanda, Macanás Jorge, Muñoz Maria, and Muraviev Dmitri N. Bifunctional polymer-metal nanocomposite ion exchange materials. In *Ion exchange technologies*, Ayben Kilislioğlu (ed.), pp. 35–72. Barcelona: InTech, 2012.

Eastoe Julian, Hollamby Martin J., and Hudson Laura. Recent advances in nanoparticle synthesis with reversed micelles. *Adv. Colloid. Interface. Sci.* 128–130 (2006): 5–15.

El-Noura Kholoud M.M. Abou, Eftaihab Ala'a, Al-Warthanb Abdulrhman, and Ammar Reda A.A. Synthesis and applications of silver nanoparticles. *Arabian J. Chem.* 3 no. 3 (2010): 135–140.

Emamifar Aryou. *Applications of antimicrobial polymer nanocomposites in food packaging.* Vienna: InTech, 2011.

Faraday Michael. The bakerian lecture: Experimental relations of gold (and other metals) to light. *Philos. Trans. Roy. Soc. Lond.* 147 (1857): 145–181.

Giannazzo Filippo, Eyben Pierre, Baranowski Jacek, Camassel Jean, and Lányi Stefan. Advanced materials nanocharacterization. *Nanoscale. Res. Lett.* 6 no. 107 (2011): 1–2.

Gomez-Romero Pedro. Hybrid organic inorganic materials-in search of synergic activity. *Adv. Mater.* 13 no. 3 (2001): 163–174.

Guo Qingchuan, Ghadiri Reza, Weigel Thomas, et al. Comparison of in situ and ex situ methods for synthesis of two-photon polymerization polymer nanocomposites. *Polymers.* 6 no. 7 (2014): 2037–2050.

Issa Bashar, Obaidat Ihab M., Albiss Borhan A., and Haik Yousef. Magnetic nanoparticles: Surface effects and properties related to biomedicine applications. *Int. J. Mol. Sci.* 14 no. 11 (2013): 21266–21305.

Jia Zhiyong, Kang Shishou, Nikles Davi E., and Harrell James W. Synthesis of FePtAu nanoparticles in high-boiling-point solvents. *Magn. IEEE Trans.* 41 no. 10 (2005): 3385–3387.

Jin Rongchao. Quantum sized, thiolate-protected gold nanoclusters. *Nanoscale.* 2 no. 3 (2010): 343–362.

Kameo Azusa, Yoshimura Tomokazu, and Esumi Kunio. Preparation of noble metal nanoparticles in supercritical carbon dioxide. *Colloids. Surf. A Physicochem. Eng. Asp.* 215 no. 1–3 (2003): 181–189.

Karttunen Mikko, Ruuskanen Pekka, Pitkänen Ville, and Albers Willem M. Electrically conductive metal polymer nanocomposites for electronics applications. *J. Electron. Mater.* 37 no. 7 (2008): 951–954.

Klabunde Kenneth J. *Nanoscale materials in chemistry.* New York: Wiley-Interscience, 2001.

Koole Rolf, Groeneveld Esther, Vanmaekelbergh Daniel, Meijerink Andries, and de Mello Donegá Celso. Size effects on semiconductor nanoparticles. In *Nanoparticles*, Celso de Mello Donegá (ed.), pp. 13–51. Berlin: Springer-Verlag, 2014.

Kreibig Uwe, and Vollmer Michael. *Optical properties of metal clusters.* Berlin: Springer, 1995.

Li Kai, and Liu Bin. Polymer-encapsulated organic nanoparticles for fluorescence and photoacoustic imaging. *Chem. Soc. Rev.* 43 (2014): 6570–6597.

Love J. Christopher, Estroff Lara A., Kriebel Jennah K., Nuzzo Ralph G., and Whitesides George M. Self-assembled monolayers of thiolates on metals as a form of nanotechnology. *Chem. Rev.* 105 (2005): 1103–1169.

Lubick Naomi, and Betts Kellyn. Silver socks have cloudy lining. *Environ. Sci. Technol.* 42 no. 11 (2008): 3910–3910.

Luckham Paul F. Manipulating forces between surfaces: Applications in colloid science and biophysics. *Adv. Colloid. Interf. Sci.* 111 no. 1–2 (2004): 29–47.

Ma Chun Yan, Mu Zhen, Li Jin Jun, et al. Mesoporous Co_3O_4 and Au/Co_3O_4 catalysts for low-temperature oxidation of trace ethylene. *J. Am. Chem. Soc.* 132 no. 8 (2010): 2608–2613.

Macanás Jorge, Ruiz Patricia, Alonso Amanda, Muñoz María, and Muraviev Dmitri N. Ion exchange-assisted synthesis of polymer stabilized metal nanoparticles. In *Ion exchange and solvent extraction*, Arup K. Sengupta (ed.), pp. 1–44. Boca Raton: CRC Press, 2011.

Murray Christopher B., Sun Shouheng, Doyle Hugh, and Betley Theodore. Monodisperse 3d transition-Metal (Co,Ni,Fe) nanoparticles and their assembly into nanoparticle superlattices. *MRS Bull.* 26 no. 12 (2001): 985–991.

Nicolais Luigi. *Metal–polymer nanocomposites.* Italy: Willey, 2004.

Okumura Mitsutaka, Haruta Masatake, Kitagawa Yasutaka, and Yamaguchi Kizashi. Theoretical study of H_2O and O_2 adsorption on Au small clusters. *Gold Bull.* 40 no. 1 (2007): 40–44.

Pineda Eduart Gutieŕrez, Alcaide Francisco, Rodríguez Presa Mariá Jose, Bolzań Agustıń E., and Gervasi Claudio A. Electrochemical preparation and characterization of polypyrrole/stainless steel electrodes decorated with gold nanoparticles. *ACS Appl. Mater. Interf.* 7 no. 4 (2015): 2677–2687.

Pomogailo Anatolii D. Polymer-immobilised clusters of the platinum group metals. *Platin. Met. Rev.* 38 no. 2 (1994): 60–70.

Pomogailo Anatolii D., and Dzhardimalieva Gulzhian I. Thermolysis of metallopolymers and their precursors as a way for synthesis of nanocomposites. In *Nanostructured materials preparation via condensation ways*, Anatolii D. Pomogailo, Gulzhian I. Dzhardimalieva (eds.), pp. 287–388. Netherlands: Springer, 2014.

Pomogailo Anatolii D., and Kestelman Vladimir Nikolaevich. *Metallopolymer nanocomposites.* New York: Springer, 2005.

Radha Boya, and Kulkarni Giridhar U. Micro-and nanostripes of self-assembled Au nanocrystal superlattices by direct micromolding. *Nano Res.* 3 no. 8 (2010): 537–544.

Rao C. N. Ramachandra, Thomas P. John, and Kulkarni Giridhar U. *Nanocrystals: Synthesis, properties and applications.* Berlin: Springer, 2007.

Reboul Julien, Furukawa Shuhei, Horike Nao, et al. Mesoscopic architectures of porous coordination polymers fabricated by pseudomorphic replication. *Nat. Mater.* 11 no. 8 (2012): 717–723.

Roncali Jean. Electrogenerated functional conjugated polymers as advanced electrode materials. *J. Mater. Chem.* 9 (1999): 1875–1893.

Sajanlal Panikkanvalappil R., Sreeprasad Theruvakkattil S., Samal Akshaya K., and Pradeep Thalappil. Anisotropic nanomaterials: Structure, growth, assembly, and functions. *Nano Rev.* 2 (2011): 1–62.

Sarathy K. Vijaya, Kulkarni Giridhar U., and Rao C. N. Ramachandra. A novel method of preparing thiol derivatised nanoparticles of gold, platinum and silver forming superstructures. *Chem. Commun.* 6 (1997): 537–538.

Sattler Klaus. The energy gap of clusters nanoparticles, and quantum dots. In *Handbook of thin films materials,* H. S. Nalwa (ed.), pp. 61–97. Hawai: Academic Press, 2002.

Spagnoli Dino, Banfield Jillian F., and Parker Stephen C. Free energy change of aggregation of nanoparticles. *J. Phys. Chem. C.* 112 no. 38 (2008): 14731–14736.

Starowicz Maria, Stypuła Barbara, and Banas Jacek. Electrochemical synthesis of silver nanoparticles. *Electrochem. Commun.* 8 no. 2 (2006): 227–230.

Sun Lei, Zhang Zhijun, and Dang Hongxin. A novel method for preparation of silver nanoparticles. *Mater. Lett.* 57 no. 24–25 (2003): 3874–3879.

Suo Liumin, Hu Yong-Sheng, Li Hong, Armand Michel, and Chen Liquan. A new class of solvent-in-salt electrolyte for high-energy rechargeable metallic lithium batteries. *Nat. Commun.* 4 no. 1481 (2013): 1–9.

Thøgersen Annett, Bonsak Jack, Fosil Carl Huseby, and Muntingh Georg. Size distributions of chemically synthesized Ag nanocrystals. *J. Appl. Phys.* 110 no. 4 (2011): 1–7.

Thomas P. John, and Kulkarni Giridhar U. From colloids to nanotechnology: Investigations on magic nuclearity palladium nanocrystals. *Curr. Sci.* 85 no. 12 (2003): 1760–1766.

Turkevich John. Colloidal gold. Part I. Historical and preparative aspects, morphology and structure. *Gold Bull.* 18 no. 3 (1985): 86–91.

Visser Jan. The concept of negative Hamaker coefficients. 1. History and present status. *Adv. Colloid Interface Sci.* 15 no. 2 (1981): 157–169.

Vollath Dieter. An introduction to synthesis, properties and application. *Nanomaterials.* 7 no. 6 (2008): 865–870.

Wang Wei, Banerjee Sarbajit, Jia Shengguo, Steigerwald Michael L., and Herman Irving P. Ligand control of growth, morphology, and capping structure of colloidal CdSe nanorods. *Chem. Mater.* 19 no. 10 (2007): 2573–2580.

Wolf Michael O. Transition-metal–polythiophene hybrid materials. *Adv. Mater.* 13 no. 8 (2001): 545–553.

Yang Guohua, Tan Li, Yang Yiyun, Chen Shaowei, and Liu Gang-Yu. Single electron tunneling and manipulation of nanoparticles on surfaces at room temperature. *Surf. Sci.* 589 no. 1–3 (2005): 129–138.

Zhang Jian, Langille Mark R., Personick Michelle L., Zhang Ke, Li Shuyou, and Mirkin Chad A. Concave cubic gold nanocrystals with high-index facets. *J. Am. Chem. Soc.* 132 no. 40 (2010): 14012–14014.

18 Thin Metal Films on Polymer Substrate

Raluca Marinica Albu

CONTENTS

18.1 INTRODUCTION

Thin films represent thin material layers with thickness sizes ranging from a fraction of a nanometer to several micrometers. Applications of thin films may be mentioned in several important domains, such as microelectronics, optical and magnetic devices, electrochemistry, protective and decorative coatings, and catalysis. Equally, polymer materials are widely used in various areas due to their advantages of low weight, low cost, and good mechanical properties. Nevertheless, polymeric materials also present some drawbacks, particularly related to surface properties, conductivity, durability, and decorativeness. For improving these aspects, metallic or ceramic films are preferred for deposition on such polymer substrates. Regardless, the metallization process of polymer substrates is of considerable technological importance and presents significant interest from a fundamental point of view. These multiphase composites systems of metal/polymer substrate and polymer/metal/polymer types are widely used in a variety of applications, such as in food packaging, on-chip interconnection, reflectors for car lights, optical data storage (compact discs), electrically shielded computer cases, decorative coatings, optical ultra-thin color filters, and substrates for biomolecules. Generally, the properties of thin metal/polymer interfaces are determined by the metal–polymer interaction and also by the polymer and by the morphology of the interface. The cohesive energy of metals is typically two orders of magnitude higher than that of polymers. Furthermore, the adhesion between moderately reactive metals and polymers is generally much weaker than the strong metal–metal binding forces.

For this reason, the polymer substrate treatment represents a very important step in the technology of metallic or ceramic coating on the polymer substrate. Therefore, noble metals and other low reactivity metals do not wet the untreated polymer surfaces, forming three-dimensional spherical clusters growing in a Volmer–Weber mode. Surface modification of the hydrophobic polymer surface onto a hydrophilic one can be achieved by wet (acid, alkali), dry (plasma), and radiation treatments (ultraviolet radiation and laser), without affecting the bulk properties. Consequently, application of different pretreatment methods represents an efficient way to improve wettability and thin metal adhesion.

18.2 FUNDAMENTAL ASPECTS OF THIN FILM GROWTH MECHANISMS

Material layers with thickness values below several micrometers typically are used for the development and modifications of different material surfaces or for designing devices such as light-emitting diodes. It is known that selection of the material influences the surface properties and their area of applicability. Moreover, the microstructure of the substrate can significantly modify the properties of films (Ohring 2002). Therefore, obtaining thin films requires a thorough understanding of the processes occurring in time, along with film microstructure evolution. During their growth, even thin films go through stages responsible for their final microstructure and physical properties (Veprek 2013). The main steps in obtaining thin films include the emission of particles from the source (heat, high voltage), transport of particles to the substrate, and condensation of particles on the substrate. The dynamic processes developed on the material substrate, which influence the initial microstructure of thin films growing in a three-dimensional fashion, are island nucleation, growth, and coalescence of the deposited atoms (Tame et al. 2013). Some of the research present the chronology of obtaining thin film, the various stages of growth, and the diffusion processes occurring on the material's surface, as shown here.

18.2.1 Adsorption and Surface Diffusion

Among the main techniques through which the thin films can be deposited mention should be made of:

- Physical or vapor deposition (PVD) is based on the thermodynamical/mechanical processes applied to obtain thin films, such as pulsed laser deposition, sputtering, and molecular beam epitaxy; the source material is placed in an energetic environment so that its particles may escape and condense on the substrate.
- Chemical vapor deposition (CVD) uses chemical processes to obtain thin films; in this case, the substrate is exposed to more volatile precursors, which react and/or decompose on the substrate's surface.
- The (electro-)chemical reaction among precursor liquids on immersed substrates.

Therefore, the incoming atoms begin to interact with the support material at a distance of a few Ångströms, causing their reflection or desorption on the interaction surface. These phenomena are related to the incoming flux, the trapping probability, and the sticking coefficient. The values of potential energy of adatoms vary, depending on the atomic structure and chemistry of the surface, which contains adsorption sites of different depth, or on their stability. Therefore, these atoms spread among the adsorption sites in a random walk process with the jump rate, G, depending on the thermal motion, as described by Equation 18.1 (Michely and Krug 2003):

$$\Gamma = \nu e^{-\frac{E_d}{k_B T}} \tag{18.1}$$

where ν is the attempt frequency; E_d represents the diffusion barrier; T is the substrate temperature; and k_B is Boltzmann's constant.

If the diffusion barrier is the same in all directions and if the adsorption sites are found at a distance, a, one can calculate the surface diffusion coefficient, D, or diffusivity, with Equation 18.2. This coefficient is a temperature-dependent constant describing the diffusion of one material into another.

$$D = \frac{1}{4}a^2\Gamma = \frac{1}{4}a^2\nu e^{-\frac{E_d}{k_B T}} = D_0 e^{-\frac{E_d}{k_B T}} \tag{18.2}$$

where factor 1/4 is due to the two-dimensional nature of surface diffusion (Michely and Krug 2003), and D_0 and E_d are the experimental diffusion coefficient and the experimental diffusion barrier, respectively.

Due to the variation of the structure and surface orientation, the diffusion barrier of the adatoms differs along a certain direction (Brune 1998). In addition, the diffusion barrier is also influenced by the presence of other adatoms, nuclei, and impurities on the surface. Thus, according to Equation 18.2, small modifications in the diffusion barriers cause an exponential increase of changes in the diffusion rates, which is an important aspect in the morphology of film growth.

18.2.2 Nucleation

The processes occurring in the first growth stage of thin films consist of the random diffusion of the adatoms stored onto a perfectly flat surface until they collide with another adatom forming a dimer or are desorbed from the surface. Furthermore, the dimer can grow by merging with another adatom, or it can unfold in two separate parts. The growth of larger cluster atoms occurs in a similar manner through the attachment and detachment of atoms. The clusters formed by two or more atoms are usually stable because the probability of accepting an atom is higher than that of losing one atom. Nucleation is a spontaneous process that starts with the appearance of small embryonic clusters that increase until a critical size, when equilibrium between their vapor pressure and the environmental pressure is attained. In some cases, the temperature during layer deposition symmetry surface can be varied, leading to critical cluster with sizes larger than unity (Michely and Krug 2003).

From a theoretical point of view, the probability of forming dimers and the nucleation density depend on the probability of the mutual meeting between two atoms and also on the increase of the density of adatom number. Moreover, an increase in nucleation density over the kinetic limits can be practically realized by modification of the nucleation behavior of the growing films by adding impurities or by creating defects on the surface to obtain extra nucleation sites.

In this context, some studies have stated that the nucleation density of Ni and Mo films deposited on highly ordered pyrolytic graphite (HOPG) can be modified by treating the substrates with various doses (number of incident ions per unit area) of 100 eV Ar^+ ions (Durand et al. 1998). The nucleation density of the deposited films was found to scale with the ion dose. A larger nucleation density, comparative to thermal evaporation, was found when the films were deposited using 100 eV Ni^+ ions as the depositing species. These ions have sufficient power to modify the first atomic layers of the substrate, causing the appearance of different types of near-surface defects (Durand et al. 1998). These defects appear as nucleation centers for adatoms, which lead to increases in nucleation density. Kalff et al. (1997) found out the same increase of island density for Pt was deposited on Pt(111) by ion beam-assisted deposition comparative to the same amount of material deposited by thermal evaporation. The processes occurring during the action of ion beam deposition are determined by energetic particles causing significant modifications on the surface such as the occurrence of additional adatoms and vacancies. These additional atoms create other nucleation that may lead to an increase of nucleation density by a factor of 27. Moreover, the impurities from residual gas can also determine the creation of auxiliary nucleation centers. These undesired processes can be diminished by choosing adequate deposition conditions, such as high deposition rates, in order to reduce the action of the residual gases on the surface (Michely and Krug 2003).

18.2.3 The Island Growth

During nuclei growth and their change into grains, multiple simultaneous processes occur. Thus, the atoms are continuously attached to and detached off the grains at a positive net growth rate. The binding energy of surface atoms is inversely proportional to the curvature of the surface grains, so that the adatom density around small grains will be higher than around large grains. Thus, the gradient of adatom density between two or more grains with different sizes creates a driving force for adatom diffusion from the small to the large island. This spontaneous phenomenon (coarsening), known as Ostwald ripening, increases the grain size difference due to the growth of larger grains at the expense of smaller particles (Lorenz and Peter 2002). In the initial growth process, many small grains are formed, with a crystallographic—more or less random—orientation, when the substrate acts as a crystallographic template (Thompson 2000). In this stage, the orientation of grains is energetically unfavorable for the equilibrium grain structure; in such cases, surfaces with high energy will be restructured. In the case of kinetic restrictions, the structures can develop into a more convenient formation (Sree Harsha 2006). At every step during the coarsening processes, the equilibrium between the grain and the mean matrix substrate is reached when their sizes attain a critical radius, R_c; grains $R > R_c$ will grow and grains with $R < R_c$ will decrease.

In contrast with Ostwald ripening, another mechanism responsible for the growth of small particles on the substrate surface is Smoluchowski ripening. In this case, entire islands migrate until they collide with other clusters, thus entering into coalescence. This process occurs when the adatoms diffuse on the cluster surface, determining the shift of the mass center. The small clusters are moving faster than the larger ones, causing formation of energetically more stable clusters.

18.2.4 Coalescence and Percolation in Thin Film Growth

During atoms' evaporation onto the material substrate, their diffusion onto the support initially occurred until collision with other atoms, nucleation, and formation of compact islands. Then the islands grow with further deposition and become elongated, percolate, fill their holes, and finally are transformed into a thin film state. When the distance between the islands is almost sufficient to interact, they will begin to coalesce to minimize the energy present in the systems (del Muro et al. 2008). This process, which leads to the reduction of island density on the substrate surface, will end when the boundaries between single-crystalline islands (i.e., grains) become immobile and coalescence (formation of polycrystalline islands) cannot be completed, provoking the formation of a continuous film (Petrov et al. 2003). The diffusion of atoms takes place mainly along the grain surface and, subsequently, into the space between grains, in order to decrease their chemical potential by moving from places with a positive curvature to a space that has a negative curvature. This surface diffusion modifies the shape of grains until they obtain an equilibrium shape. As the size of the coalescing grains varies, the extensive grains will grow, following the larger curvature of the smaller grains. An increase in island coalescence will decrease the areal fraction of the surface covered by the deposited material. These newly created surfaces are frequently sufficiently large to promote the nucleation of new generations of islands, known as secondary nuclei.

Some studies (Iijima and Ajayan 1991; Warrender and Aziz 2003) have analyzed the island coalescence of Au islands obtained on a SiO_2 substrate by transmission electron microscopy, their examination revealing that the small islands (1 nm in diameter) coalesce like liquid droplets, forming a new equiaxed island within 1/60 s. Then, the 2–3 nm in diameter islands are able to complete coalescence within 1/20 s. The bigger particles (with diameter > 5 nm) exhibit a different coalescence behavior because they are crystalline. When the crystallographic orientation of grains differs, a channel or hole-like formation, between the clusters with a grain boundary in the middle, appears. The channel is slowly completed, creating the possibility of diffusing out through the smaller grain in the coalescing pair. Therefore, for big particles, the coalescence process is much slower than for small liquid-like particles.

The growth of thin films involves some important morphological transitions, which depend on the interplay between the coalescence time and the time scales connected to film deposition. Thus, the elongation transition appears when the average grain exclusively shows an equiaxed in-plane shape. Percolation transition occurs while the films of conductive materials become conductive at macroscopic scale (the electrons percolate through the films) forming a continuous pellicle. Further on, during deposition, recording of a coalescence time longer than the time between island impingement processes may appear. Thus, before ending the coalescence process, another island may grow, affecting one of the coalescing islands and causing formation of clusters, which remain elongated on the surface and kinetically frozen (Warrender and Aziz 2003). Moreover, elongated clusters may also appear if a new material is added to the coalescing islands at a rate that increases the coalescence time. Coalescence can be better controlled by means of pulsed deposition processes that allow an independent storage of the amount of material per pulse and pulsing frequency (Warrender and Aziz 2003). Further deposition leads to joining of the elongated clusters among them, determining the formation of a percolator film with a sinuous network of channels. Subsequent deposition will continue to fill the channels until the whole substrate is covered by a continuous and complete film (Ruffino and Grimaldi 2010). Consequently, understanding and controlling the process of layer formation implies, in each case, detailed knowledge of the nucleation and growth modes.

18.3 DYNAMICS OF THIN FILM GROWTH

Considering the fact that thin films are widely used in the electronic industry, their growth mode and surface morphology have been investigated since the 1970s. Among the various methods used by researchers for studying the behavior of the material during the growth of thin films, one should mention TEM, scanning electron microscopy (SEM), low-energy electron microscopy (LEEM), reflection high-energy electron diffraction (RHEED), medium-energy electron diffraction (MEED), spot profile analysis low-energy electron diffraction (SPALEED), and scanning tunneling microscopy (STM) (Sheldon et al. 2011). Many theoretical and computational research have been focused on the elucidation of the fundamental phenomena controlling thin film growth. In this way, it discovered that the main processes that occur during thin film formation are nucleation and growth. Moreover, the initial nucleation and growth stages are influenced by the thermodynamic parameters of the deposit and by the substrate surface.

Furthermore, some studies reported in the literature state that the thin film nucleation process depends first and foremost on the nature of the deposited material. On the other hand, from the point of view of crystallinity, thin films can be classified into three classes:

- Amorphous: These are common insulator films without the presence of crystalline structures and crystalline defects, such as amorphous SiO_2.
- Polycrystalline: These are composed of lots of grain boundaries of elemental metals, which are grown at near room temperature.
- Epitaxial (single crystalline): They require high temperatures and slow growth rates; these are high-quality thin films, such as III-V semiconducting films and complex oxides.

Based on the surface energy theory, the different modes of crystal growth are synthetically illustrated in Scheme 18.1. The atomic structure of the substrate plays

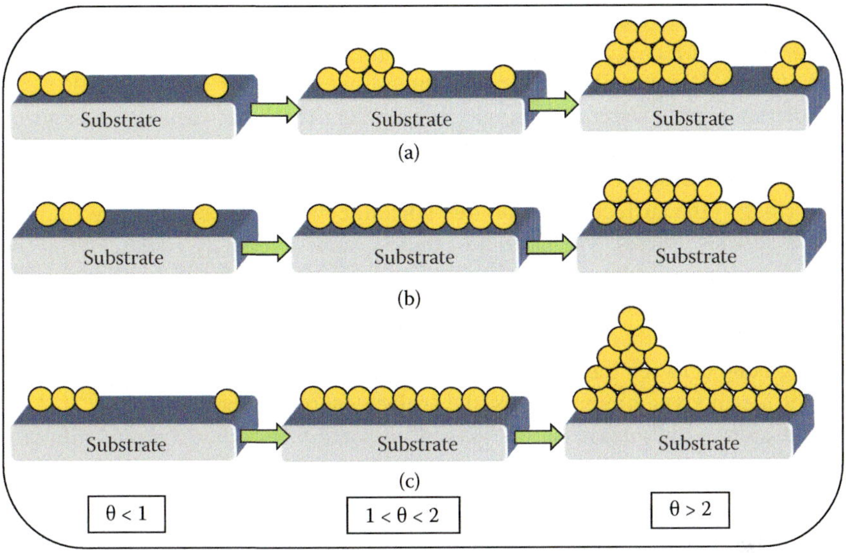

SCHEME 18.1 Three different growth modes of adatoms on a substrate: (a) Volmer–Weber; (b) Frank–Van der Merwe; and (c) Stranski–Krastanov; θ = coverage in the monolayer. (Adapted from Kintzel J. Edward et al. *J. Cryst. Growth.* 289 no. 1 (2006): 345–350.)

a considerable role in positioning the incoming adatoms during deposition onto the solid substrate in a highly symmetrical position. The process that determines the oriented growth of a crystalline film on a single-crystal surface is called epitaxial growth because the overlayer determines the underlying surface ordering. Based on this aspect, it is possible not only to use the substrate as a support for the overlayer but also as a template to obtain materials with metastable crystalline structures and outstanding properties not found in bulk phases. When both the film and the substrate are made of a similar material, it is named homoepitaxy (A on A). When the film and the substrate are made of different materials, it is called heteroepitaxy (A on B).

Taking into account the thermodynamic equilibrium, the growth morphology may be calculated from the balance of the different surface free energies involved (Bauer 1958). Therefore, the differences among these energies can cause various structures of the substance to be deposited on a surface, ranging from two- and three-dimensional clusters to a completely closed layer. In addition, this criterion states that, at equilibrium, the growth mode is determined by the energy difference expressed in Bauer's equation (Wuttig and Liu 2004):

$$\Delta\gamma_{\text{total}} = \gamma_{\text{f}} + \gamma_{\text{s/f}} - \gamma_{\text{s}} \tag{18.3}$$

where γ_s is the surface energy of the substrate, γ_f is the surface energy of the film, and $\gamma_{s/f}$ is the surface energy of the substrate–film interface.

In this context, the mode of thin film growth is mainly governed by the surface free energies of the substrate, films, and interfaces. Usually, the mode of growth of a thin film can be classified into three different growth modes: Volmer–Weber, Frank–Van der Merwe, and Stranski–Krastanov (Zinke-Allmang et al. 1992).

In Volmer–Weber (VW) mode (i.e., three-dimensional or island growth mode), the adatom–adatom interactions are stronger than those between the adatom and the surface, leading to the appearance of three-dimensional clusters that nucleate directly on the substrate surface and then grow into island in all directions, including the normal to the surface direction (the adatom cohesive force is stronger than the surface adhesive force). The growth of these islands ultimately coalesces and then forms a continuous film on the substrate surface or rough multi-layer films. This mode is displayed by several systems growing on insulators (such as mica or organic substrates) because there are no unsaturated dangling bonds on the surface to form a strong bond between the impinging molecules and the substrate. From the point of view of thermodynamic properties, the surface energy of the substrate, γ_s, is lower than the sum of the surface energy of the film γ_f and that of the substrate–film interface $\gamma_{s/f}$:

$$\gamma_s < \gamma_f + \gamma_{s/f} \tag{18.4}$$

Therefore, the total surface energy is positive because the system tries to keep the substrate surface uncovered by minimizing the exposed area of storage. These growth modes are schematically depicted in Scheme 18.1a.

In the Frank–van der Merwe (FVDM) mode (two-dimensional or layer-by-layer growth), represented in Scheme 18.1b, the adatoms are more strongly bound to the substrate sites than to each other, leading the growth of the film in a layer-by-layer manner. Thus, the surface energy of the substrate is higher than the sum of the surface energy of the film and that of the substrate–film interface (the surface adhesive force is stronger than the adatom cohesive force):

$$\gamma_s \geq \gamma_f + \gamma_{s/f} \tag{18.5}$$

The layer-by-layer growth is two-dimensional, denoting that the formation of complete films occurs prior to that of the subsequent layers' growth. This growth type commonly appears when the substrate and the film are either homogeneous materials or some particular dissimilar materials, such as the epitaxial growth of semiconductor and oxide materials.

The Stranski–Krastanov (SK) mode (initially layer-by-layer growth, then two-dimensional islands), known as layer-plus-island growth, is an "intermediate" case. Initially, the process starts with two-dimensional growth, namely formation of the first monolayer, up to a few monolayers on the substrate. Then, at a critical layer thickness, the subsequent layer growth becomes unfavorable and three-dimensional islands are formed on top of this substrate. The growth continues through nucleation and coalescence of adsorbate "islands" (Scheme 18.1c).

One may conclude that thin films first grow two-dimensionally in a layer-by-layer mode, and then three-dimensionally in an island mode. This latter type of growth, characterized by two-dimensional layer and three-dimensional island growth, is highly dependent on the chemical and physical properties, such as surface energies and lattice parameters, of both the substrate and film. As a function of the monolayer coverage, θ, two cases occur during thin film growth, as follows:

$$\gamma_s \geq \gamma_f + \gamma_{s/f}, \text{ when } \theta \leq \theta_{critic} \quad (18.6)$$

$$\gamma_s < \gamma_f + \gamma_{s/f}, \text{ when } \theta > \theta_{critic} \quad (18.7)$$

Therefore, the evolution mode of film growth depends on the ability of atoms on the surface and from the grains to move along the grain surface and among grains. When the deposition does not occur under thermodynamic equilibrium, the kinetic aspects should consider the growth mode. Therefore, it can be concluded that the evolution of film growth includes the following steps: nucleation, island growth, coalescence of islands, formation of polycrystalline islands and channels, development of a continuous structure, and thickness growth.

18.4 THIN METAL NANOLAYERS DEPOSITED ON POLYMER SUBSTRATES

Polymers represent an important class of materials that present several advantageous features—low weight, low cost, and good mechanical properties, along with a number of disadvantages referring to surface properties, conductivity, and durability, which sometimes limit their area of applicability. Therefore, improvement of these aspects has attracted a great deal of attention in much of the research, with one of the best solutions being the deposition of metallic or ceramic films on a polymeric substrate (Kim et al. 2005). Consequently, the adhesion between the polymer substrate and the film should be stronger and in no way affected by physical or chemical factors. Moreover, the polymeric substrates that possess the capacity to form complexes with various metal or ceramic films have been preferred for many applications, due to the special properties of the new metal–polymer composite. Because the chemical nature and properties of the metal and polymer are different, adhesion at this interface is very poor. For this reason, an important step in the technology of metallic or ceramic coating on the polymer substrate sometimes consists of obtaining a bond formation at the interface, by functionalization of the polymer surface, through different surface treatment techniques. The adhesion between the metal and polymer substrate strongly depends on the type of metal and on the functional sites present on the surface. Thus, for increasing adhesion at the metal/polymer substrate interface, the most suitable methods recommended are as follows:

- Oxidation of the polymer surface
- Chemical modification of the substrate with polar groups, such as hydroxyl, carbonyl, and carboxylic acid moieties
- Physical modification of the substrate through its plasma treatment prior to the deposition of the metal

For example, the low-temperature plasma treatment is a technique that does not negatively affect the environment, so that it is widely applied for modifying the polymer substrate surface. The main benefits of this technique are (Kim et al. 2005):

- It is an elementary process where high activation energies of some electron volts can appear in low-pressure plasmas without elevated gas (ion, neutral) temperatures. Sensitive polymer materials show almost no thermal load during the low-pressure plasma treatment.
- The bulk properties are not influenced by the major modifications of the chemical structure of surface layers.
- Variation of low-pressure plasmas induces numerous modifications at the substrate surface.
- The amount of toxic products is low, compared to wet chemistry.
- Utilization of low-pressure plasmas without high gas (ion, neutral) temperatures leads to the appearance of basic processes with high activation energies of some electron volts.

Therefore, some results of the research showed the influence of the sputtering cleaning of the polymer substrate on the adhesion of titanium films and modification of surface properties, which affect adhesion (Hoa et al. 2006). In this context, Hoa et al. investigated the effect of low-pressure Ar gas plasma treatment on the surface films of polyamide/propylene (PP/PA) and polyamide/polyethylene (PP/PE) blends prior to the deposition of a titanium thin film on this surface. Coating of the polymer substrate with a titanium film was performed by the RF sputtering technique, a film with a thickness about 0.15 μm resulted. The surface morphology of the polymer substrate analyzed by SEM indicated that the surfaces of both PA/PP and PA/PE untreated polymer substrates are smooth. After plasma treatment, the polymer surface becomes rougher because the positive ions of plasma interact with the polymer substrate causing various phenomena, such as breakage of the covalent bonds, cross-linking, grafting, and incorporation of chemical groups in the plasma. Moreover, modification of roughness depends on the treatment conditions, nature of the polymer substrate, and the homogeneity of the two polymer types from the polymer substrate. It may be observed that the effect of plasma treatment on the PA/PP substrate under various conditions shows a slight change in roughness and contact angle (the solvent used for measurements was ethanol), as is shown in Figure 18.1. These small variations of surface properties with modification of the treatment conditions are attributed to the treatment mechanism, which is quite difficult to explain in detail, especially in relation to the variation in surface properties.

On the other hand, investigation of the influence of plasma treatment on titanium film adhesion on the PA/PA substrate reveals a considerable increase in adhesion after plasma action. Table 18.1 shows that adhesion of the titanium film on the untreated PA/PP substrate is only 12%, which presents the utilization of the metal-PA/PP composite in any type of application.

For plasma conditions of 30 W power and 60 s, a higher adhesion was obtained, which shows that adhesion increases with the increase of plasma power and treatment time. The obtained results demonstrated that, by using the plasma treatment

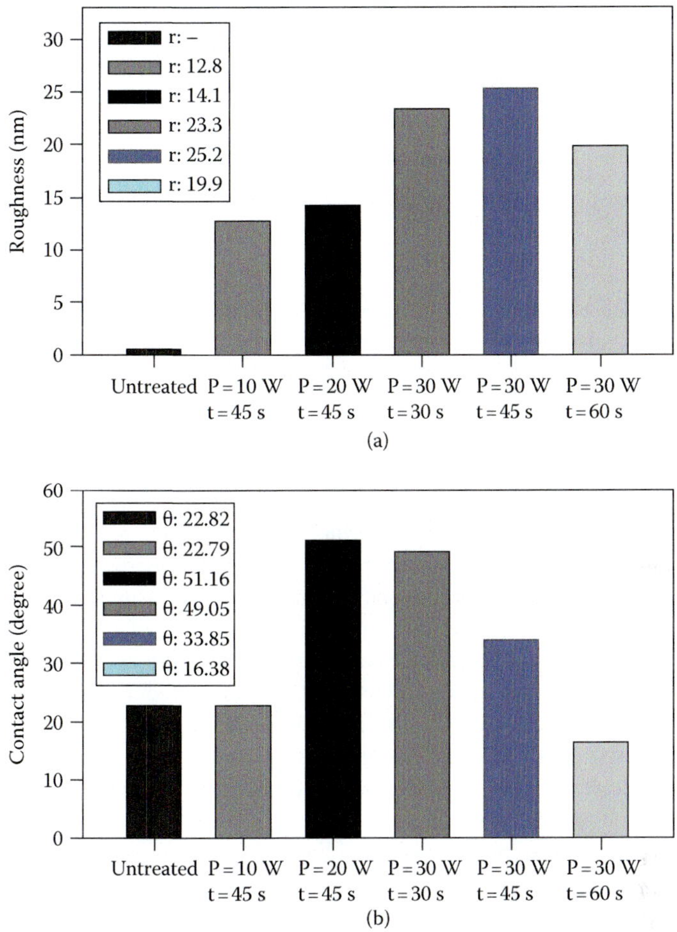

FIGURE 18.1 Effect of plasma treatment on the surface properties of a PA/PP substrate: (a) on roughness and (b) on variation of the contact angle (θ). (Adapted from Hoa Ta Thi Phuong et al. Improvement of the adhesion of titanium film on polymer substrate by argon plasma treatment. *Proceedings of the International Conference on Engineering Physics*, pp. 397–400, 2006.)

technique, adhesion of the titanium film on the PA/PP substrate increases as a function of the treatment conditions.

Other conclusions of the investigation evidenced that most metalized polymers are used in a wide range of applications, from food packaging to decorative coatings. Metal–polymer interaction types, as well as interface morphology, essentially induce variations in the properties of metal–polymer systems (Putz et al. 2015). Generally, the cohesive energy of polymers is lower than that of reactive metals, which determines a low adhesion between them. More than that, metals with low reactivity do not wet the untreated polymer surface, as the moderately reactive metals forming

TABLE 18.1
Influence of Plasma Treatment Conditions on the Adhesion of Titanium Films Deposited on PA/PP Substrate

	Different Plasma Treatment Conditions				
Untreated sample	P = 10 W, t = 45 s	P = 20 W, t = 30 s	P = 30 W, t = 45 s	P = 30 W, t = 45 s	P = 30 W, t = 60 s
12%	24%	40%	52%	60%	80%

Source: Adapted from Hoa Ta Thi Phuong et al. Improvement of the adhesion of titanium film on polymer substrate by argon plasma treatment. *Proceedings of the International Conference on Engineering Physics*, pp. 397–400, 2006.

three-dimensional spherical clusters so. One of the most preferred materials used for metallization of the polymer substrate is gold because it possesses numerous favorable properties required for applications in optics, electronics, and chemical catalysis (Daniel and Astruc 2004). The most attractive properties of gold required in the manufacture of structures and devices are chemical stability and large atomic mass (electromigration resistance). An efficient hole-injection occurs when using gold as an anode in organic LEDs and solar cells. However, if gold is used as a cathode, n-type doping will facilitate electron injection into organic semiconductors (Walzer et al. 2007). Realization of a structurally continuous gold thin film is often difficult to obtain on amorphous and crystalline surfaces, such as native silicon oxide, glass, Si_3N_4, TiO_2, and some transparent polymers. In this respect, some studies have looked at obtaining epitaxial single-crystal gold islands or sheets on a crystalline substrate, such as LiF, KCl, silicon mica, or sapphire (Fedotov et al. 2012). Other studies with a significant impact on thin-film electronics, fabrication of transparent gold contacts, biosensing, metamaterials, plasmonic superlenses, and optical antennas made use of transparent optical polymers (CYCLOTENE3022—derived from B-staged bis-benzocyclobutene-based monomers and ORMOCER—ORganically MOdified CERamics) as supports for obtaining thin gold films (Švorčík et al. 2010; Leosson et al. 2013). Electrical, optical, and structural analysis of the gold films deposited on CYCLOTENE and ORMOCER substrates confirmed that an ultra-thin gold film had been obtained. Electrical resistance measurements and SEM confirmed that the gold film presents continuity thresholds and lower coalescence, when compared with those deposited on SiO_2. In addition, X-ray reflection and diffraction measurements showed that the gold films deposited on the polymer surface have a high density and a thickness less than 2 nm.

Thin films with thickness values below 20 nm may be obtained on the polymer substrate by a traditional lift-off process, without the use of adhesion layers that limit optical performance. The maximum internal transmission measured at 4.2 nm deposition thicknesses and at 530 nm wavelengths was 75%. Surface characterization methods, such as X-ray photoelectron spectroscopy, provide a better analysis of the wetting chemistry of gold on different polymer surfaces. Therefore, by using the CYCLOTENE and ORMOCER transparent substrates, smooth and patternable gold films can be fabricated by traditional deposition techniques, at room temperature, in the absence of adhesion layers. Some studies report obtaining continuous and homogeneous 20-nm gold layers sputtered on the polyethylene terephthalate (PET) (Švorčík et al. 2010). Sputtering of the gold layer on PET has considerably influenced both surface morphology and physical properties. Thus, an AFM investigation of PET surface and gold-coated PET shows that, after deposition of 20-nm gold layers onto the substrate, surface roughness increases from 0.5 to 1.9 nm and surface morphology is modified. Measurement of the zeta potential confirms the increase of surface conductance of Au/PET, while the higher values of the zeta potential of pristine PET over an alkaline pH range show a more hydrophilic surface, compared with the Au/PET structure. Some investigations focused on the increase of adhesion between the metal layer and the polymer substrate by different methods, such as chemical changes, plasma charge, and laser and ion beam irradiation (Hollenberg et al. 2006; Yeager et al. 2008).

In addition, thin metal films on flexible substrates with high chemical and thermal stability, such as polyimides, may be used in more flexible electronic devices. These applications often use gold as a thin film because it is inert and easy to pattern through lithographic methods (Hollenberg et al. 2006). Thus, an increase of adhesion between the gold film and polyimide may be achieved when a variety of substrate treatments are applied prior to gold deposition (Yeager et al. 2008). Some investigations on the effect of oxygen plasma treatments prior to gold deposition on a polyimide substrate demonstrate an increase in the interfacial adhesion energy from 0.8 to 1.5 J/m^2. In other cases, deposition of an adhesive titanium layer on the polyimide substrate may increase adhesion up to 4 J/m^2 (Yeager and Bahr 2010; Albu et al. 2014). Moreover, for a better understanding of the process–structure–properties relationship in these systems, a microstructural investigation of the gold deposited on the treated and untreated polyimide was performed.

This analysis is especially important because it provides information about the dependence between the elastic properties of systems and the orientation of grains. The size and shape of grains can induce diffusion, fracture, and some thermal, electrical, and mechanical properties. In this context, microstructural examination reveals that the gold film deposited on untreated polyimides presents a strong {111} texture, and both pretreated systems exhibit a random texture. This denotes that the pretreatment modifies the surface energy of the polymer substrate and influences the manner of gold film growth. Therefore, texture differences can be responsible for the differences in adhesion measured among the samples. The plasma-etching pretreatment is shown to have no statistical effect on grain size; however, it does randomize the texture. Inclusion of a titanium interlayer after plasma etching slightly enhances the grain size without additional modification of the random texture. Moreover, other experimental data presented in the literature have also investigated the effect of plasma treatment on the morphological aspect and adhesion between the thin gold layer and some quaternized polysulfones containing different tertiary amines, such as *N,N*-dimethylethylamine and *N,N*-dimethyloctylamine (Albu et al. 2014). Investigation of the influence of substrate nature, plasma treatment, and gold sputtering times on the gold adhesion to quaternized polysulfone enables the optimal parameters for obtaining continuous thin hold layer to be found. Therefore, the sample that contains an –*N*,*N*-dimethyloctylamine group is more rigid than the one with an ethyl radical. This leads to a higher roughness and a larger average grain diameter of the gold layer at a higher deposition time, with all these aspects determining a better adhesion, particularly after plasma treatment (Albu 2015). The reported results confirm that, for longer deposition times in plasma, adhesion of the gold layer onto the quaternized polysulfone substrate is higher. Moreover, understanding how various methods may change the microstructure of the thin gold film could offer new possibilities for future development of other sputtered thin film systems of interest.

Another metal that is widely deposited on polymer substrates for biomedical utilizations, in the food packaging industry, or for space applications, is silver due to its antibacterial and anti-inflammatory properties (Brandon et al. 2011). Some techniques, such as sputtering, ion exchange, and sol–gel methods, are presented in the literature for providing antibacterial properties to the thermally sensitive polymer parts used in many medical devices (Balagna et al. 2012). Ion beam–assisted treatments and magnetron sputtering are used for the deposition of silver coatings at lower (10^{-3} Pa) and higher pressure (0.0–0.1 Pa). Because medical devices contain thermally sensitive polymers, some experimental studies were necessary that focused on silver coating at low temperature. Also, the combination of magnetron sputtering with the atom beam (Saddle Field) plasma source for silver deposition onto polyurethane, silicone, tubes of polyvinyl chloride, polyurethane pellethane, and cyclic-aliphatic polyurethane was realized at a maximum deposition temperature of 70°C (Dowling et al. 2001). A series of mechanical tests, such as the pull test, shows that the rigidity of the polymer substrate influences coating adhesion considerably.

For example, coating on a less-rigid polymer, such as polyurethane, shows good adhesion and a 2–3 log reduction in bacterial adhesion when using *Staphylococcus epidermidis* and manifests no cytotoxicity when using fibroblast cells. The obtained results highlight that the combination of magnetron sputtering with the atom beam source is suitable for silver deposition at higher pressure and low substrate temperatures.

18.5 GENERAL REMARKS

The goal of this study is to summarize the research devoted to the deposition of thin material layers with thickness values below several micrometers onto a polymer substrate to provide new properties that are unattainable in the individual substrates. Thin films play an essential role in enhancing the properties and performance of tools and machines because their special properties differ significantly from the bulk form. In order to obtain the desired properties, preparation of nanostructured materials in a controlled manner implies exploration of morphology and understanding of the growth mechanism. Such studies are necessary for correlating the structural changes with the various process parameters, such as deposition characteristics (i.e., rate, time). These new multiphase systems, thin film–polymer substrates, could be used for potential nanotechnology applications. Therefore, control of the physical processes allows control of the chemical and structural characteristics of the functional nanomaterials. From this point of view, the present chapter explains the evolution and the phenomena occurring during every step in the growth of thin films and the importance of metallic or ceramic film deposition on the polymer substrate.

Multiphase composites formed of polymer substrates coated with thin metal films show special properties and are in great demand for various applications. The metal–polymer substrate interaction and the morphological structure at the interface influence the final properties of the composites of thin metal–polymer substrates. Thus, modifying the properties of the polymer substrate by wet (acid, alkali), dry (plasma), and radiation treatments (ultraviolet radiation and laser) appears as a significant step for increasing adhesion of thin film onto polymeric substrates.

The main stages of thin film formation are diffusion of the atoms deposited on the substrate around until they encounter other atoms, nucleation, and then formation of islands. Subsequent deposition determines the increase of islands until they come in contact with one another and coalescence appears, which determines formation of continuous films. Therefore, understanding the nucleation, the growth of ultrathin films on a polymer surface, and minute exploration of the new surface properties thus obtained is very important in engineering polymers for future research and development in plasmonics, metamaterials, light-emitting and light-harvesting devices, and other related areas.

REFERENCES

Albu Raluca Marinica. Functionalized polysulfone-metal complexes. In *Functionalized polysulfones. Synthesis, characterization, and applications*, Silvi Ioan (ed.), pp. 201–232. Boca Raton: CRC Press, 2015.

Albu Raluca Marinica, Stoica Iuliana, Avram Ecaterina, Ioanid Emil Ghiocel, and Ioan Silvia. Gold layers on untreated and plasma-treated substrates of quaternized polysulfones. *J. Solid State Electrochem.* 18 no. 10 (2014): 2803–2813.

Balagna Cristina, Perero Sergio, Ferraris Sara, et al. Antibacterial coating on polymer for space application. *Mater. Chem. Phys.* 135 no. 2–3 (2012): 714–722.

Bauer Ernst. Phaenomenologische theorie der kristallabscheidung an oberflaechen I. *Z. Krisallogr.* 110 no. 1–6 (1958): 372–394.

Brandon J. Erik, Vozoff Max, Kolawa A. Elizabeth, et al. Structural health management technologies for inflatable/deployable structures: Integrating sensing and self-healing. *Acta Astronaut.* 68 no. 7–8 (2011): 883–903.

Brune Haralad. Microscopic view of epitaxial metal growth: Nucleation and aggregation. *Surf. Sci. Rep.* 31 no. 3–4 (1998): 125–229.

Daniel Marie-Christine, and Astruc Didier. Gold nanoparticles: Assembly, supramolecular chemistry, quantum-size-related properties, and applications toward biology, catalysis, and nanotechnology. *Chem. Rev.* 104 no. 1 (2004): 293–346.

del Muro Montserrat Garcia, Konstantinovic Zorica, Varela M., Batlle Xavier, and Labarta Amílcar. Metallic nanoparticles embedded in a dielectric matrix: Growth mechanisms and percolation. *J. Nanomater.* 2008 (2008): 1–5.

Dowling P. Denis, Donnelly Kevin, McConnell M.L., Eloy Rosy, and Arnaud M.N. Deposition of anti-bacterial silver coatings on polymeric substrates. *Thin Solid Films.* 398–399 (2001): 602–606.

Durand Hervé-André, Sekine Kazuhiko, Kazuyuki Etoh, Kazuhiko Ito, and Izumi Kataoka. Relation of initial thin film formation to defects induced by low energy ions. *Thin Solid Films.* 336 no. 1–2 (1998): 42–48.

Fedotov A. Vassili, Uchino Toshiyuki, and Ou Jun-Yu. Low-loss plasmonic metamaterial based on epitaxial gold monocrystal film. *Opt. Express.* 20 no. 9 (2012): 9545–9550.

Hoa Ta Thi Phuong, Mai Nguyen Thi Thanh, An Nguyen Hoang, Tuan Nguyen Anh, Hung Le Thanh, and Song Young Sik. Improvement of the adhesion of titanium film on polymer substrate by argon plasma treatment. *Proceedings of the International Conference on Engineering Physics*, pp. 397–400, October 9–12, 2006.

Hollenberg A. Brian, Richards D. Cecilia, Richards Robert, Bahr F. David, and Rector M. David. A MEMS fabricated flexible electrode array for recording surface field potentials. *J. Neurosci. Meth.* 153 no. 1 (2006): 147–153.

Iijima Sumio, and Ajayan M. Pulickel. Substrate and size effects on the coalescence of small particles. *J. Appl. Phys.* 70 (1991): 5138–5140.

Kalff Matthias, Breeman Marinus, Morgenstern Markus, Michely Thomas, and Comsa George. Effect of energetic particles on island formation in sputter deposition of Pt on Pt (111). *Appl. Phys. Lett.* 70 no. 2 (1997): 182–184.

Kim Soo Hong, Cho Su Hyeun, Lee Nae-Eung, Kim Hoon Mo, Nam Yun Woo, and Kim Young-Ho. Adhesion properties of Cu/Cr films on polyimide substrate treated by dielectric barrier discharge plasma. *Surf. Coat. Tech.* 193 no. 1–3 (2005): 101–106.

Kintzel J. Edward, Smilgies Detlef Matthias, Skofronick G. James, Safron A. Safron, and Winkle Van H. David. Ultrathin film growth of p-phenylene oligomers on alkali halide substrates. *J. Cryst. Growth.* 289 no. 1 (2006): 345–350.

Leosson Kristjan, Ingason S. Arni, Agnarsson Bjorn, Kossoy Anna, Olafsson Sveinn, and Gather C. Malte. Ultra-thin gold films on transparent polymers. *Nanophotonics.* 2 no. 1 (2013): 3–11.

Lorenz Ratke, and Peter W. Voorhees. *Growth and coarsening: Ostwald ripening in material processing.* Berlin: Springer, 2002.

Michely Thomas, and Krug Joachim. *Islands, mounds and atoms.* Berlin: Springer, 2003.

Ohring Milton. *Materials science of thin films.* San Diego: Academic Press, 2002.

Petrov Ivan, Barna B. Peter, Hultman Lars, and Greene E. Jenny. Microstructural evolution during film growth. *J. Vac. Sci. Technol. A.* 21 (2003): S117–S128.

Putz Barbara, Schoeppner L. Rachel, Glushko Oleksandr, Bahr F. David, and Cordill J. Megan. Improved electro-mechanical performance of gold films on polyimide without adhesion layers. *Scripta Mater.* 102 (2015): 23–26.

Ruffino Francesco, and Grimaldi Maria Gratia. Island-to-percolation transition during the room-temperature growth of sputtered nanoscale Pd films on hexagonal SiC. *J. App. Phys.* 107 (2010): 074301-1–074301-6.

Sheldon W. Brian, Nicholas D. Jason, and Mandowara Sunil. Tensile stress evolution during the early-stage constrained sintering of gadolinium-doped ceria films. *J. Am. Ceram. Soc.* 94 no. 1 (2011): 110–117.

Sree Harsha S. Karnamadakala. *Principles of physical vapor deposition of thin films.* Oxford: Elsevier, 2006.

Švorčík Vaclav, Kolská Zdenca, Luxbacher Thomas, and Mistrík Jan. Properties of Au nanolayer sputtered on polyethyleneterephthalate. *Mater. Lett.* 64 no. 5 (2010): 611–613.

Tame S. Mark, McEnery K.R., Özdemir Şahin Kaya, Lee Jinhyoung, Maier A. Sebastian, and Kim Mun Dae. Quantum plasmonics. *Nat. Phys.* 9 (2013): 329–340.

Thompson C. V. Structure evolution during processing of polycrystalline films. *Ann. Rev. Mater. Sci.* 30 (2000): 159–190.

Veprek Stan. Recent search for new superhard materials: Go nano!. *J. Vac. Sci. Technol. A.* 31 no. 5 (2013): 050822-1–050822-33.

Walzer Karsten, Maennig Bert, Pfeiffer Martin, and Leo Karl. Highly efficient organic devices based on electrically doped transport layers. *Chem. Rev.* 107 (2007): 1233–1271.

Warrender M. Jeffrey, and Aziz J. Michael. Morphological evolution of Ag/mica films grown by pulsed laser deposition. *Mat. Res. Soc. Symp. Proc.* 749 (2003): W3.1.1–W3.1.6.

Wuttig Matthias, and Liu Xiangdong. *Ultrathin metal films: Magnetic and structural properties*. New York: Springer, 2004.

Yeager D. John, and Bahr F. David. Microstructural characterization of thin gold films on a polyimide substrate. *Thin Solid Films* 518 (2010): 5896–5900.

Yeager D. John, Phillips D. Jonathan, Rector M. David, and Bahr F. David. Characterization of flexible ECoG electrode arrays for chronic recording in awake rats. *J. Neurosci. Meth.* 173 no. 2 (2008): 279–285.

Zinke-Allmang Martin, Feldman C. Leonard, and Grabow H. Marcia. Clustering on surfaces. *Surf. Sci. Rep.* 16 no. 8 (1992): 377–463.

Index

E

F

G

H

I

K

L

M

N

O

P

Q

R

S

T

U

V

W

X

Y

Z

For Product Safety Concerns and Information please contact our EU representative GPSR@taylorandfrancis.com Taylor & Francis Verlag GmbH, Kaufingerstraße 24, 80331 München, Germany

Batch number: 10745645

Printed by Printforce, the Netherlands